BIOCHEMISTRY ILLUSTRATED

Dedication to the first edition
We would like to dedicate this book to our wives, Mollie and Shirley and our children, Alastair (a paediatrician), Julia (a Health Visitor) and Corinne (a budding biochemist), all of whom have been affected by, and sometimes afflicted by, our love for biochemistry.

Dedication to the second edition
In addition to the above one of us as an indicator of progress would like to add his four grandchildren, at least one of whom already finds the book interesting.

BIOCHEMISTRY ILLUSTRATED

AN ILLUSTRATED SUMMARY OF THE SUBJECT FOR MEDICAL AND OTHER STUDENTS OF BIOCHEMISTRY

PETER N. CAMPBELL
PROFESSOR OF BIOCHEMISTRY

ANTHONY D. SMITH
READER IN BIOCHEMISTRY

University College and Middlesex School of Medicine, London

ILLUSTRATOR: **SUE HARRIS**

SECOND EDITION

CHURCHILL LIVINGSTONE
EDINBURGH LONDON MELBOURNE AND NEW YORK 1988

CHURCHILL LIVINGSTONE
Medical Division of Longman Group UK Limited

Distributed in the United States of America by
Churchill Livingstone Inc., 1560 Broadway,
New York, N.Y. 10036, and by associated companies,
branches and representatives throughout the world.

First edition
Second edition 1988

ISBN 0 443 03454 0

British Library Cataloguing in Publication Data

Campbell, P.N.
Biochemistry illustrated : an illustrated summary of the subject for medical and other students of biochemistry.—2nd ed.
1. Biological chemistry
2. Title II. Smith, Anthony D. (Anthony Donald)
574.19'2 QP514.2

Library of Congress Cataloging in Publication Data

Campbell, P. N. (Peter Nelson)
Biochemistry illustrated.

Bibliography: p.
Includes index.
1. Biological chemistry — Outlines, syllabi, etc.
I. Smith, A. D. (Anthony David) II. Title.
[DNLM: 1. Biochemistry — outlines. QU 18 C189ba]
QP518.3.C36 1987 574.19'2 87-9013

Produced by Longman Singapore Publishers Pte Ltd
Printed in Singapore

PREFACE TO THE SECOND EDITION

The first edition of this book published in 1982 seems largely to have met its objective and so we have been invited to prepare a second edition. After much consideration we decided to retain the original objectives and to resist the temptation to greatly expand the text. Each page has been re-assessed in the light of the advance in knowledge and the many suggestions from our readers all over the world. In the process we have expanded the text by some 50 pages. Once again we would welcome further suggestions and hope that the new edition is still considered to fulfill our objectives.

1988

P.N.C.
A.D.S.

PREFACE TO THE FIRST EDITION

The purpose of this book is to provide a survey of biochemistry in an easily assimilable form. We had in mind firstly students who might appreciate a succinct summary of some of the tenets on which a more advanced study of biochemistry is based. We hoped we might even encourage some students to develop their thirst for more extensive texts. Secondly we were aware of the needs of students in developing countries, who may not be too conversant with English. We thought it might be helpful to the students if the teachers could refer to clear diagrams in a modest book. Thirdly we wished to assist all those who, while not specializing in biochemistry, want to be aware of the major trends in the subject. In this respect we hope the book may prove useful to school teachers, physical scientists and medical doctors.

In making our choice of subject matter we have aimed at the student of medicine and the biochemistry student who is primarily interested in animals. We are aware that our choice has been arbitrary and we would be delighted to have suggestions concerning the subject matter and how we could improve the presentation.

We have of course worked within certain constraints. Firstly we have confined ourselves to about 200 pages, and secondly we have used only one colour in addition to black and white. As authors we consider we have been very fortunate in our collaborators: the artist, Sue Harris, worked closely with us, producing clear diagrams from our rough outlines; and the publishers have helped to keep us on sensible lines. In addition we thank Professor Maharani Chakravorty, Professor A. D. Patrick and Dr Ton So Har for their critical reading of the manuscript, and assisting us to ensure accuracy. We are more than grateful to all the numerous people and organizations who have given permission for the use of figures which we have redrawn from previous publications. We have a special debt to all our colleagues in the Courtauld Institute and elsewhere on whom we have called for expert advice on many occasions.

1982

P.N.C.
A.D.S.

CONTENTS

1 CELLULAR BASIS OF BIOCHEMISTRY

2 THE PROTEINS

3 STRUCTURE AND FUNCTION OF ENZYMES

4

NUCLEIC ACIDS AND PROTEIN BIOSYNTHESIS

5 COENZYMES AND WATER-SOLUBLE VITAMINS

6 CARBOHYDRATE CHEMISTRY AND INTERCONVERSIONS OF MONOSACCHARIDES

9 MEMBRANE STRUCTURE AND FUNCTION

INDEX

A NOTE ON THE LAYOUT

The left hand column contains the section headings and a very brief summary of the subject matter. Further details are contained in the right hand column which is also where the illustrations are presented.

Each page is numbered so that it seemed unnecessary to number separately the illustrations. Where more than one illustration is presented they are lettered A, B, C, etc. It should therefore be easy for a lecturer to refer the students to a particular illustration.

At the first mention of an important word it is presented in italics. The word will also appear in the index together with other entries.

Rather than acknowledge each figure separately we have grouped the acknowledgements together. The division of the text into frames does not in any way imply that the book can be used as a teaching programme. It is intended that it be used in conjunction with more complete textbooks and a course of lectures.

The arrangement of carbohydrate and fat metabolism is unusual, but is based on experience of teaching medical students over a number of years. The division of pathways involved in the synthesis and degradation of energy-storage compounds into the fed and fasting states of metabolism has been found to give students a lasting understanding of the functions of these pathways. It also leads to a ready analysis of many clinical metabolic problems. Structural lipids and carbohydrates are described under the subject of membranes, to which they are functionally related.

Whilst some of the diagrams may seem to contain more information than appears necessary in a short summary (e.g. detailed structures of active sites), we felt that such additional detail does not in any way obscure the essential concept being conveyed, and gives an authentic basis for a more precise examination of the concept by students who wish to think more deeply about the subject.

ACKNOWLEDGEMENTS

The authors are pleased to acknowledge the source of those illustrations in the book which have been based on illustrations published elsewhere and to acknowledge the assistance of those who have helped in the preparation of various illustrations. Our acknowledgements have been grouped as follows:

a. The authors and publishers of books and original reports in journals. This list may also be of value to those who wish to extend their reading.
b. Those persons and companies who have provided material from their own sources.
c. Those who have offered their expert opinion.

In all cases the numbers in [ ] refer to the number of the illustration in the present book. The other numbers refer to the original source.

a.

General textbooks

Basic Biochemistry for Medical Students (Campbell, P.N. & Kilby, B.A. eds.)
Academic Press, London and New York. Apart from the editors the authors were J.B.C. Findlay, H. Hassall, R.P. Hullin, A.J. Kenny and J.H. Parish, Figs 2.8 [11B], 2.10 [12A], 2.15 [11A], 4.10 [19A], 4.16 [14], 4.18 [15A], 4.21 [30A], 5.24 [70A], 6.7 [191A], 12.3 [89C], 12.12 [96B], 12.13 [100A], 12.16 [99B], 12.17 [102B], 12.23 [101A], 12.25 [103A], 12.26 [103B], 12.27 [104A]. Tables 2.8 [15B], 4.2 [15C].

Biochemistry: A Functional Approach (3rd Edn.) (McGilvery, R.W. and Goldstein, G.W.)
W.B. Saunders Co., Philadelphia.
Fig. 3.12 [21B].

Biochemistry with Clinical Correlations (2nd Edn.) (Devlin, T.M. ed.)
J. Wiley & Sons, New York.
Figs 3.6 [67B], 6.39 [124C], 6.50 [175B], 6.51 [179B], 15.4(a) [68B].

Biochemistry (Stryer, L.) (2nd Edn.)
W.H. Freeman, Oxford and San Francisco.
Figs 4.2 [24A], 4.3 [24B], 4.6 [24C], 4.20 [23B], 30.21 [91B].

Principles of Biochemistry (6th Edn.) (White, A., Handler, P., Smith, E.L., Hill, R. L. & Lehman, I.R.)
McGraw-Hill Book Co., New York.
Figs 36.1, 36.2, 36.5, 36.6, [49A,B,C,D], 36.4, 36.9 [50A, B].

Molecular Biology (Freifelder, D.)
Jones and Bartlett Publishers, Inc. Boston USA.
Fig. 4.32 [83A], Fig. 21.27 [93B].

Biochemistry for the Medical Sciences (Newsholme, E.A. & Leech, A.R.)
J. Wiley & Sons, Chichester.
Figs. 7.9 [193A], 11.5 [200B].

Books on special topics

Open University Course Book (S322 Units 1–2) (1977)
Open University Press, Milton Keynes.
Fig. 3, p. 14 [18].

Cells and Organelles (2nd Edn.) (Novikoff, A.B. & Holtzmann, E.)
Holt, Rinehart & Winston, New York.
Figs. 1.23 [3], 1.25 [4B].

Structure and Action of Proteins (Dickerson, R.E. & Geis, I.)
Benjamin Inc., New York.
Pages 47 [21A], 56 [22A].

Enzyme Structure and Mechanism (2nd Edn.) (Fersht, A.)
W.H. Freeman, Oxford and San Francisco.
Figs. 1.12 [60A, B], 1.13 [62], 12.13 [59].

Chance and Necessity (Monod, J. Trans. by Wainhouse, A.)
A.A. Knopf, New York, and William Collins, London.
p. 74, Fig. 4 [106].

The Structure and Function of Animal Cell Components (Campbell, P.N.)
Pergamon Press, Oxford.
Figs 2.2 [180], 5.2 [153B]. Table 1.1 [4A].
Advancing Chemistry (Lewis, M. & Waller, G.)
Oxford University Press.
p. 311, [174].
Supplement to DNA Replication (Kornberg, A.)
W.H. Freeman and Co. San Francisco 1982.
Frontispiece [92B].
Molecular Basis of Antibiotic Action (Gale, E.G., Cundliffe, E., Reynolds, P.E., Richmond, M.E. & Waring, M.J. eds.)
Wiley Interscience, New York and Toronto
Cundliffe, E., p. 278 [93C].
Waring, M.J., p. 173 [14B].
Membranes and their Cellular Functions (2nd Edn.) (Finean, J.B., Coleman, R. & Michell, R.H.)
Blackwell Scientific Publications, Oxford.
p. 37 [253B].
Structure of Mitochondria (Munn, E.A. ed.,)
Academic Press, London and New York.
Kroger, A., & Klingenberg, M., p. 2820 [177A].
A Guided Tour of the Living Cell (Christian de Duve)
Scientific American Library.
Illustration Copyright (C) 1982 Neil Hardy p. 272 [109A], p344 [44A].
Molecular Biology of the Cell (Alberts, B., Bray, D., Lewis, J. Raff, M., Roberts, K., Watson, J.D. eds)
Garland Publishing Inc. New York & London.
Fig. 8–24 [86], 8–59 [6A].
Immunology (Eisen, H.N.)
Harper & Row, New York.
p. 132, Fig. 2 [41A].
Principles of Gene Manipulation 3rd Edn. (Old, R.W. & Primrose, S.B.) Blackwell Scientific Publications, Oxford.
Figs. on 1.3, 1.4 [117 A & B].
From Cells to Atoms (Rees, A.R. & Steinberg, M.J.E.)
Blackwell Scientific Publications, Oxford.
Fig. 10.1 [51A], Fig. 11.2 [37A], Fig. 26.1 [86A], Fig. 40.2 [43A].
Recombinant DNA. A Short Course. (Watson, J.D., Tooze, J. & Kurtz, D.T.)
Scientific American Books.
Fig. 5–3 [118], Fig. 5–4 [119].
Immunology (Roitt, I.M., Brostoff, J., Male, D.K.).
Churchill Livingstone, Edinburgh and Gower Medical Publishing, London & New York.
Fig. 5.15 [42B], Fig. 7–8 [43B].
Separation of Plasma Proteins (Curling, J.M. ed.)
Pharmacia Fine Chemicals AB, Uppsala, Sweden.
Fig. 39 [32B].
Albumin, an Overview and Bibliography
Miles Laboratories Inc. IN 46515, USA.
Physiological transport functions of albumin [34A].
The Ultrastructural Anatomy of the Cell (Allen, T.D.)
Cancer Research Campaign, London SW1Y 5AR [5 & 6].

Reviews

Companion to Biochemistry (Bull, A.T., Lagnado, J.R., Thomas, J.O., Tipton, K.F., eds.)
Longman, London and New York.
Campbell, P.N. (1979) *2*, Fig. 8.1 [107A].
FEBS Symposium Vol. 53 (Rapoport, S. & Scherve, T. eds.)
Pergamon Press, Oxford.
Grant, M.E., p. 29–41 [109B].
The Plasma Proteins (Putnam, F.W. ed.)
Academic Press, London and New York, Putnam, F.W., Vol. III, p. 14 [39A].
The Enzymes 3rd Edn. (Boyer, P.D. ed.)
Academic Press, London and New York

Dickerson, R.E. & Timkovich, R., Vol. XI, p. 441, Fig. 8 [61A].
Rossman, M.G., Liljas, A., Brinder, C-I. & Banaszak, L., Vol. XIA, p. 68, Fig. [63].

Essays in Biochemistry (Campbell, P.N., Greville, G.D. & Dickens, F. eds.)
Academic Press, London and New York
Hales, C.N. (1967) *3*, p. 75, Fig. 1 [193A]
Grant, P.T. & Combs, T.L. (1970) *6*, p. 76, Fig. 3 [78A].
Williamson, A.R. (1982) *18*, p. 24, Fig. 13 [116B].

Biochemical Society Symposia (Grant, J.K. ed.)
Cambridge University Press, Cambridge
Whittaker, V.P. (1963) *23*, p. 120, Fig. 4. [176A].

Atlas of Protein Sequence and Structure (Dayhoff, M.O. ed.)
National Biomedical Research Foundation, Washington D.C. (1972). Dayhoff, M.O., Park, C.M. and McLaughlin, P.J., Vol. 5, p. 8 [61B].

Current Topics in Cell Regulation (Horecker, B.L. & Stadtman, E.R. eds.)
Academic Press, London and New York.
Masters, C.J. (1977) *12*, p. 77, Fig. 2 [63B].

Trends in Biochemical Sciences
Elsevier/North-Holland Biomedical Press, Amsterdam
Benesch, R. (1978) *3*, N 126 [25B]
Huber, M. (1979) *4*, p. 271, Fig. 7 [42A].

Seminars in Hematology
Grune and Strutton, New York
Rachmilewitz, E.A., (1974) *11*, p. 453, Fig. 5 [26B].

Annual Reviews of Medicine
Annual Reviews Inc., California
Stamatoyannopoulos, G., Bellingham, A.J., Lenfant, C. & Finch, C.A. (1971) *22*, p. 224 Fig. 1 [27A].

Annual Reviews of Biochemistry
Annual Reviews Inc. California
Bennett, V. (1985) *54*, p. 283, Fig. 1 [253C].
McIntosh, J.R. & Snyder, J.A. (1976), *45*, p. 706 Fig. 1 [254A].

Scientific American
W.H. Freeman, Oxford and San Francisco
Grobstein, C. (1977) July, p. 30 [112B].
Brown, M.S. & Goldstein, J.L. (1984), Nov. p. 55 [224] p. 56, [261A].
Hinkle, P.C. & McCarty R.E. (1978), March, p. 110 [178], p. 106, [179A].
Rothman, J.E. (1985), Sept. p. 86 [247].
Dunant, Y. & Israel, M. (1985) April p. 42 [258A, B, C].
Lodish, H.F. & Dautry-Varsat A. (1984) May p. 51 [261A], p. 52 [263].

Biochimica et Biophysica Acta
Elsevier/North Holland, Biomedical Press, Amsterdam
Lotan, R. & Nicolson, G.L. (1979) *559*, p. 329 [251B].
Kagawa, Y. (1978) *505*, p. 47 [256A, B].
Small, D.M. Penkett S.A. & Chapman D. (1969) *176*, p. 178 Fig 7 [227B].

Biomedicine
Springer International, Berlin, Heidelberg, New York
Maclouf, J., Sors, H. & Rigaud, M. (1977) *26*, 362 [241].

Haemoglobin and Red Cell Structure and Function (Brewer, G.J. ed.)
Plenum Press, London and New York (1972).
Brenna, O., Luzzana, M., Pace, M., Perrella, M., Rossi, F., Rossi-Bernardi, L. & Roughton, F.J.W. p. 20, Fig. 1 [25C].

Advances in Protein Chemistry
Academic Press, N. York & London (1981)
Richardson, J.S. *34*, p. 254, Fig. 71 [52], p. 262, Fig. 73, p. 263, Fig 74, p. 266, Fig. 77 [51].

Papers in journals

Journal of Molecular Biology
Academic Press, London and New York
Josephs, R., Jarosch, H.S. & Edelstein, S.J. (1976) *102*, p. 409, Fig. 6d [26A].
Valentine, R.C. & Green, N.M. (1967) *27*, p. 615 [41B].
Sigler, P.B., Blow, D.M., Matthews, B.W. & Henderson, R. (1968) *35*, p. 143 Fig. 6 [59A].
Rich, A. (1961), *3*, p. 483 Fig. 2 [37A].

Biochemical Education
International Union of Biochemistry and Pergamon Press, Oxford
Hall, L. & Campbell, P.N. (1979) 7, p. 57 [114A, 113A, 115B].
Henderson, J.F. (1979) 7, p. 52, Fig. 2 [129B].
Smith, I. (1980) *8*, p. 1 [38].
Science
American Association for the Advancement of Science
Britten, R.J. & Kohne, D.E. (1968) *161*, p. 530, Figs. 1 and 2 [84B & C].
Histochemistry & Cytochemistry
Elsevier North Holland Inc. New York
Keller, E.B., Zamecnik, P. & Loftfield, R.B. (1954) *2*, p. 378 [95A].
FEBS Letters
Elsevier/North Holland Biomedical Press, Amsterdam
Campbell, P.N. & Blobel, G. (1976) 72, p. 216, Fig. 1 [108B].
Cell
M.I.T. Press, Cambridge, Mass.
Lai, E.C., Stein *el al.* (1979) *18*, p. 834, Fig. 6 [116A].
European Journal of Biochemistry
Springer Verlag, Berlin
Krisman, C.R. & Barengo, R. (1975) *52*, p. 122, Fig. 7 [212B].
Proceedings of the National Academy of Sciences, U.S.A.
The National Academy of Sciences, Washington
Palade, G.E. (1964) *52*, p. 617, Fig. 2 [177A].
Silverton, E.L. (1977), *74*, p. 5142, Fig. 3 [39B].
Philosophical Transactions of the Royal Society B.
The Royal Society, London.
Evans, P.R., Farrants, G.W. and Hudson, P.J. (1981) *293*, p53 Fig. 2B [64B].
Journal of Cell Biology
The Rockefeller University Press, New York
Fernandez-Moran, H., Oda, T., Blair, P.V. & Green, D.E. (1964) *22*, p. 73, Figs., 6 and 7 [177B].
Osborn M., Webster, R.E. & Weber, K. (1978) 77, R. 29, [254B].
Nature, London
Macmillan Journals Ltd. London
Arnone, A. (1972) *237*, p. 148 [23B].
Williams, A.F. (1984) *308*, p. 12 [45B].
Poorman, R.A. *et al.* (1984), *309*, p. 468 [64A].
Ungewickell, E. & Branton, D. (1981) *289*, p. 420, Fig. 3 [265B].
Mishina, M. *et al* (1985) *313*, p. 364 Fig. 1 [267A].
Fig. 3. Table 1 [268A].
Biochemical Journal
The Biochemical Society
Andrews, P. (1964) *91*, p. 222 [29C].
Journal of Biological Chemistry
American Society of Biological Chemists Inc.
Rosenberg, L., Hellmann, W. & Kleinschmidt, A.K. (1975) *250*, p. 1877, Fig. 1 [138B].
Molecular & Cellular Endocrinology
Elsevier North Holland, Amsterdam.
Pilkis, S.J. *et al.* (1982) *25*, p. 245 Fig. 10 [168B].
Immunology Today
Elsevier North Holland Inc. Amsterdam
Brodsky, F.M. (1984) *5*, p. 350 Figs. 2 & 4 [264, 265A].

b.

We wish to thank the following for detailed help with the diagrams indicated, or for allowing us to use diagrams already prepared.

1 Pharmacia Fine Chemicals AB, Box 175, S-7514, Uppsala 1, Sweden. [29A, B; 32A]
2 Boehringer Mannheim, P.O. Box 310120, 6800 Mannheim 31, W. Germany. [72A, B]
3 Dr Elinor M. Steen and Dr. Alberto N. Dutra [2]
4 Dr A.L. Miller and Miss Laura Worsley [33]
5 Professor P. McLean and Dr. M.R. Ball [184B]

6 Dr G.L. Mills [223B]
7 Dr G. Allt [257A]
8 Dr J.L.H. O'Riordan and Dr L.M. Sandler [239B]
9 Bio-Rad Laboratories, Watford, Herts, England [31B]
10 Dr B.R.F. Pearce, Lab. Mol. Biol. Cambridge, England [265B]
11 Professor R. Carrell, Dept. of Clin. Biochem. Christchurch Clin. Sch. Otago, New Zealand, [34C, 35]
12 Dr A. Moshtaghfard [30B]

c.
We also thank Dr D.A. Bender for reading and criticising the sections on nitrogen metabolism and coenzymes, Professor R.K. Craig for his helpful comments on the molecular biology sections, Professor R.P. Ekins for his views on thermodynamics and equilibra and the many other colleagues who have offered opinions at various times.

1

CELLULAR BASIS OF BIOCHEMISTRY

A

Types of living cells

Living cells may be subdivided into two groups.

Prokaryotes e.g. bacteria (shown on the left) and *Eukaryotes* e.g. animals, plants, fungi, protozoa shown on the right.

Note absence of mitochondria/chloroplasts and other organelles in bacteria. Chloroplasts are confined to plants.

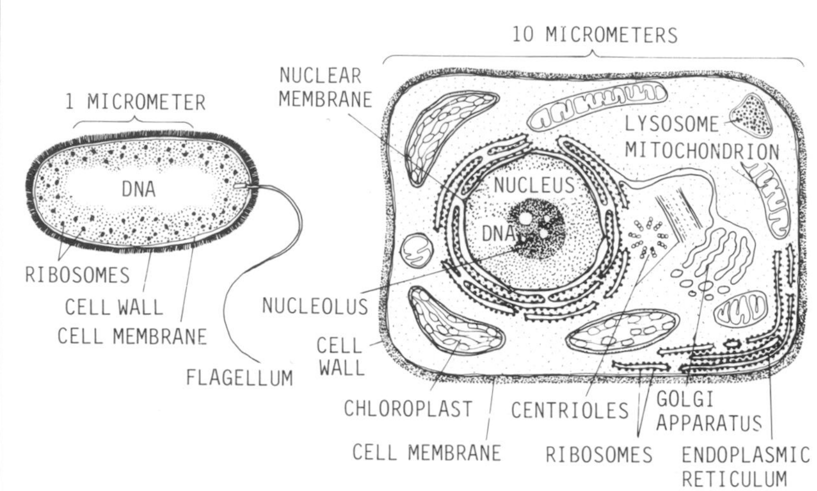

B

Types of biochemical preparations

Living cells may be studied at various levels of organization.

In assessing the physiological significance of a biochemical finding there is usually no substitute for the so-called intact animal. It is, however, difficult to control the various parameters in the whole body and so biochemists resort to other systems such as the incubation of single cells in carefully defined media. Organ explants (tiny pieces of tissue) have an advantage in that they can be effectively aerated with oxygen. The perfusion of intact organs represents a half-way stage.

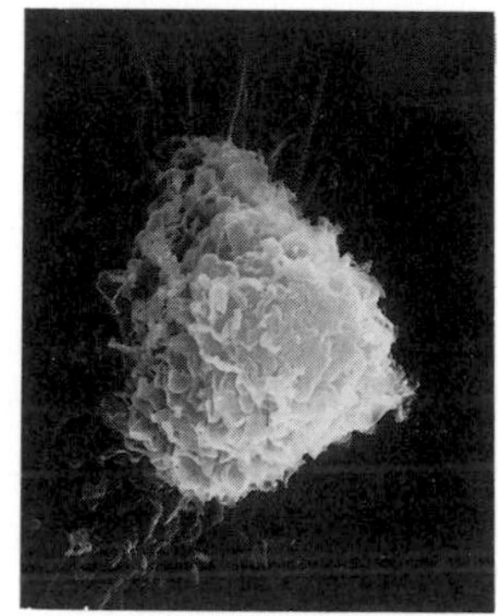

Single cell, eukaryotic cells in culture

Scanning electron micrograph of a mouse peritoneal macrophage.

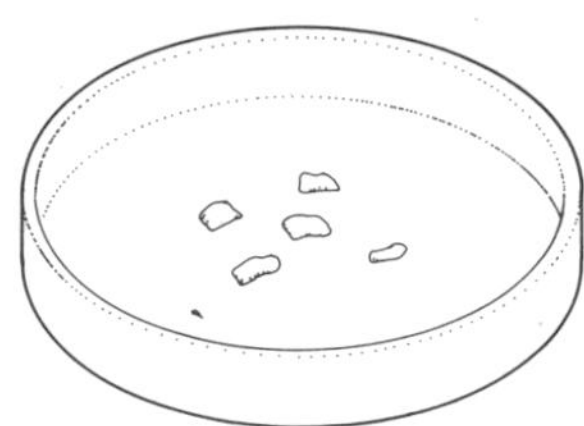

Organ explants, and slices of tissue

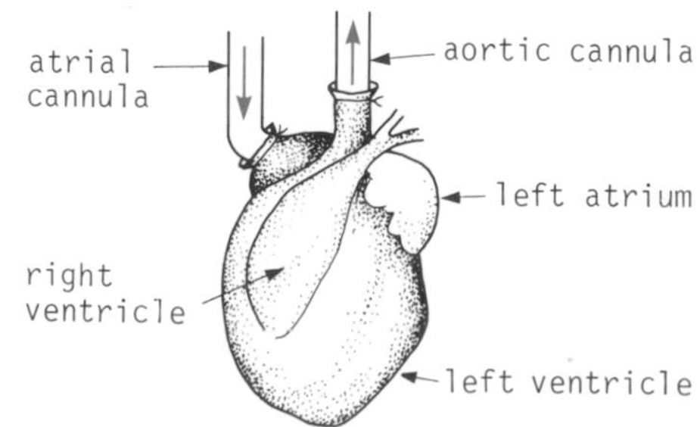

Intact organs

Perfusion: It is often convenient to study the metabolism of an intact organ under controlled conditions, e.g. a heart or liver.

A

Subcellular fractionation

Cells can be disrupted and the constituents separated by differential centrifugation.

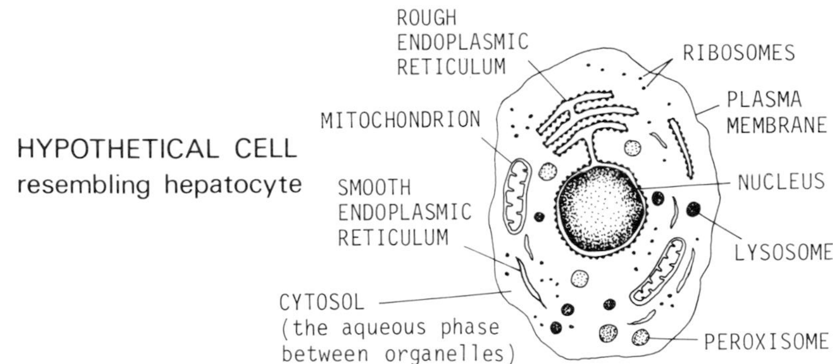

B

The disruption is usually carried out in an isotonic medium (this medium may be a salt solution but is often 0.25 M sucrose).

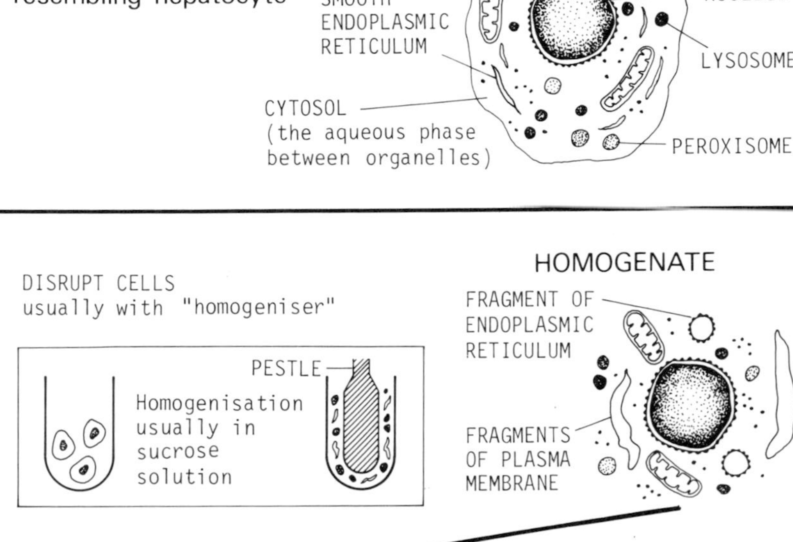

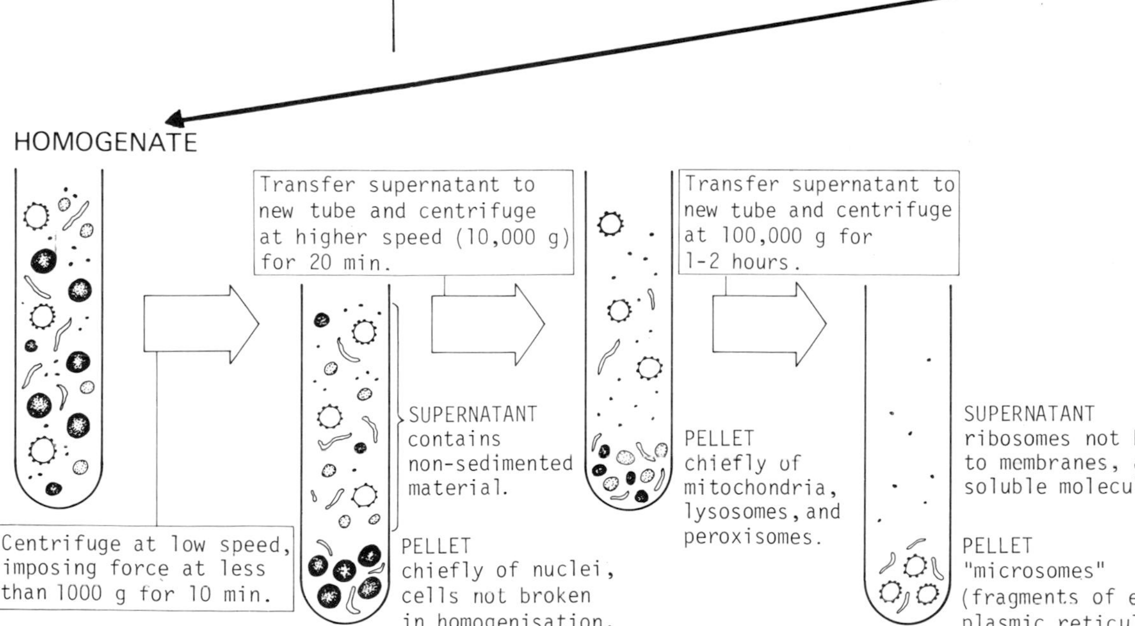

A

Biochemical functions of organelles

Chemical analysis of cells and tissues has revealed that the molecules of which they are composed can be grouped into the classes

1. Nucleic acids
2. Proteins
3. Carbohydrates
4. Lipids

The metabolic pathways that are concerned with the synthesis and breakdown of these molecules are characteristically located in certain cell organelles.

The rat liver cell
(Calculations made by Professor Drabkin)

Liver cell	Diameter	19.3 μm
	Volume	3081 μm^3 (as tetrakaidecahedron)
	Number per gram	3.25×10^8
Nuclei	Diameter	7.7 μm
	Volume	239 μm^3
	Nuclear fraction	7.76%
	Number per cell	1.0 (assumed)
Mitochondria	Diameter	0.35 μm to 1.2 μm (mean = 0.83 μm)
	Volume	Volume, as prolate spheroid for mean of 0.83 μm = 0.483 μm^3
Lysosomes	Mitochondrial fraction	21.1%
	Number per cell	Mean = 1343
Microsomes	Fraction of Cell	10.85%
Ribosomes	Diameters	0.023 μm = 230 Å and 0.0140 μm (140 Å) major and minor axes respectively of oblate spheroid
	Volume	$3.871 \times 10^{-6} \mu m^3$
	Fraction of cell	1.04%
	Number per cell	
	Calc. by e.m.	8.65×10^6
	Calc. from RNA	8.19×10^6

B

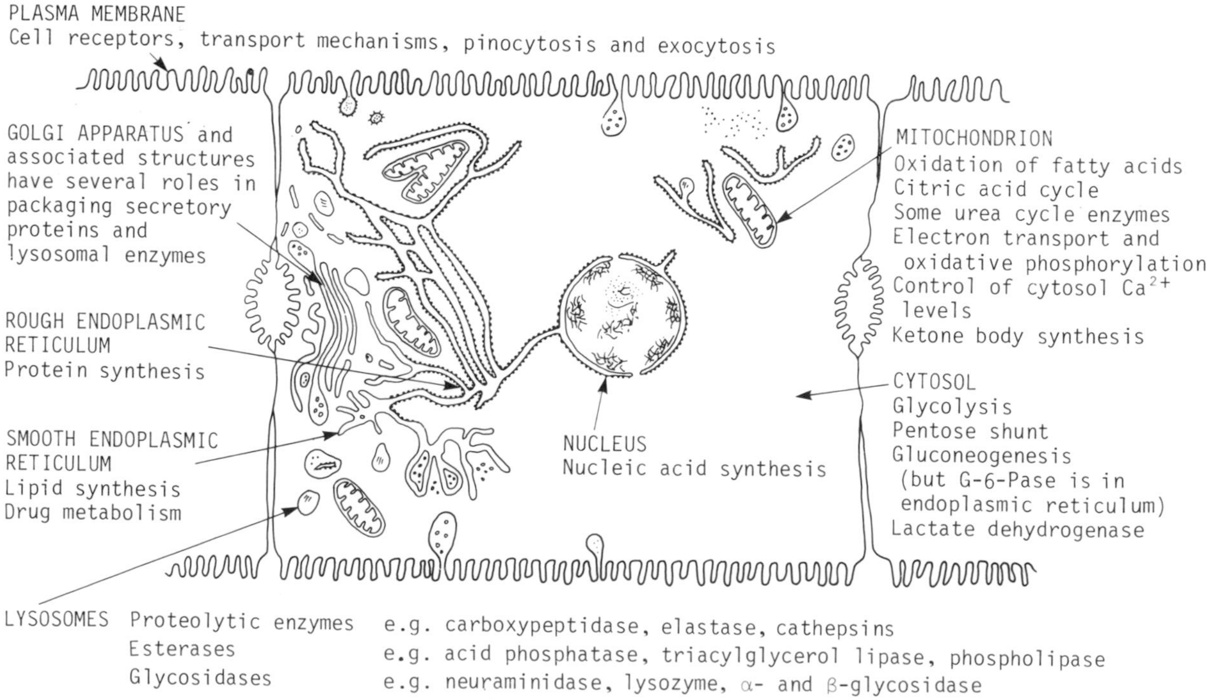

A

The definition of cell organelles and subcellular fractions

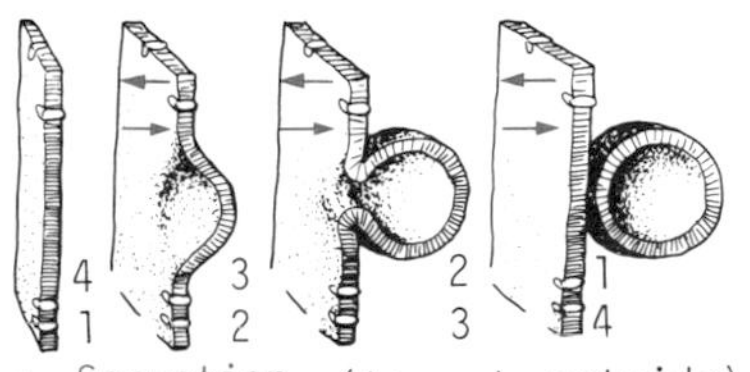

Plasma membrane. This refers to the outer membrane of the cell in contact with the extracellular fluid. It has a particular role in secretion.

Cytoplasm. Strictly this consists of all the components of a cell apart from the nucleus. It is often used, however, to mean all the components apart from the nucleus and the mitochondria.

Cytosol. This is that part of a cell in which we cannot readily detect organized components (organelles). It is not the same as the soluble supernatant obtained by disrupting a cell and centrifuging out all the particulate components (also often referred to as the cytosol). Such a soluble fraction will contain various soluble components extracted from the various organelles.

B

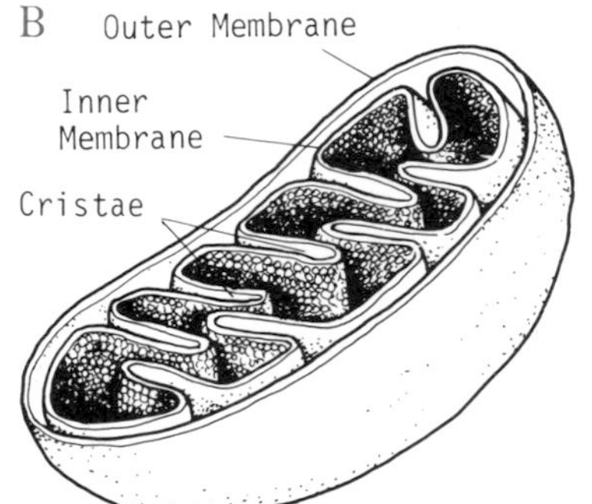

Mitochondria

Mitochondria are unique organelles within the cytoplasm, lacking any direct structural relationship with other organelles and containing their own DNA. Besides producing energy, they also help to control the level of calcium in the cytoplasm.

Mitochondrial membranes are about 6.5 nm thick with the inner membrane folded to form cristae. The inner surfaces of the cristae are closely packed with 8.5 nm particles, the sites of oxidative phosphorylation (see p. 176).

C

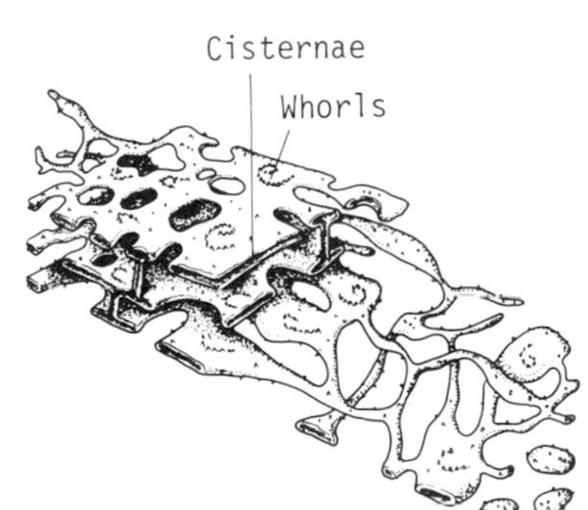

The endoplasmic reticulum

The endoplasmic reticulum is a network of membranes throughout the matrix of the cytoplasm. These membranes form a complex arrangement of connecting vesicles and tubules or large flattened sacs. The membranes run parallel to each other creating channels which are called cisternae. Large areas of membrane are thus created and it is estimated that the surface area of the endoplasmic reticulum in 1 ml of cytoplasm is 11 m^2. The surface may bear ribosomes (rough surfaced endoplasmic reticulum) or may not (smooth surfaced endoplasmic reticulum). The ribosomes are often characteristically arranged as 'whorls'.

Microsomes

The microsome fraction is obtained by disrupting a cell and separating the components by differential centrifugation. That fraction which sediments more slowly than the mitochondrial fraction but is particulate is defined as the microsomal fraction. In liver extracts it consists predominantly of fragments of the endoplasmic reticulum. This is not necessarily the case with other types of cell and in all cases it is necessary to determine the morphological components of the microsome fraction before meaningful comparisons may be made.

A

The nucleus

The nucleus contains the DNA organized into separate chromosomes. In a human diploid cell there are 23 pairs of chromosomes giving a diploid number of 46, with half as many in a haploid cell. The nucleolus is a region within the nucleus in which the genes for three of the four ribosomal RNA molecules are located. The nuclear membrane is double-layered with a perinuclear space. Transfer of substances between the cytoplasm and nucleus is through the nuclear pore which is a complex of proteins arranged in an octagonal array with a central hole.

The lamina is a structure that surrounds the inner nuclear membrane and contains three proteins the lamins A, B, C. The lamina interconnects with the chromosomes.

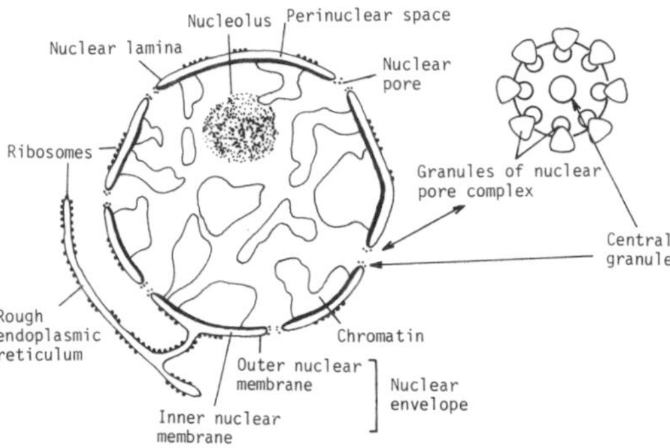

B

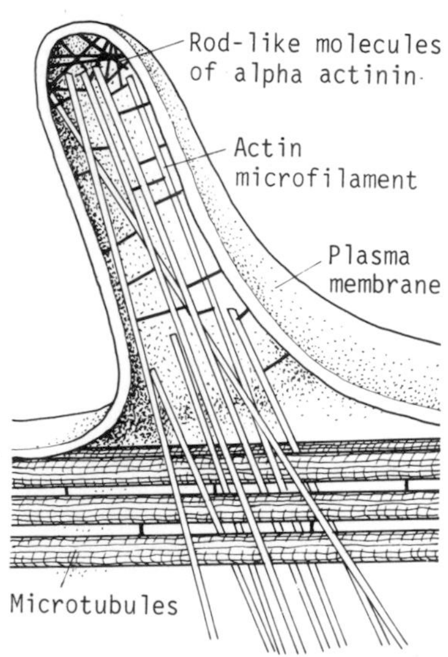

A microvillus

Microtubules and microfilaments

The high resolution that has been achieved with the high-voltage electron microscope has revealed that there is a structural lattice even within the cytosol. This lattice involves the *microtubules* which traverse the cytoplasm, the *microfilaments* that are adjacent to the plasma membrane and a web of finer filaments called the *microtrabecular lattice*. In the diagram of a *microvillus*, such as might occur on the lining of the small intestine, the membranes and fibrillar components involved in the structure are shown.

Lysosomes

Lysosomes are membrane-bound vesicles containing a wide range of hydrolytic enzymes. They are central components in the *intracellular digestive system* so that, for example, substances and components of the cell that are to be degraded form *phagosomes* which fuse with the *lysosomes* to become *secondary lysosomes*.

Marker enzymes

Confirmation of the nature of the morphological constituents of the various subcellular fractions isolated by differential centrifugation can be obtained either by electron microscopy of the pellets or by biochemical analysis. Thus one can determine the DNA or RNA/protein ratio. Enzymic analysis is also useful, based on the principle that one enzyme is only to be found associated with one particular morphological constituent of the cell (usually but not always true) and that the enzymic make up of a particular constituent is unique (e.g. all mitochondria are identical, probably true). Such enzymes are known as *marker enzymes*. Examples are glucose 6-phosphatase for endoplasmic reticulum and succinic dehydrogenase for mitochondria.

2

THE PROTEINS

1. THE CHEMISTRY OF AMINO ACIDS AND PEPTIDES

Structure of amino acids

Amino acids have the general structure

$$^{+}H_3N\overset{\displaystyle R}{\overset{|}{C}}HCOO^{-}$$

The internationally approved three letter and single letter abbreviations for each amino acid are indicated.

The α-carbon is optically active in α-amino acids other than glycine. The two possible isomers are termed D and L. All naturally occurring amino acids found in the proteins are of the L configuration (see page 10).

All the common amino acids except for proline have the same general structure, in that the α-C bears a COOH- and an NH_2-group, but differ with respect to their 'R' groups. The R groups confer on the amino acids their respective characteristic properties. The amino acids are grouped according to the nature of their R groups as follows.

1. Non-polar or hydrophobic R Groups

L-Alanine (Ala) A

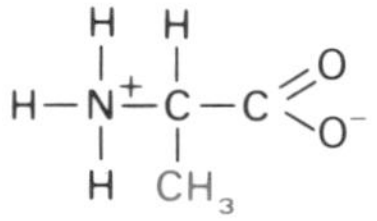

L-Valine (Val) V

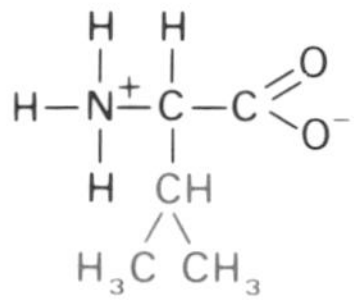

L-Leucine (Leu) L

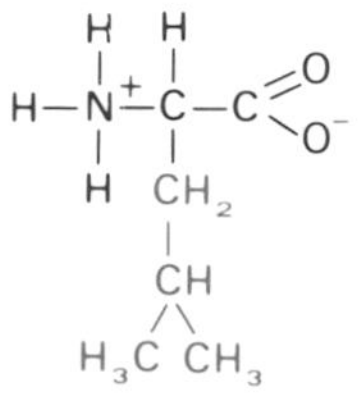

L-Isoleucine (Ile) I

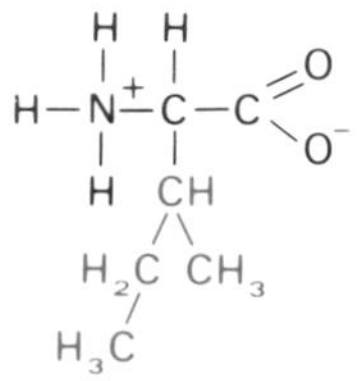

Methionine (Met) M

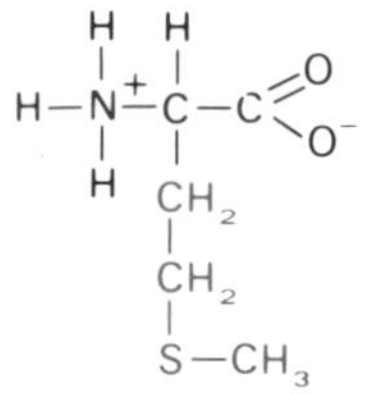

L-Proline (Pro) P

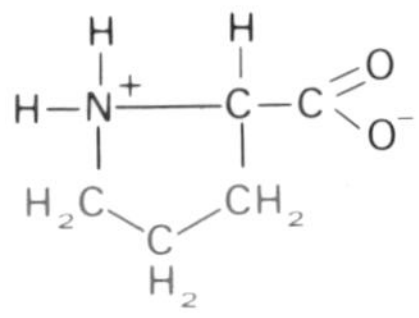

L-Phenylalanine (Phe) F

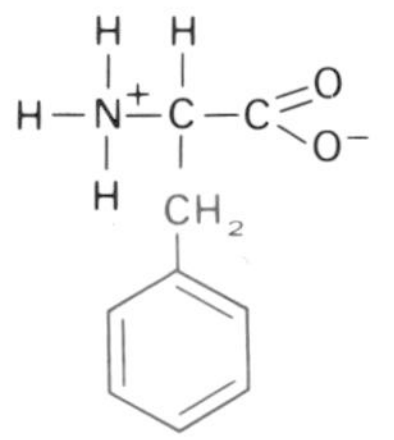

L-Tryptophan (Trp) W

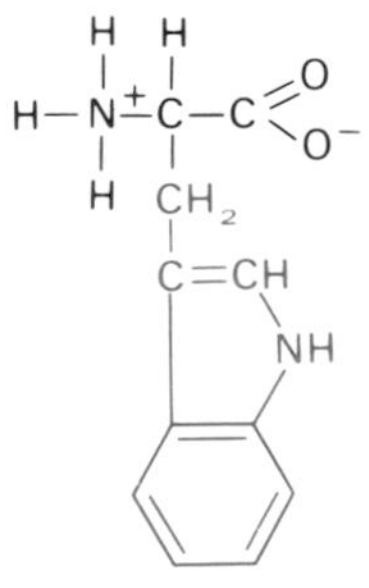

2. Negatively charged R groups at pH 6–7

L-Aspartic acid (Asp) D

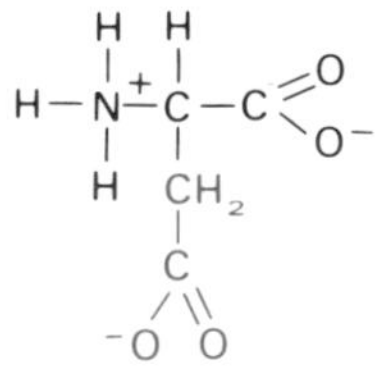

L-Glutamic acid (Glu) E

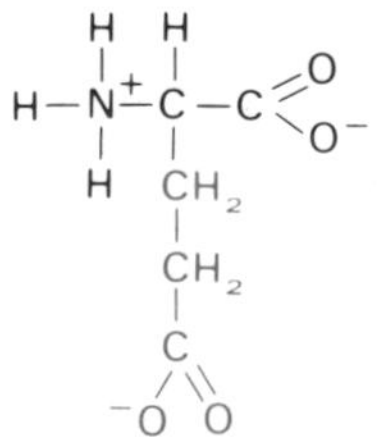

3. Uncharged or hydrophilic R groups

L-Asparagine (Asn) N

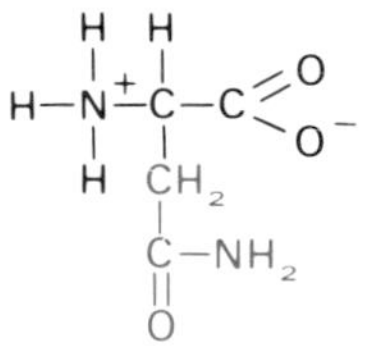

L-Glutamine (Gln) Q

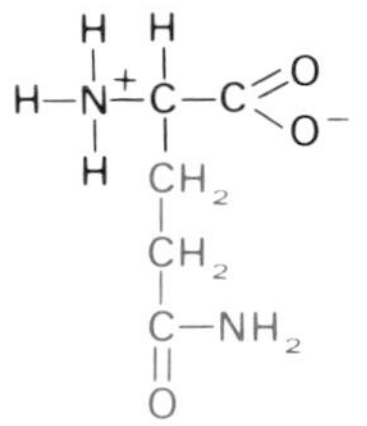

Glycine (Gly) G

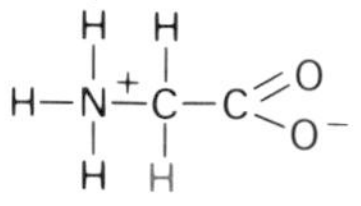

L-Serine (Ser) S

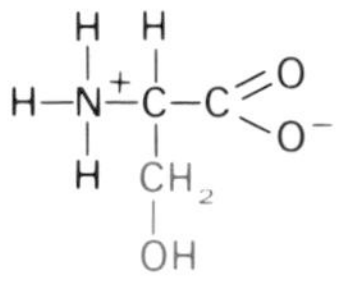

L-Threonine (Thr) T

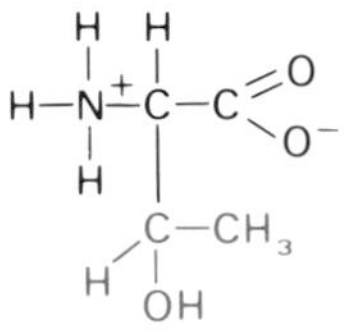

L-Tyrosine (Tyr) Y

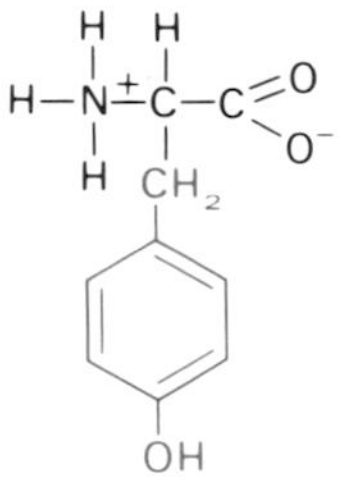

L-Cysteine (Cys) C

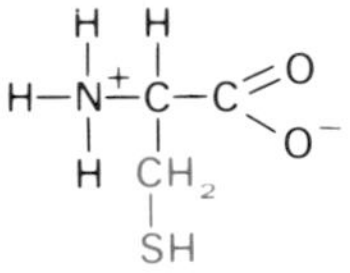

4. Positively charged R groups at pH 6–7

L-Lysine (Lys) K

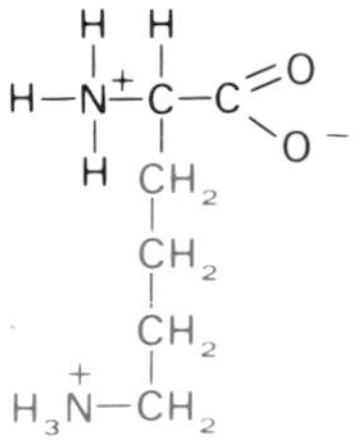

L-Arginine (Arg) R

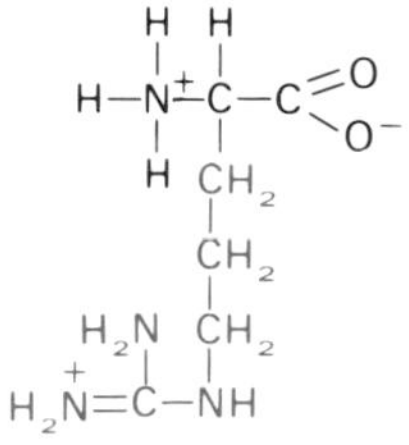

L-Histidine (His) H

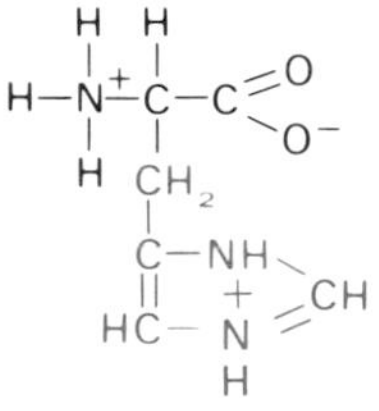

The term 'acid' and 'base' as applied to amino acids.

Acids are defined as proton donors and bases as proton acceptors. It follows that at pH 6–7 as shown, amino acids in group 2 are present as a free base (an anion) and those in group 4 as a free acid (a cation). The terms 'acidic' and 'basic', as applied to amino acids, should therefore be used with caution, since they refer to the protonated forms of group 2 or the unprotonated forms of group 4. A compound such as an amino acid which carries both basic and acidic groups is referred to as *amphoteric*.

A

Asymmetry in Biochemistry

Chirality
(from the Greek word 'cheir' for 'hand' — the left and right hands are mirror images of each other). Asymmetry in molecular structure is of great importance in biochemistry.
A chiral molecule possesses at least one asymmetric centre, that is, a carbon atom to which are joined 4 groups different from each other.

The amino acid alanine may exist in two forms, denoted D-alanine and L-alanine.

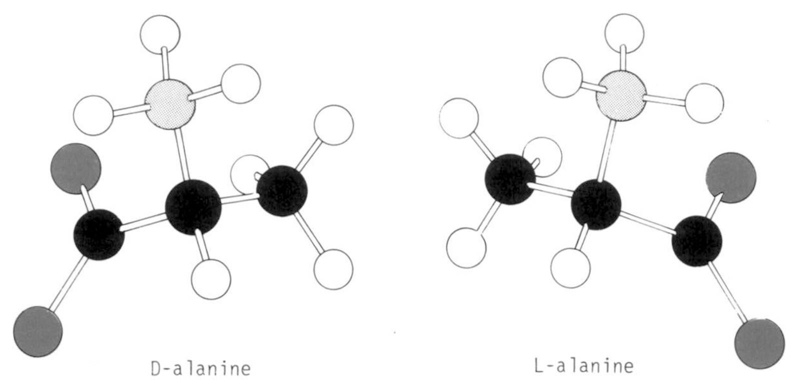

The amino acids contained in mammalian proteins are of the L form. Sugars also are chiral molecules. D-Sugars predominate in mammalian carbohydrates (see p. 134). Red denotes carboxyl group oxygen atoms, grey is the amino group nitrogen.

B

Non-chiral asymmetry

Even if a molecule is not chiral, it may contain identical groups which are nevertheless sterically distinguishable. The classic biochemical example is citric acid. Although this has a plane of symmetry, the centre carboxyl group and the hydroxy group can be held in such a way that the two $-CH_2COOH$ groups can be distinguished (see p.169). This can be seen in a simplified representation of a hypothetical molecule, which interacts with an enzyme with specific binding sites for different groups in the molecule.

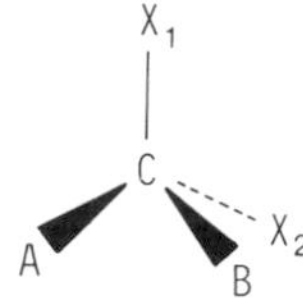

If A and B are held in space on a surface, then X_1 and X_2 can be distinguished. Such a molecule is termed 'pro-chiral' in that it can be made chiral by changing a group on only one of the central carbon bonds.

C

R and S convention
A chiral centre can be denoted R or S.

> The order of priority of some biochemically important groups is $-SH$ (highest), $-OH$, $-NH_2$ $-COOH$, $-CHO$, $-CH_3$, $-H$ (lowest).

The method for ascribing the R or S designation to a centre is as follows.

1. List functional groups in order of priority (see examples of the priority sequence in the margin). Then orientate the molecule so that the group of the lowest priority points away from the observer.
2. If the order of priority (high to low) of the remaining groups is clockwise, the centre is R. If anti-clockwise, it is S.

Thus, L-alanine (above) has the S configuration.

A

Ionic properties of amino acids

The amphoteric* properties of amino acids account for their separation on electrophoresis on paper at pH 6.0.

*Property of behaviour either as an acid or a base.

The amphoteric nature of α-amino acids determines that, in the absence of other acids or bases, the carboxyl and amino groups are both ionized fully, giving rise to the term zwitterion (German zwitter = hybrid, or hermaphrodite).

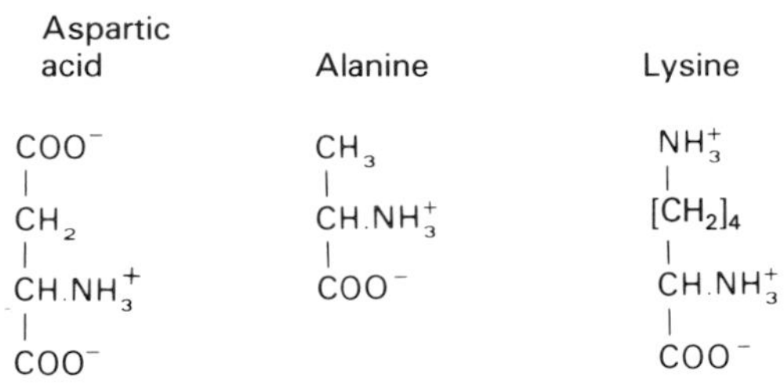

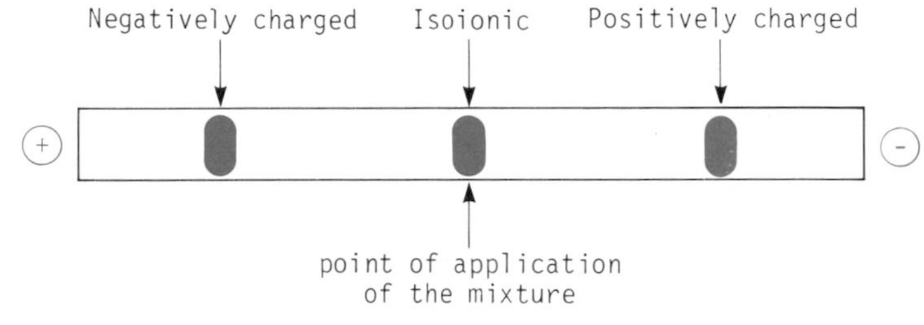

B

The buffering capacity of alanine is shown by titration with acid and alkali.

The term isoionic implies that there is no net electric charge on the molecule.

pK is defined on page 66

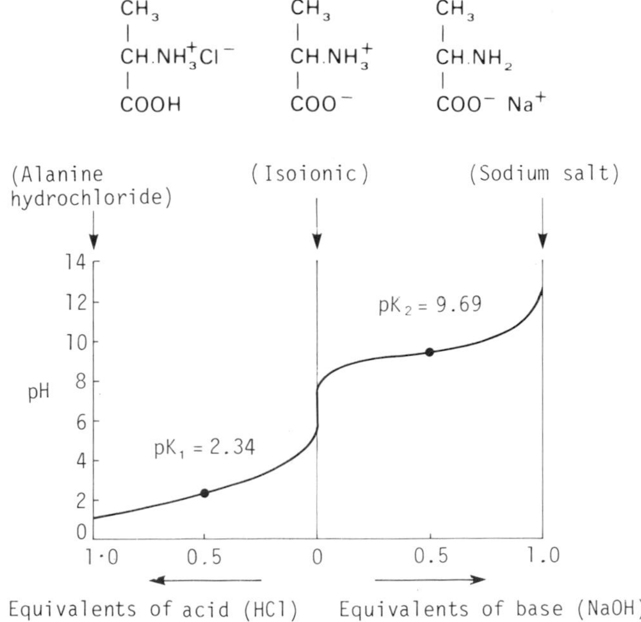

The titration curve of alanine

A

Histidine is an important amino acid in proteins since it contributes to their buffering capacity.

The basic groups of histidine may partially ionize at physiological pH.

The imidazole group of histidine is only weakly basic having a pK_a of 6.00 and therefore exists as a mixture of the protonated and dissociated forms in solutions at physiological pH.

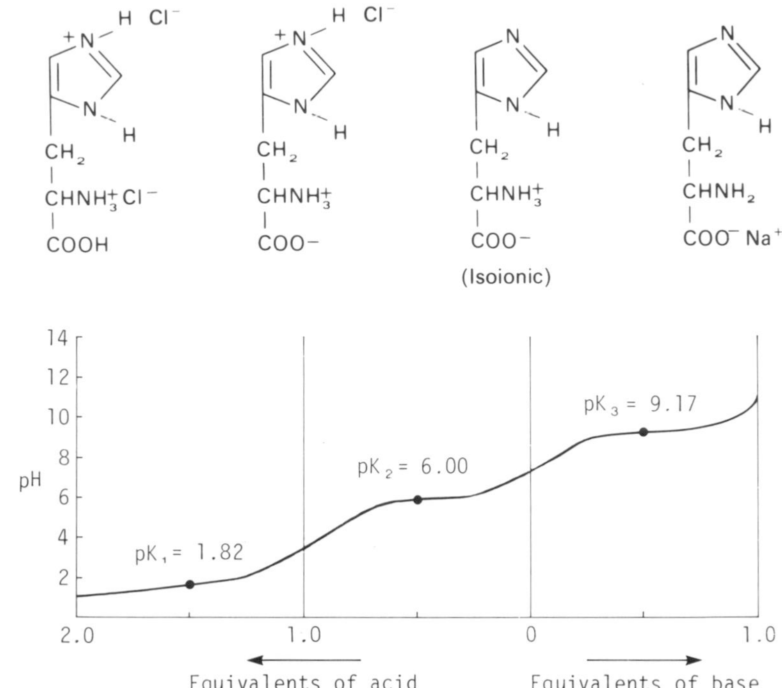

The titration curve of histidine

B

Amino acids can be separated by ion exchange chromatography.

In addition to permitting the separation of amino acids by electrophoresis, their ionic properties also permit their separation by ion-exchange chromatography. Ion-exchange is performed using a resin to which positively-charged groups (anion-exchange resin) or negatively-charged groups (cation-exchange resin) are covalently bound and thus immobilized. Ions passed down a column of such a resin bind competitively to the charged groups.

At pH 3

$$\boxed{\text{RESIN}}\text{—}SO_3^- \ldots {}^+H_3NR_1 + {}^+H_3NR_2 + Cl^-$$

Amine 1 can be displaced by amine 2 :—

$$\boxed{\text{RESIN}}\text{—}SO_3^- \ldots {}^+H_3NR_2 + {}^+H_3NR_1 + Cl^-$$

or by other cations, such as Na^+ :—

$$\boxed{\text{RESIN}}\text{—}SO_3^- \ldots Na^+ + {}^+H_3NR_1 + Cl^-$$

(Sulphonic acids are strong acids, and thus remain ionized at pH 3.)

As the various ions differ in their affinity for the resin, some are displaced more readily than others by an ionic solution (e.g. NaCl) passed down the column (eluent) and so appear more rapidly in the solution flowing out of the column (eluate).

A

The peptide bond

The structure of peptides.

The peptide bond is formed by the interaction of two amino acids with the elimination of water between the NH_2 and COOH groups.

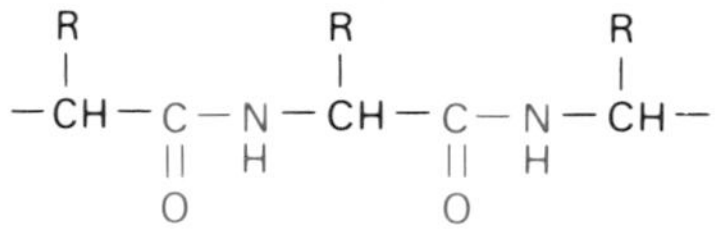

```
  R           R           R
  |           |           |
—CH—C—N—CH—C—N—CH—
     ||  H       ||  H
     O           O
```

Proline can also participate in a peptide bond

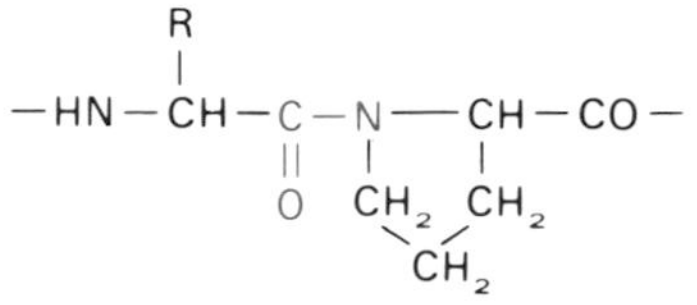

```
          R
          |
—HN—CH—C—N——CH—CO—
          ||  |     |
          O  CH2  CH2
                \  /
                 CH2
```

but there is no H available on the $\underset{\underset{O}{\|}}{C}\text{–}\underset{|}{N}\text{–}$ for hydrogen bonding.

Sharing of electrons between the —C=O and —C—NH— bonds confers rigidity on the peptide bond.

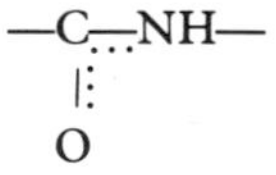

```
—C⋯NH—
 |:
 O
```

This has important implications for protein structure (see p. 19 B).

B

The Biuret reaction

Biuret has the formula NH_2 CONHCONH_2 and, therefore, is a simple substance containing a peptide bond. When Biuret is treated with $CuSO_4$ in alkaline solution a purple colour is produced. This is known as the Biuret Reaction and proteins give a strong reaction.

C

Notation used for peptides

H_2N-Tyr-Gly-Gly-Phe-Met-COOH

Enkephalin

In writing the primary structure of a peptide one starts with the NH_2-terminus so that in the three letter code it is abbreviated as above or in the single letter code YGGFM.

Non-covalent bonds in proteins

The R-groups interact in a *polypeptide chain* to create the tertiary structure of a protein.

Proteins are composed of chains of amino acids linked by peptide bonds. These are termed polypeptide chains.

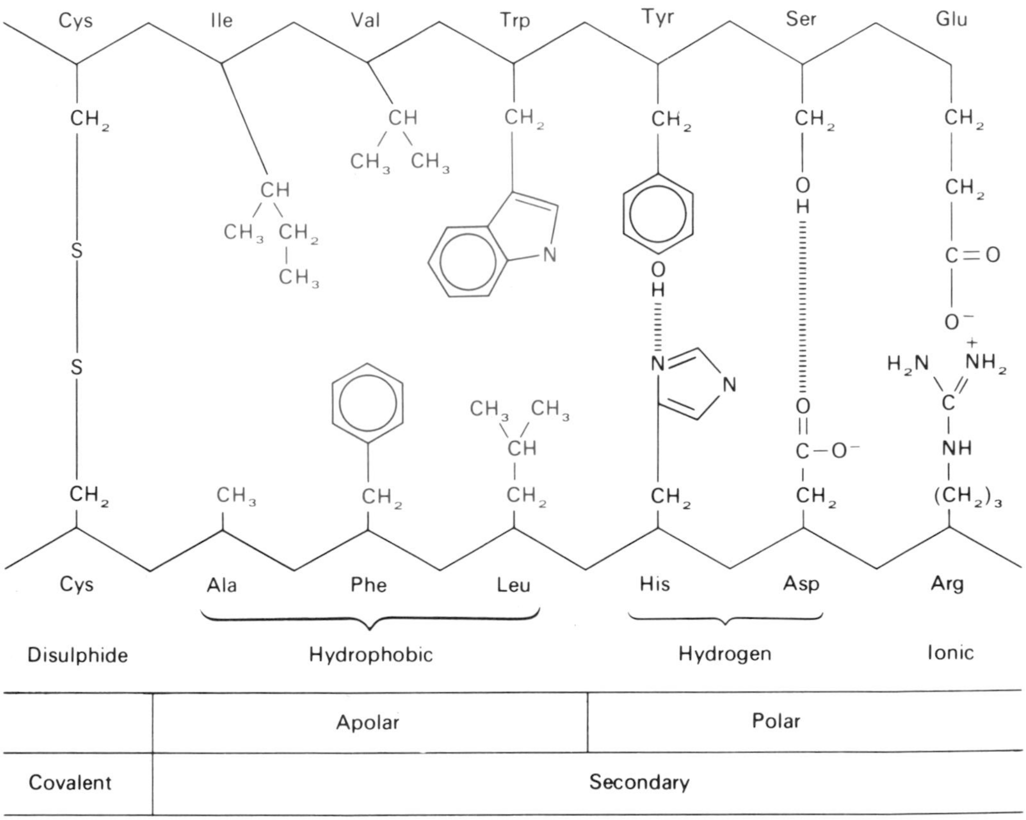

Disulphide	Hydrophobic		Hydrogen	Ionic
	Apolar		Polar	
Covalent	Secondary			

The variety of bonds or interactions which stabilize the tertiary structure of protein molecules is shown. The S–S bonds formed by the oxidation of two sulphydryl groups are *covalent*. The other interactions are *non-covalent*. These may be either *Apolar* i.e. hydrophobic, or *Polar* i.e. hydrogen-bonding and ionic.

Hydrophobic interactions may be due to (a) Van der Waal's interactions which arise from an attraction between atoms due to fluctuating electric dipoles originating from the electronic cloud and positive nucleus; (b) the hydrophobic effect which refers to the tendency of non-polar groups to associate with one another rather than to be in contact with water.

Hydrogen bonds. These arise because when H is linked to O or N there is a shift of electrons that leads to a partial negative charge on the other atom. This produces an electric dipole that can interact with dipoles that exist elsewhere. The commonest H bond is between N-H and C = O as in the α-helix or β-pleated sheet (see pages 19 and 20) but other bonds are possible as shown.

Ionic interaction or salt bridges are formed by the close approach of two atoms of opposite charge e.g. between Glu and Arg as shown.

A

Ionic properties of peptides

A 'basic' protein with a predominance of group 4 amino acids is a cation at pH 7 and an 'acidic' protein with a predominance of group 2 amino acids is an anion at pH 7.

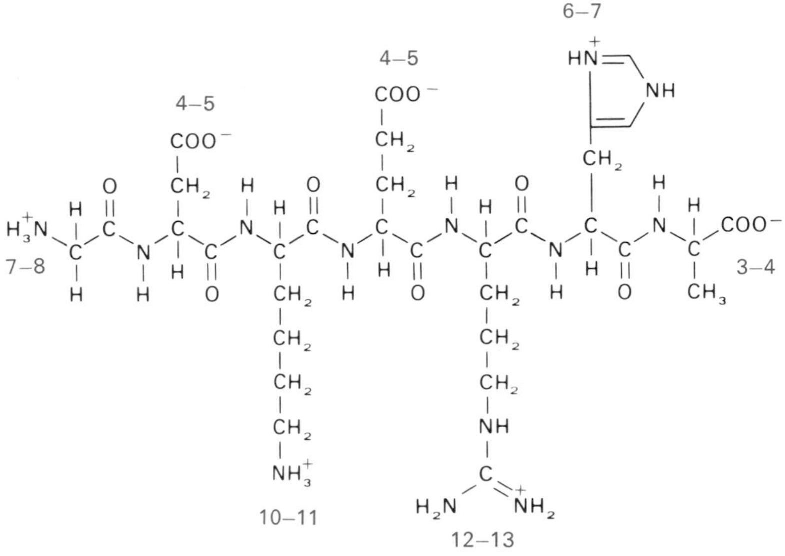

glycyl-aspartyl-lysyl-glutamyl-arginyl-histidyl-alanine

Illustrated is a hypothetical polypeptide containing all the R-groups that normally contribute charges to proteins. The numbers represent the p*K* range of each dissociating group. Histidine is very important since its charge may vary over the physiological range.

B

Physical properties and separation of proteins

Proteins as electrolytes

Isoelectric points of some common proteins

Blood proteins		Miscellaneous Proteins	
Protein	Isoelectric point	Protein	Isoelectric point
α_1-Globulin	2.0	Pepsin	1.0
Haptoglobin	4.1	Ovalbumin	4.6
Serum albumin	4.7	Insulin	5.4
γ_1-Globulin	5.8	Histones	7.5–11.0
Fibrinogen	5.8	Ribonuclease	9.6
Haemoglobin	7.2	Cytochrome *c*	9.8
γ_2-Globulin	7.4	Lysozyme	11.1

The isoelectric point is that pH at which the protein carries a net charge of zero. According to the isoelectric point proteins are described as basic, neutral or acidic depending on whether their overall charge at physiological pH is positive, approximately zero, or negative.

C

The titratable groups of a protein (ribonuclease)

Group	Amino acid involved	No. of residues	p*K*
α-NH_2	N-Terminal	1	7–8
Side-chain COOH	asp, glu	10	4–5
Side-chain NH_2	lys	10	10–11
Guanidyl	arg	4	12–13
Imidazole	his	4	6–7
α-COOH	C-Terminal	1	3–4

A

Methods for sequence determination proteins

The primary structure of a protein can be determined either by the use of proteolytic enzymes or by the use of specific chemical reagents or both.

In a typical investigation, the N-terminus is determined by dansyl chloride, fluorodinitrobenzene or the Edman degradation technique. The Edman degradation can be used directly on the polypeptide chain to determine the amino acid sequence, as the amino acids are released one by one from the N-terminus. In other cases, the protein is split into peptides, either by CNBr to break the chain at methionine residues or by proteolytic enzymes acting at specific amino acid residues (see p. 208). The amino acid sequence of the resulting peptides is then determined, usually by the Edman degradation. The objective is to obtain peptides by the various methods that are overlapping in sequence, thus enabling the order in which individual peptide sequences occur in the protein to be determined. Since it is now possible to determine the base sequence of genes, the amino acid sequence may in certain cases be checked against the base sequence of the structural gene using the genetic code.

B

Chemical reagents

Only one reagent is used commonly to break a polypeptide chain at a specific amino acid and this is *cyanogen bromide* (N≡C-Br) which cleaves adjacent to methionine residues.

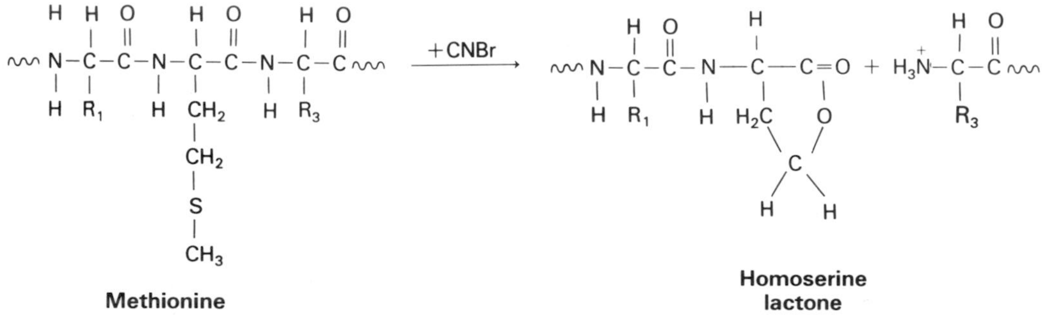

$$\sim N(H)-CH(R_1)-C(=O)-N(H)-CH(CH_2CH_2SCH_3)-C(=O)-N(H)-CH(R_3)-C(=O)\sim \xrightarrow{+CNBr} \sim N(H)-CH(R_1)-C(=O)-N(H)-CH(H_2C)-C{=}O \ (\text{lactone ring via } O-CH_2) + H_3\overset{+}{N}-CH(R_3)-C(=O)\sim$$

Methionine

Homoserine lactone

C

Reactions at the N-terminus

Another procedure is to react the N-terminal residue either with *fluorodinitrobenzene* or *dansyl chloride*. After reaction the peptide is hydrolysed with acid and the DNP amino acid or dansyl amino acid (fluorescent) identified.

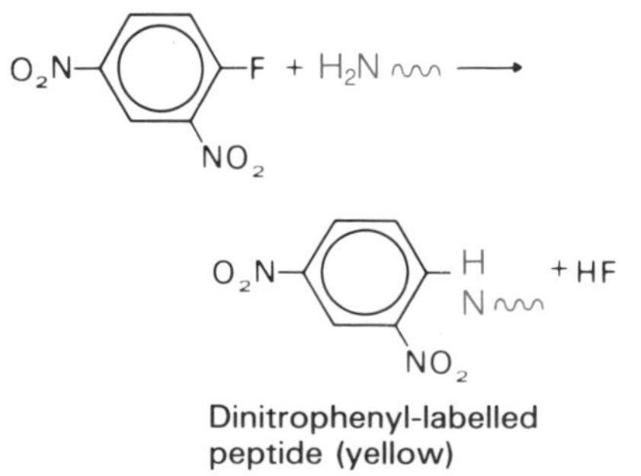

$$O_2N-C_6H_3(NO_2)-F + H_2N\sim \longrightarrow O_2N-C_6H_3(NO_2)-NH\sim + HF$$

Dinitrophenyl-labelled peptide (yellow)

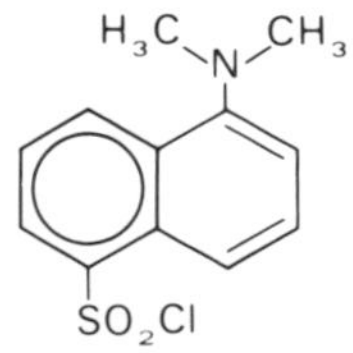

Dansyl chloride: $(H_3C)_2N-C_{10}H_6-SO_2Cl$

Dansyl chloride
(5-Dimethylaminonaphthylsulfonyl chloride)

A

The Edman degradation

In the *Edman degradation* the procedure is repeated on the ever-shortening chain and the procedure can be automated.

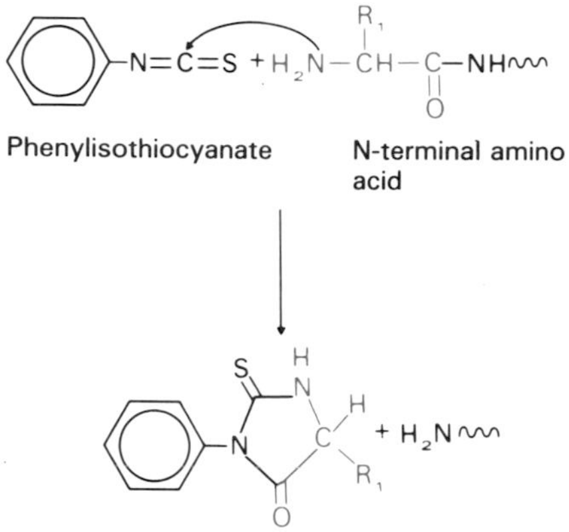

Phenylisothiocyanate + H_2N–CH(R_1)–C(=O)–NH〰

Phenylisothiocyanate — N-terminal amino acid

↓

Phenylthiohydantoin (PTH) of N-terminal amino acid + H_2N〰

The PTH of the amino acid can be identified.

Phenylisothiocyanate is used for the step-wise degradation of a peptide.

The Edman degradation. The labelled amino-terminal residue (PTH-alanine in the first round) can be released without hydrolyzing the rest of the peptide. Hence, the amino-terminal residue of the shortened peptide (Gly-Asp-Phe-Arg-Gly) can be determined in the second round. Three more rounds of the Edman degradation reveal the complete sequence of the original peptide.

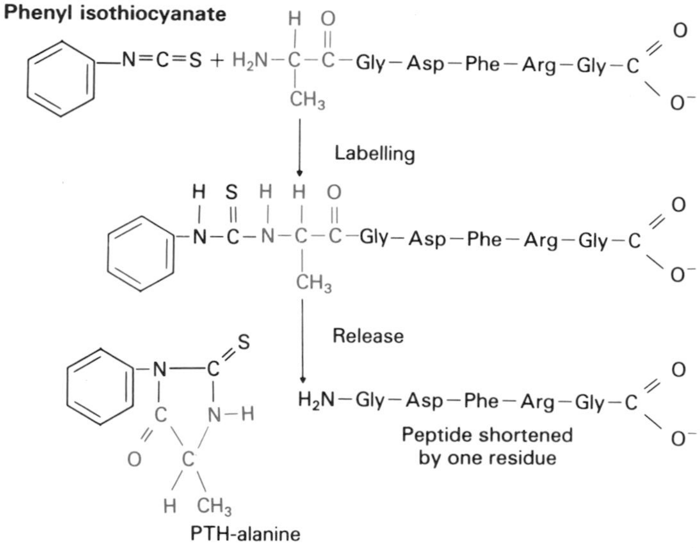

Phenyl isothiocyanate

Ph–N=C=S + H_2N–CH(CH_3)–CO–Gly–Asp–Phe–Arg–Gly–COO^-

↓ Labelling

Ph–NH–C(=S)–NH–CH(CH_3)–CO–Gly–Asp–Phe–Arg–Gly–COO^-

↓ Release

PTH-alanine + H_2N–Gly–Asp–Phe–Arg–Gly–COO^-

Peptide shortened by one residue

B

The use of proteolytic enzymes.

The specificity of the various proteolytic enzymes that may be used to degrade a protein to peptides, the structure of which may then be determined, is shown on page 208.
As an example the action of trypsin is indicated:

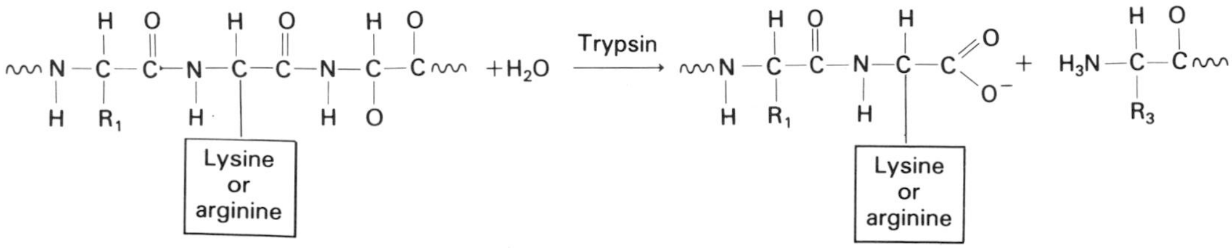

〰NH–CH(R_1)–CO–NH–CH(Lysine or arginine)–CO–NH–CH(R_3)–CO〰 + H_2O —Trypsin→ 〰NH–CH(R_1)–CO–NH–CH(Lysine or arginine)–COO^- + H_3N–CH(R_3)–CO〰

Trypsin hydrolyses polypeptides on the carboxyl side or arginine and lysine residues.

2. PROTEIN STRUCTURE, PROPERTIES AND SEPARATION

A

Hierarchies

The structure of a protein can usefully be considered in four hierarchies.

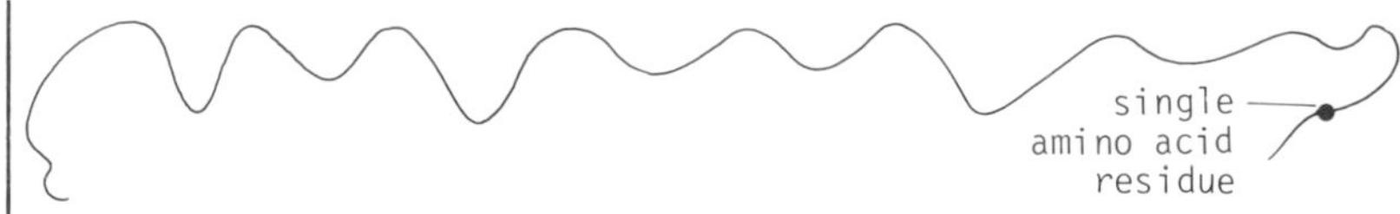

Primary structure describes the order of covalently linked amino acid residues.

B

Role of hydrogen bonds.

Secondary structure describes the way in which certain lengths of polypeptides interact through CO : NH hydrogen bonds either intramolecularly or intermolecularly.

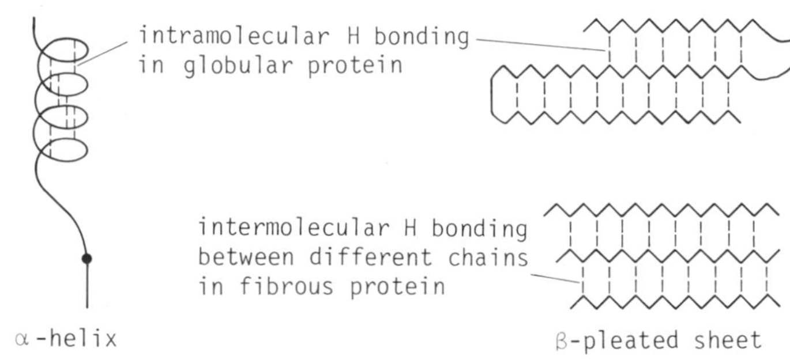

C

Both *β*-pleated sheet and *α*-helix play a role in tertiary structure.

Tertiary structure describes how the chains with secondary structure further interact through the R groups of the amino acid residues to give a 3-D shape.

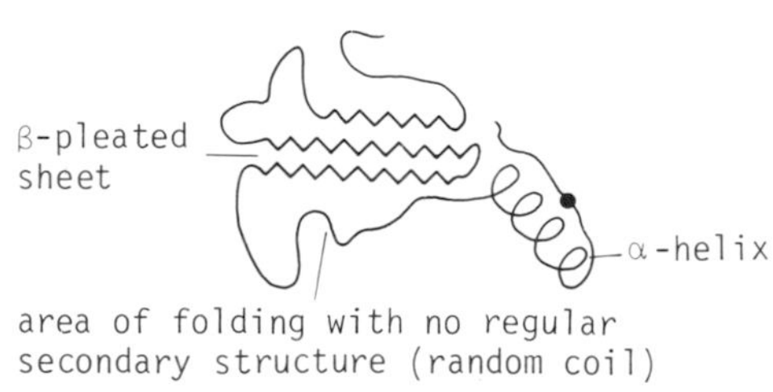

D

A protein with a quaternary structure is composed of several subunits. Such a protein is an *oligomeric* protein.

Quaternary structure describes the interaction, through weak bonds, of the polypeptide subunits. *Dimers* are associations of 2 subunits, *tetramers* are associations of 4 subunits, each subunit being a *monomer*.

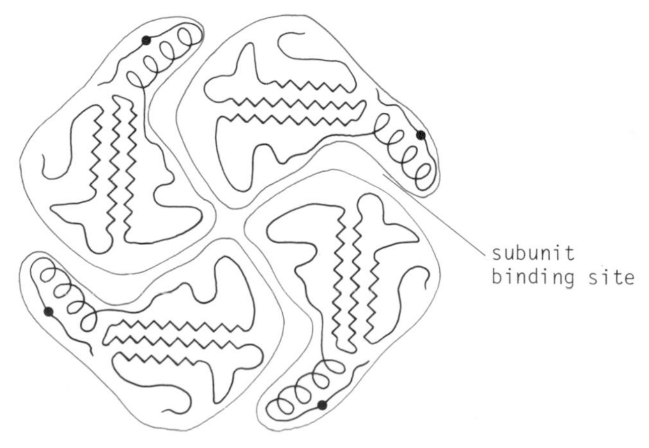

A

Nature of the peptide bond

Rotation in the peptide chain is limited.

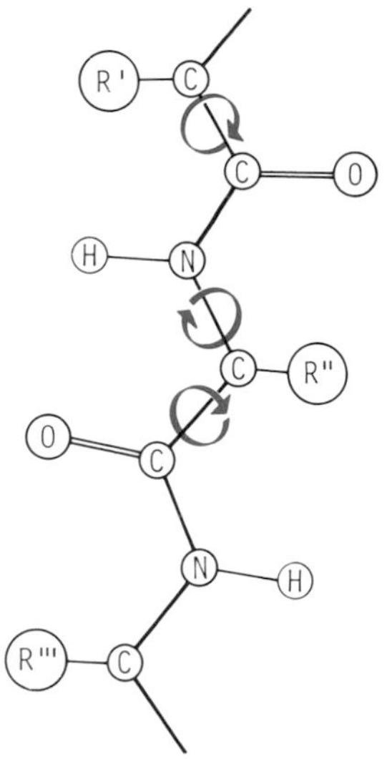

In a polypeptide chain rotation is only possible between certain atoms as indicated. The —CONHC— is planar.

B

The α-helix

Hydrogen bonds in the α-helix help to maintain its conformation.

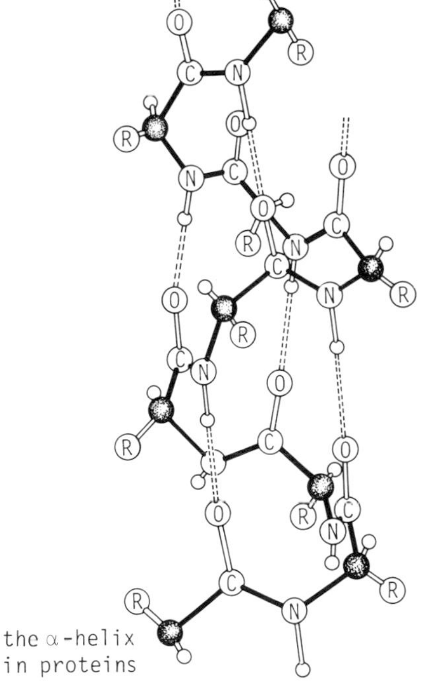

In the α-helix the maximum number of H-bonds is present so that there are 3.6 amino acid residues per turn. In order for this to be achieved the polypeptide chain is twisted into a right-handed helix.

This structure is found not only in the fibrous proteins such as *keratin* but also in the tertiary structure of the globular proteins as will be shown.

Because a proline residue in peptide linkage has no spare H to bond, proline has the effect of preventing α-helical formation.

A

The *β*-pleated sheet

Hydrogen bonds in the *β*-pleated sheet.

A polypeptide chain in which there are no intrachain H bonds is described as in the *β*-configuration. Such chains may interact as shown. The R groups are in the opposite plane to the H bonds and so do not interact. This *β-pleated sheet* structure is not only found in fibrous proteins such as silk fibroin but also in the tertiary structure of globular proteins as will be illustrated.

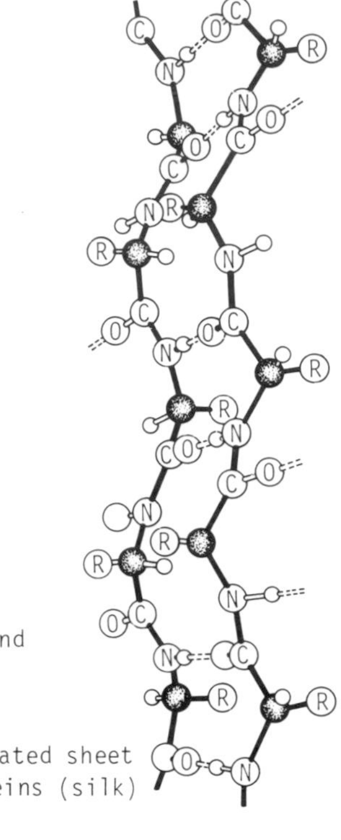

B

Myoglobin and haemoglobin structure

The organization of a globular protein maximizes association between hydrophobic residues and between hydrophilic groups.

The polypeptide chain tends to fold so that the apolar groups are located in the interior giving a hydrophobic core, whilst the polar groups lie on the hydrophilic surface of the protein. A characteristic of enzymes (see later) is that often they have a cleft extending into the protein interior which may contain some polar groupings.

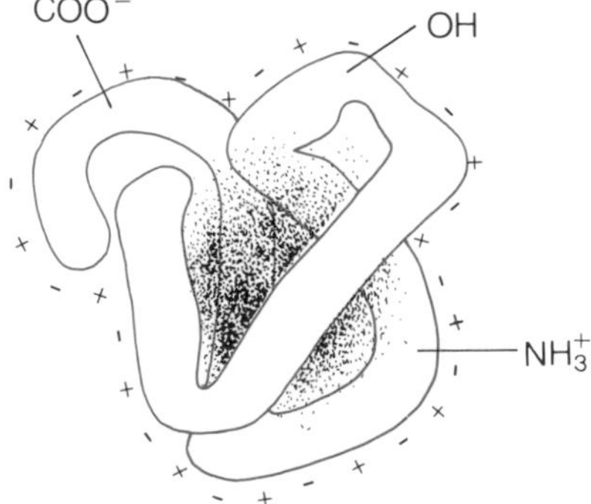

Myoglobin and haemoglobin will be used as models of the way in which primary structure affects tertiary and quaternary structure and also the function of the molecule.

A

The architecture of myoglobin illustrates the importance of tertiary structure in the function of the protein.

The haem group has a greater affinity for CO than O_2 and this accounts for the poisonous effect of carbon monoxide.

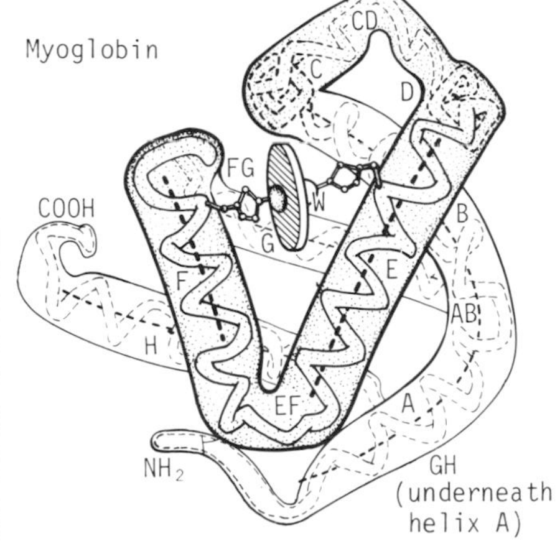

There are 8 stretches of α-helix surrounding the haem group. Histidines in helices E and F interact with haem on either side. The O_2 sits at W. Helices E and F form the walls of a box for the haem, B, G and H are the floor and the CD corner closes the open end. The haem 'pocket' consists in the main of non-polar (group 1) amino acids

Myoglobin carries charged amino acid residues at positions which are hydrophobic in haemoglobin and which are important in bonding the haemoglobin subunits together. Hence the myoglobin subunits do not readily associate to form a quaternary structure. For designation of regions see page 23.

B

An example of the importance of quaternary structure can be seen in the way that the four chains of haemoglobin interact to give a compact structure, that is essential for correct protein function.

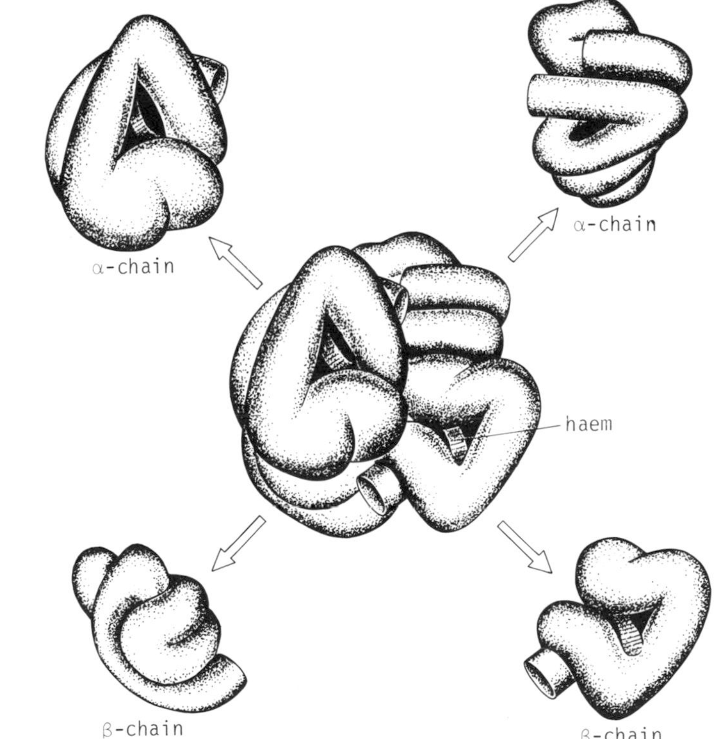

Human adult haemoglobin consists of two identical α-chains and two identical β-chains. If there is any change in the structure of the α and β-chains resulting from mutations that affect their interaction, the properties of the haemoglobin will be changed.

A

The bonding between the subunits of haemoglobin is greatest between the chains that are dissimilar ($\alpha\beta$).

The binding of oxygen to haemoglobin results in the movement of α_1- and β_1-chains as a unit relative to the α_2- and β_2-chains. Deoxyhaemoglobin is denoted T-state and oxyhaemoglobin R-state.

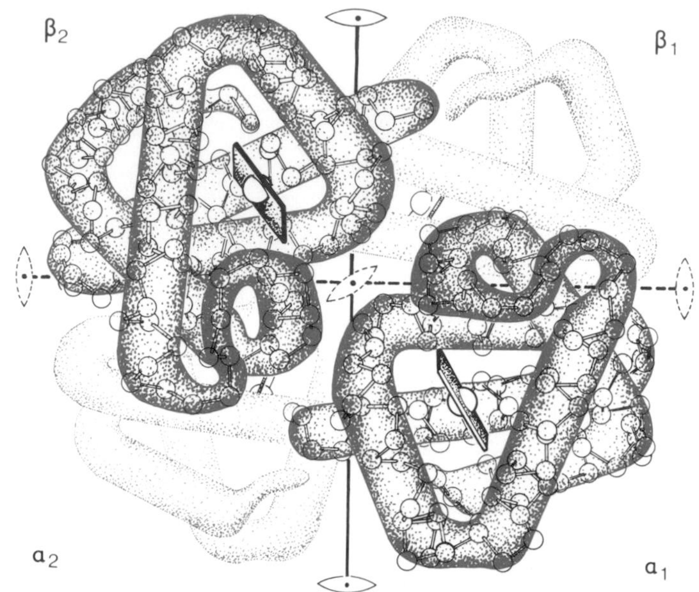

The four chains of haemoglobin are designated $\alpha_1\alpha_2\beta_1\beta_2$. There are few bonds between the two α-chains or between the two β-chains. However there are strong hydrophobic bonds betwen unlike chains, e.g. $\alpha_1\beta_1$ or $\alpha_1\beta_2$, $\alpha_2\beta_1$ and $\alpha_2\beta_2$. The bonds between α_1 and β_2 are more rigid than those between the dimers $\alpha_1\beta_2$ and $\alpha_2\beta_1$. Thus if haemoglobin is dissociated the dimers are always $\alpha\beta$.

B

Oxygen binds to Fe^{2+} and this triggers the change in conformation.

Note that oxygen does not bind when the Fe^{2+} is oxidized to Fe^{3+} as it occurs in methaemoglobin.

The binding of oxygen to the iron atom reduces the diameter of the iron atom and it moves into the plane of the porphyin ring. The oxygen binds at a site near the distal histidine on helix E (see p.21A and p.23A). The movement of the iron atom moves the proximal histidine F8 and produces a conformation change in that subunit. This is transmitted to other subunits (see above).

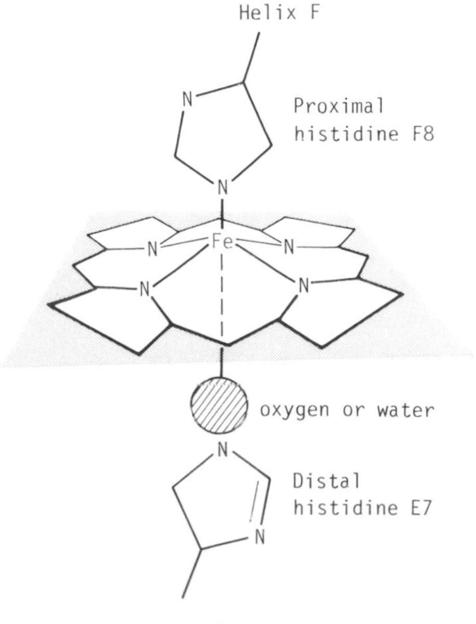

Helix F

Proximal histidine F8

N

N

N Fe N

N N

oxygen or water

N

Distal histidine E7

N

A

The regions of the haemoglobin molecule are given letters that correspond to those in myoglobin.

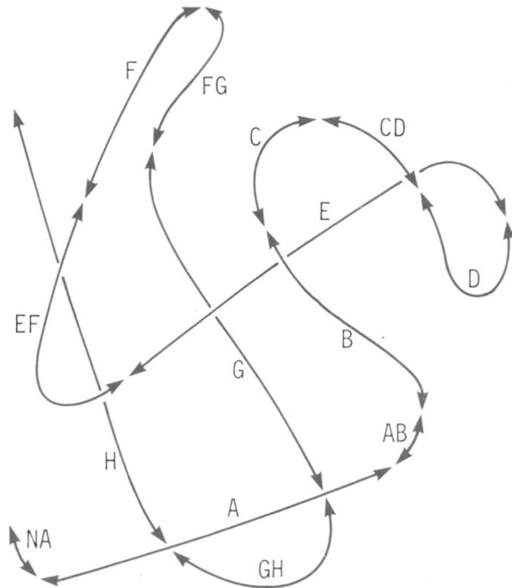

Single letters, A, B, C, etc. denote helical regions. Double letters AB, BC, CD, etc. denote non-helical regions joining the helices denoted by the letters used (AB joins A to B). Although this folding is common to the haemoglobins of every species possessing this type of molecule only about 10 amino acid residues are *invariant*.* Some changes in the primary structure make little difference to the tertiary structure whereas others have a profound effect, e.g. in *sickle-cell haemoglobin*.

The α chain of haemoglobin differs from the β chain by the deletion of one residue in the NA segment, the addition of two residues in the AB corner and the deletion of six residues in the CD segment and D helix.

* See page 27

B

Bisphosphoglycerate (BPG) binds specifically to deoxyhaemoglobin.

Bisphosphoglycerate (for its physiological role, see p. 25A) binds to the central cavity created by the four subunits of deoxyhaemoglobin. It is extruded on oxygenation because this cavity becomes too small.

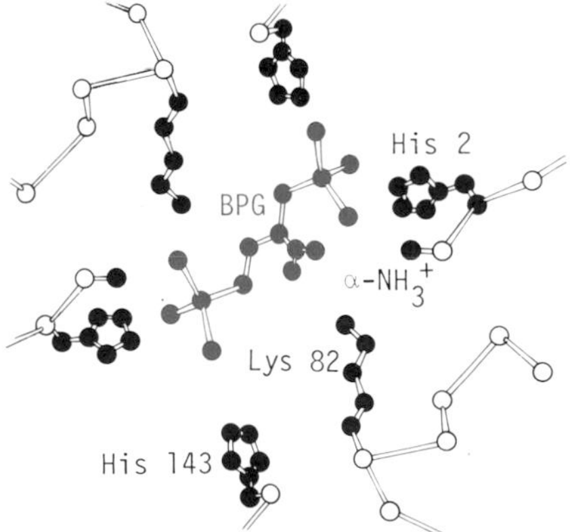

Mode of binding of BPG to human deoxyhaemoglobin. BPG interacts with three positively charged groups on each β-chain.

A

Myoglobin and haemoglobin properties

The difference in subunit structure of myoglobin and haemoglobin is reflected in their oxygen saturation curves.

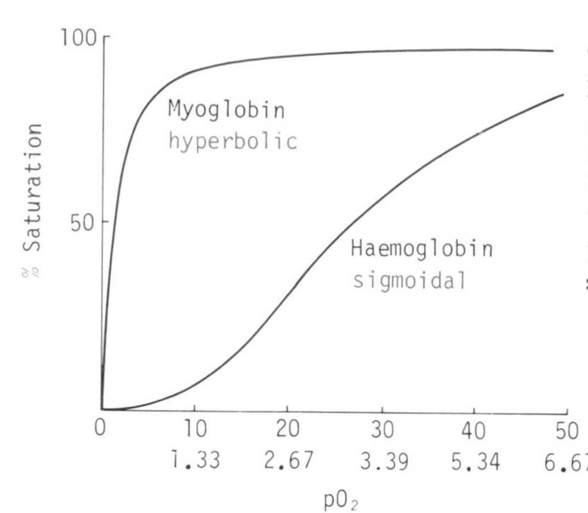

The sigmoidal curve of haemoglobin is indicative of cooperative interactions between protein subunits; the affinity for O_2 becomes progressively greater as successive O_2 sites are filled.

B

The effect of the pO_2 in capillaries and lungs is such that in the lungs haemoglobin becomes virtually fully saturated, whilst in the muscle capillaries it loses over 50% of its oxygen.

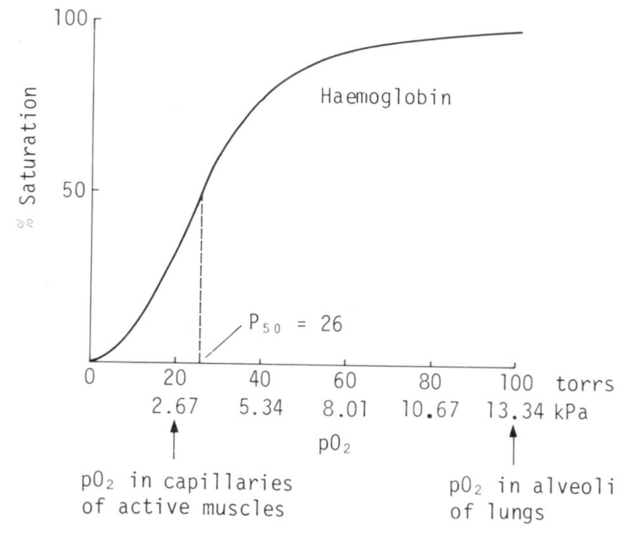

C

A lowering of the pH from 7.6 to 7.2 results in the release of O_2 from oxyhaemoglobin.

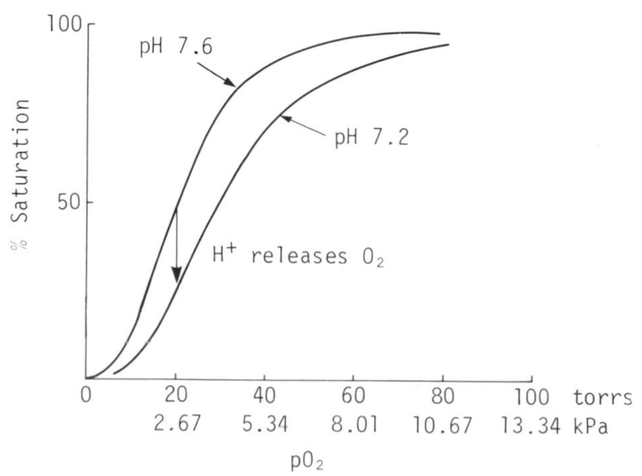

A

Bisphosphoglyceric acid (BPG) interacts with haemoglobin in the red blood cell.

Human red cells contain 2,3-bisphosphoglycerate (BPG). This substance reduces the affinity of haemoglobin for O_2. The amount of BPG varies with the physiological condition, being the same in mother and foetus. The affinity of BPG for foetal haemoglobin is much weaker than for the adult so that O_2 combines preferentially with foetal haemoglobin.

2,3-bisphosphoglycerate

B

The presence of BPG reduces the affinity of O_2 for haemoglobin to a degree dependent on the BPG/Hb ratio.

To the extent that the binding of O_2 to haemoglobin is to be likened to the interaction of a substrate with an enzyme, BPG can be regarded as an allosteric effector (*see* p63).

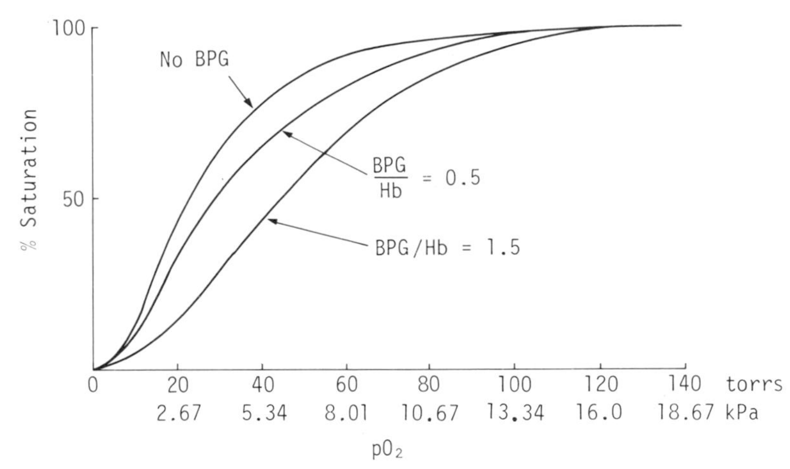

C

The effect of CO_2 is to decrease the amount of haemoglobin which is present in the oxygenated form. The effects of CO_2 and BPG are cumulative.

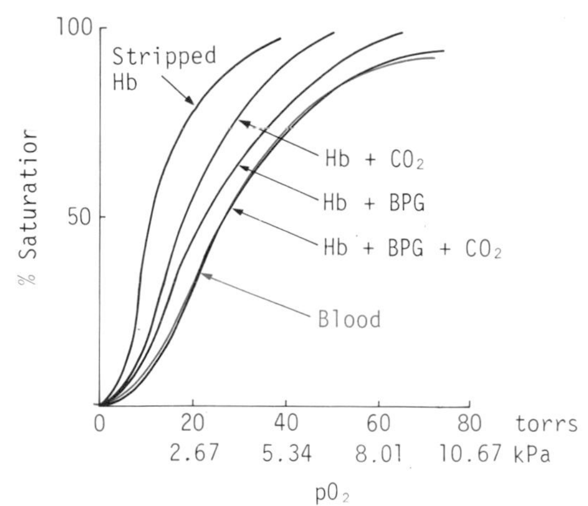

A

Abnormal haemoglobins

There are many cases in which one amino acid in one of the haemoglobin chains is replaced as a result of a single point mutation in a globin gene.

a. At normal oxygen tension.
b. At reduced oxygen tension.

Sickle-cell haemoglobin (HbS)

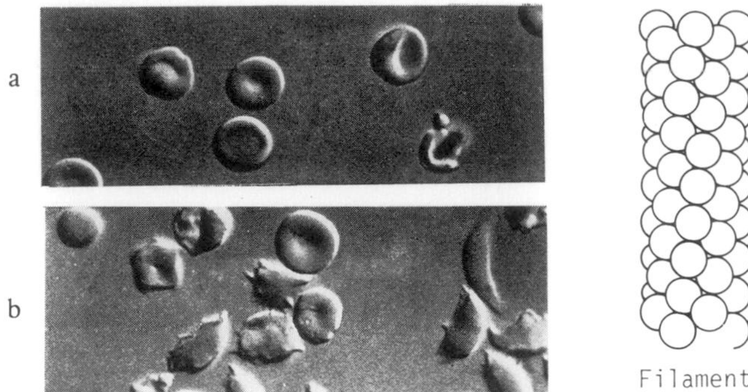

Sickle-cells

In HbS a glutamic acid in the β-chain is replaced by valine. This has little effect on the physiological properties of oxygenated HbS but deoxygenated HbS forms rod-like filaments which cause the red cells to sickle.

B

Some replacements in different haemoglobins: Residues participating in $\alpha_1\beta_1$ and $\alpha_1\beta_2$ contacts are shown in red (●). The positions of other variants which do not produce clinical symptoms are denoted half red (◐).

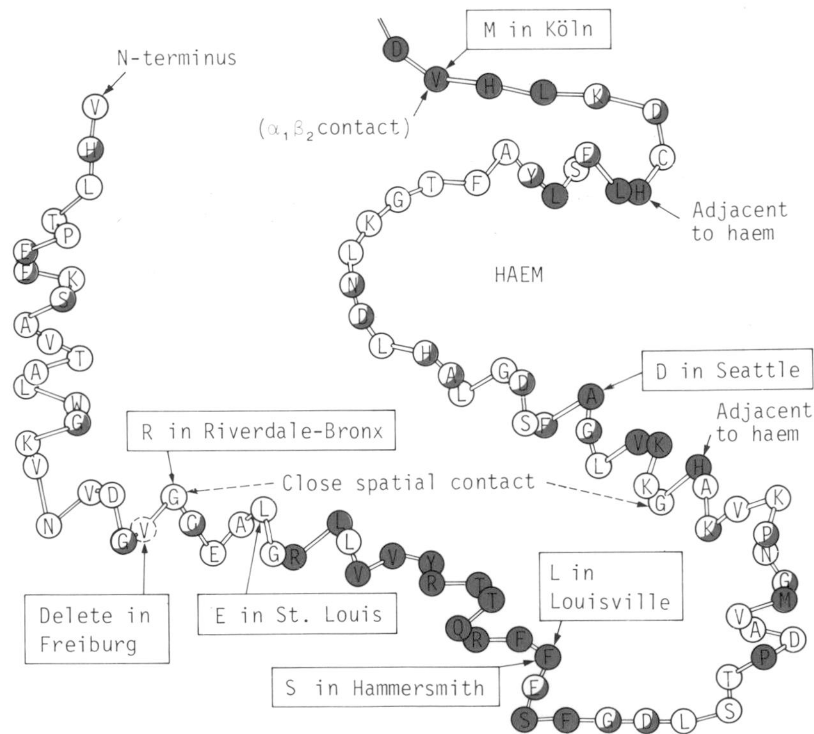

An invariant amino acid is one that has never been found to vary in any haemoglobin so far sequenced. There are only two amino-acid residues that are common to all the globins: a histidine that forms a covalent link with the haem iron; and a phenylalanine in position 1 of the loop made by helices C and D which wedges the haem into its pocket. All other residues are replaceable, but in 33 specific positions replacements are restricted to non-polar residues.

A

A few abnormal haemoglobins have reduced oxygen affinity.

The effect on the saturation curve can be very marked.

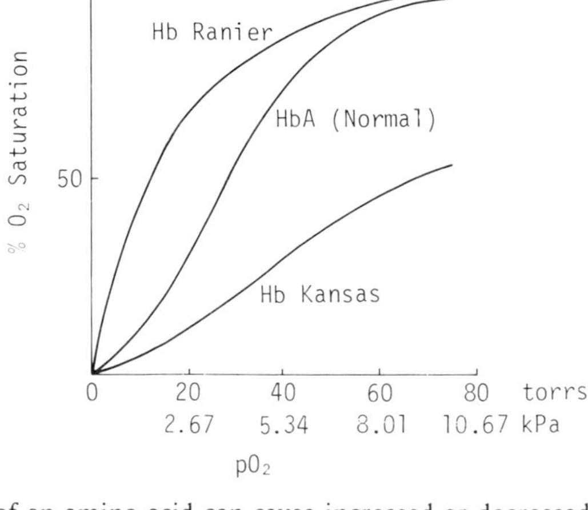

Substitution of an amino acid can cause increased or decreased oxygen affinity.

Many other abnormal haemoglobins have increased oxygen affinity.

Some substitutions causing increased oxygen affinity

Haemoglobin	Substitutions	Site affected
Hb Rainier	β 145 (Tyr — Cys)	C-terminal
Hb Denmark Hill	α 95 (Pro — Ala)	$\alpha_1\beta_2$ contact
Hb Syracuse	β 143 (His — Pro)	BPG-β contact
Hb San Diego	β 109 (Val — Met)	Haem pocket

Only a few haemoglobins are known with reduced oxygen affinity. One of these is Hb Kansas, in which threonine is substituted for asparagine at β 102.

B

Unstable haemoglobins

Note: FG5 β98 indicates amino acid 98 in the β-chain, at position 5 in the FG region.

Other substitutions can cause instability of the haemoglobin molecule, leading to formation of methaemoglobin (HbM).

The cause of the instability of some unstable haemoglobins

Haemoglobin	Substitution	Cause of instability
Hb Köln	FG5 β98 Val $\longrightarrow$ Met	Large bulky side chain of methionine distorts FG segment, breaking several haem contact amino acids.
Hb Hammersmith	CD1 β42 Phe $\longrightarrow$ Ser	Phenylalanine is an important haem contact, the side chain of serine being too short to reach haem group
Hb Bristol	E11 β67 Val $\longrightarrow$ Asp	Non-polar valine is replaced by polar aspartic acid; the internal siting of aspartic acid would result in gross distortion of E helix to neutralize its charge

A

Abnormal haemoglobins are detected either by a clinical abnormality or by screening by electrophoresis. In this regard only those changes leading to altered electrophoretic behaviour are detected. About 400 abnormal haemoglobins have been detected to date.

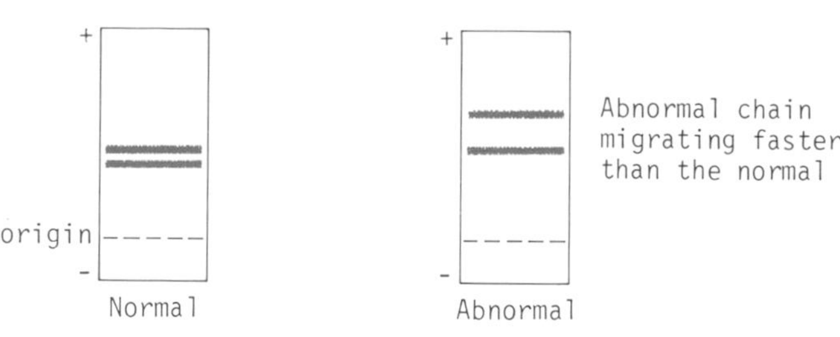

B

So-called glycosylated haemoglobin.

Glucose may react with the N-terminal amino acid residues (valine in the human) of the β-chains of HbA to produce a so-called 'fast' haemoglobin A_1. The means of attachment is through a non-enzymic reaction so that a ketoamine is formed. Glucose-CH_2-NH-βA. The reaction is a continuous process occurring slowly during the life span of the red blood cell over a period of about 120 days.

The concentration of HbA_1 is correlated to the degree of diabetic control so that a quantitative determination of HbA_1 reflects the patient's average blood glucose concentration over a long period.

HbA_1 is usually named glycosylated haemoglobin but this is a confusing nomenclature since the carbohydrate link is not glycosidic and is quite different from that of the glycoproteins. The name *glycated* haemoglobin is to be preferred.

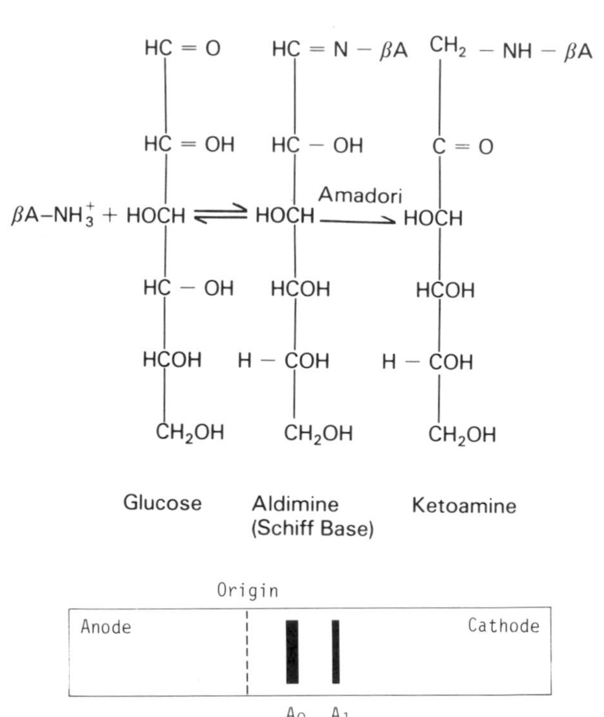

The unglycated Hb, A_0 can be separated from glycated HbA_1 either by column chromatography or by electrophoresis as shown on agar gel. The charged groups of agar interact with A_0 to a greater degree than A_1, thus retarding the migration of A_0.

C

A note on the representation of molecular weights.

The term relative molecular mass (symbol: M_r) is preferred to molecular weight. Both M_r and molecular weight are ratios and hence it is *incorrect* to give them units such as Daltons. It is therefore incorrect to say that the M_r or the molecular weight of substance X is 10^5 Daltons. The Dalton is a unit of mass equal to one-twelfth the mass of an atom of carbon-12. Hence it is *correct* to say the molecular mass of X is 10^5 Daltons or to use expressions such as the 16 000–Dalton peptide. For entities that do not have a definable molecular weight it is *correct* to say, for example, 'the mass of a ribosome is 10^7 Daltons.'

A

Separation of proteins

Gel-permeation chromatography.

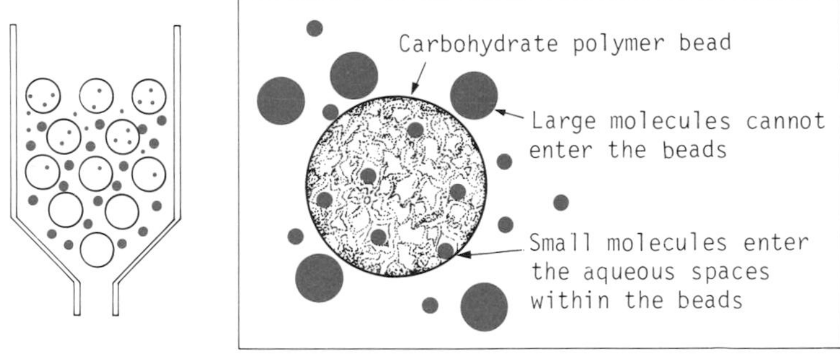

The method depends upon the use of a cross-linked dextran (Sephadex) which is usually placed in a column. Sephadex may be used to separate small molecules (e.g. salts) from large molecules (e.g. proteins) and also to separate proteins of various molecular sizes.

B

Sephadex chromatography of some serum proteins.

The red line indicates the separation of total protein in serum and the black line the separation of specific proteins (on different scales).

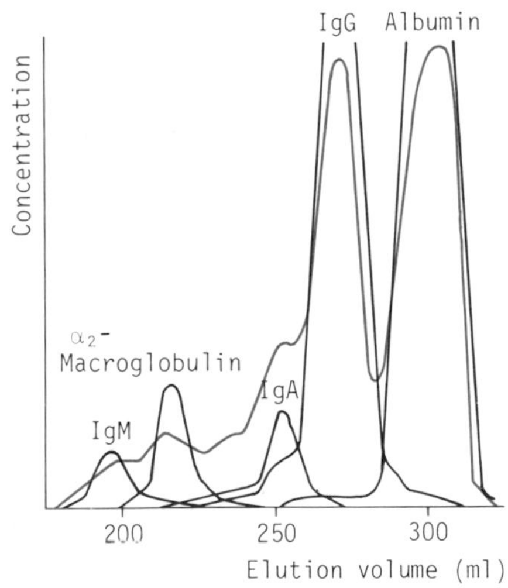

C

Gel-permeation can be used for the determination of the M_r of a protein.

Plot of the elution volume V_e of native proteins of known M_r on Sephadex G-75 (●) and G-100 (○) versus log M_r. (The grades of Sephadex differ in the extent of cross-linking of dextran).

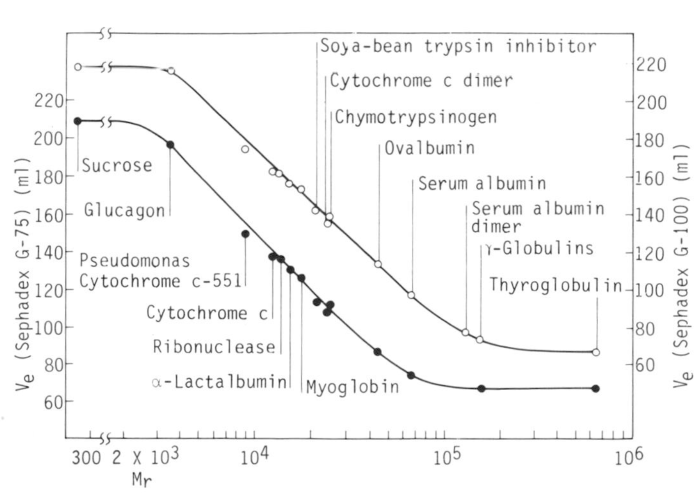

A

The separation of proteins by electrophoresis.

This may be performed on paper or cellulose acetate in buffer at pH 8.6. The proteins are stained after denaturation.

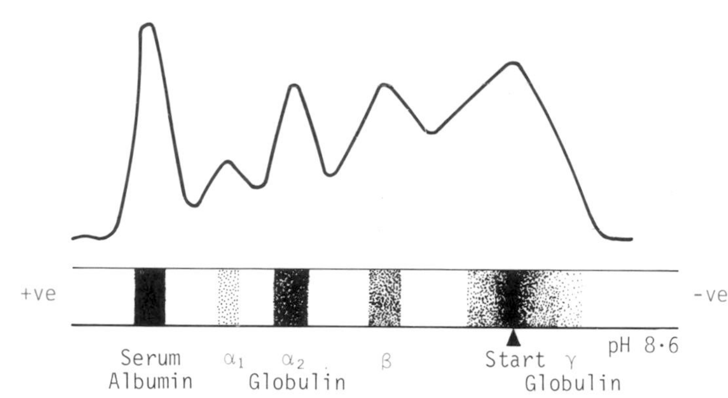

The strip is passed through a scanner which records the areas stained.

B

Polyacrylamide gel electrophoresis (PAGE)

PAGE can be used to separate native proteins according to their charge and size.

PAGE can also be carried out in the presence of sodium dodecyl sulphate (SDS). In this case oligoproteins are separated into their subunits.

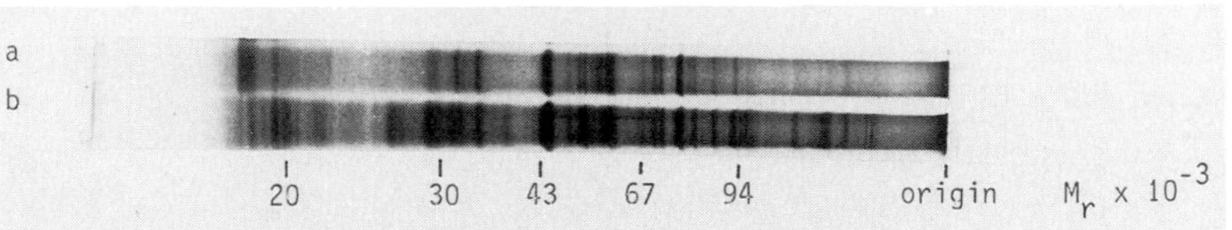

Above is shown the resolution of membrane proteins from (a) spleen lymphocytes and (b) thymocytes as a result of SDS-PAGE.

The proteins are suspended in a 1% solution of sodium dodecyl sulphate (SDS). This detergent disrupts most protein-protein and protein-lipid interactions. Very often, 2-mercapto ethanol is added to disrupt disulphide bonds. The solution is layered on an acrylamide gel containing SDS and subjected to electrophoresis. The electrophoretic mobility of most proteins, but not glycoproteins, depends on their M_r rather than their net charge in the absence of SDS. The negative charge contributed by SDS molecules bound to the protein is much larger than the net charge of the protein itself. A pattern of bands appears when the gel is stained with Coomassie Blue.

Agarose may be used in place of polyacrylamide for larger proteins or to obtain a different type of separation in the absence of SDS.

A

Affinity chromatography

The specific ligand is covalently attached to the column matrix.

Examples of types of ligand that have been used are shown.

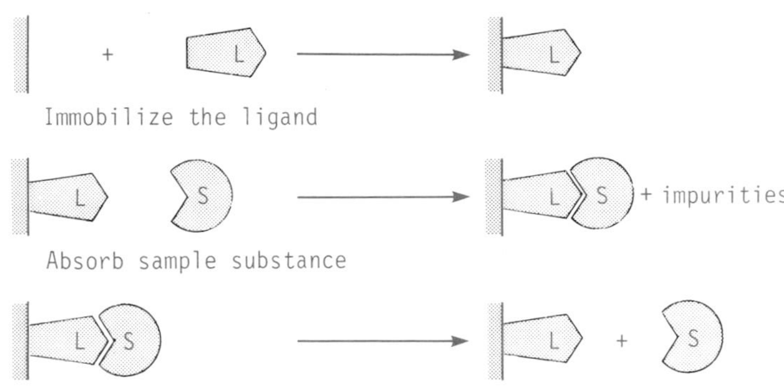

enzyme:	substrate analogue, inhibitor, cofactor
antibody:	antigen, virus, cell
lectin:	polysaccharide, glycoprotein, cell surface receptor, cell
nucleic acid:	complementary base sequence, histone, nucleic acid polymerase, binding protein
hormone, vitamin:	receptor, carrier protein
cell:	cell surface specific protein, lectin

B

High performance liquid chromatography

Peptide Separation

1. Oxytocin
2. Met-Enkephalin
3. TRH
4. α-Endorphin
5. LHRH
6. Neurotensin
7. α-MSH
8. Angiotensin II
9. Substance P
10. β-Endorphin

Conditions

Instrument:	Bio-Rad protein chromatography system
Column:	Bio-Gel TSK SP-5-PW
Sample:	2 μg each peptide in 50 μl, except LHRH (μg)
Eluant:	30 min linear gradient from A to B A. 0.02M phosphate buffer of pH 3/CH_3CN (70/30) B. 0.5M phosphate buffer of pH 3/CH_3CN (70/30)
Flow Rate:	1.0 ml/min
Temperature:	25°C
Detection:	UV at 220 nm

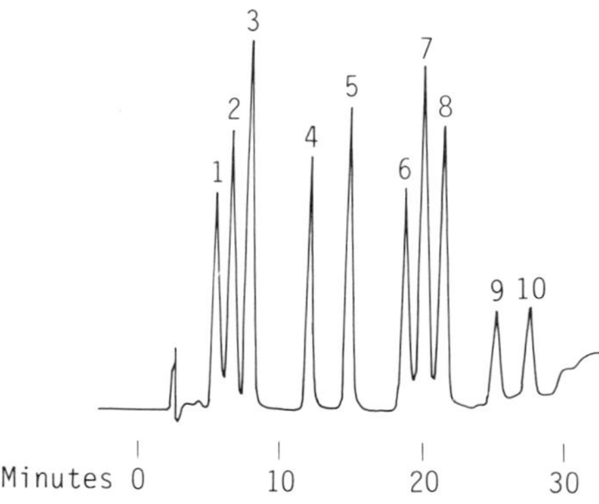

As an example of the high resolution of HPLC the separation of peptides on a cation exchange column is shown.

The technique depends on the use of microfine column matrixes to give high resolution rapidly.

In addition to ion exchange, gel filtration and reversed phase chromatography may be used.

A

Isoelectric focusing — separation, pI

Isoelectric focusing (IEF) is an electrophoretic technique that separates proteins according to their isoelectric point (pI) in a stable pH gradient generated by carrier ampholytes. These carrier ampholytes migrate under the influence of an electric current to generate the pH gradient, which increases from the anode to the cathode.

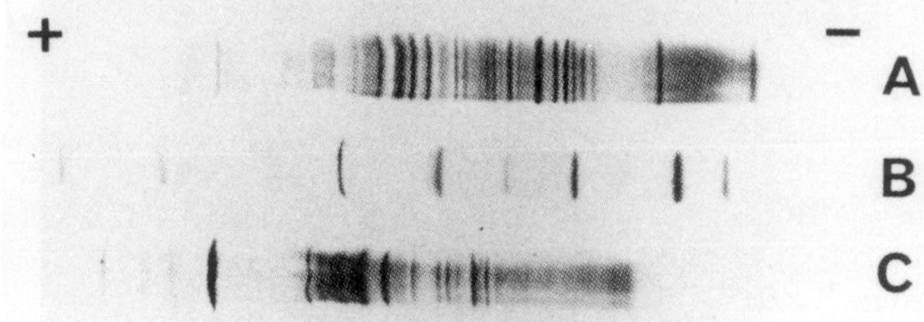

Separation of proteins by horizontal, flat bed isoelectric focusing using an ampholyte range from pH 3–10 (Pharmalyte 3–10).
A. Rainbow trout muscle extract B. Marker proteins pI range 3–10 C. Pike muscle extract. The sample is applied near the middle of the gel.

B

Large-scale separation of proteins.

The scheme for plasma protein shows how many different methods are used for the separation of some of the many plasma proteins. The products are analyzed by gel electrophoresis.

Cryoprecipitation depends on the lesser solubility of some proteins in the cold.

A chromatographic procedure for plasma protein fractionation

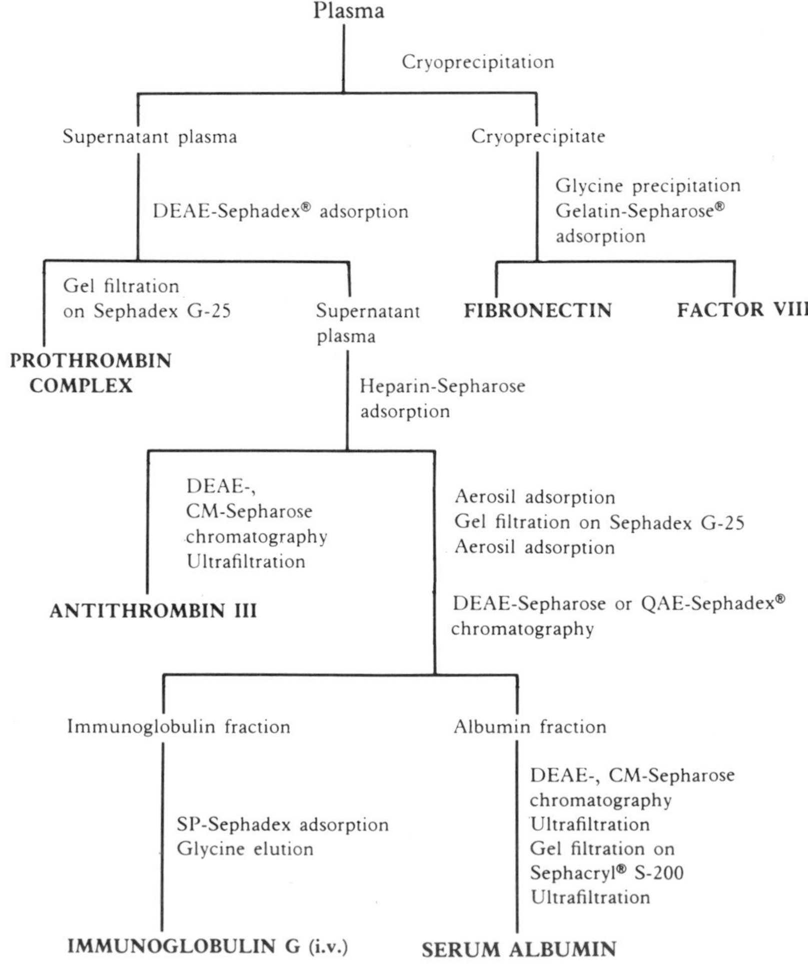

The pattern of serum proteins on electrophoresis may be used in the diagnosis of disease.

As explained previously either paper, cellulose acetate, or polyacrylamide gel may be used as the medium

Normal Pattern

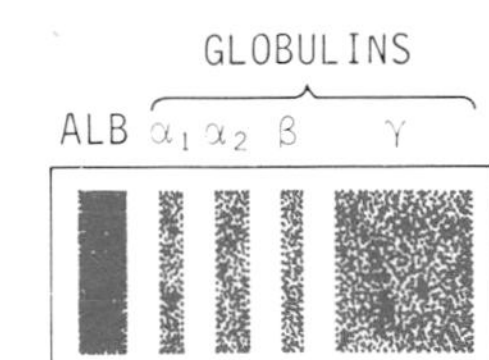

Primary Immune Deficiency

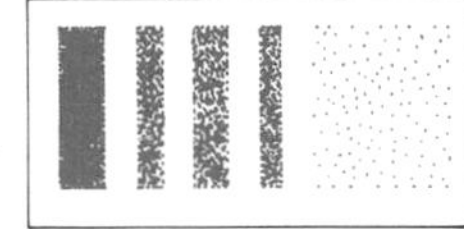

Impaired synthesis of immunoglobulins. Usually familial.

Multiple Myeloma

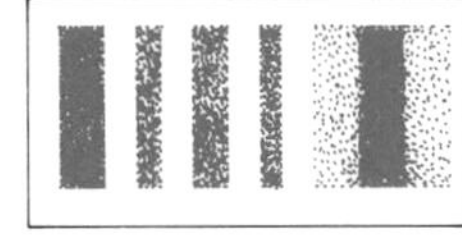

Paraprotein band between α_2 & end of γ region. Normal γ-globulin often decreased. Paraproteins are also found in other diseases.

Nephrotic Syndrome

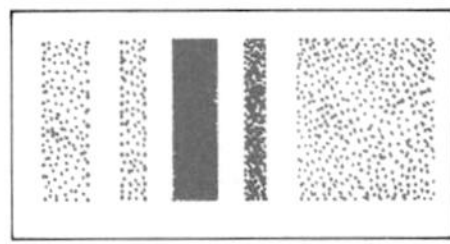

Albumin lost into urine, and sometimes γ-globulin. Increase in α_2-globulin.

Cirrhosis of liver

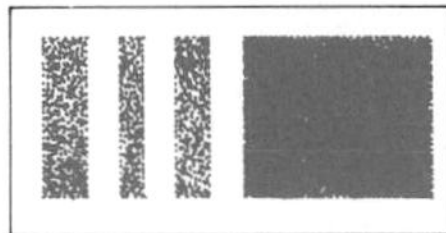

Decreased albumin, increased production of other unidentified proteins which migrate in $\beta\gamma$ region, causing impaired $\beta\gamma$ resolution.

Infection

Elevated α_1 and α_2 proteins. Usually decreased albumin.

Chronic Lymphatic Leukaemia

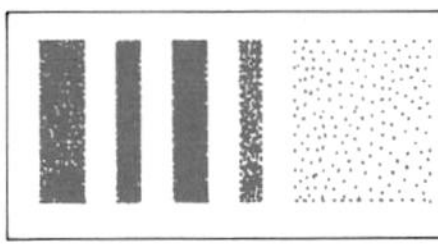

Quite often accompanied by decreased γ-globulin.

Plasma should not be used!

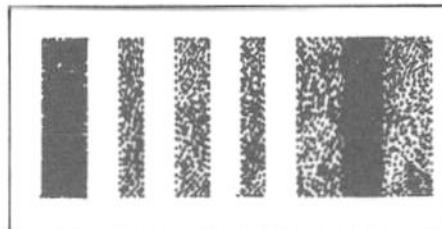

If plasma is used instead of serum, fibrinogen band gives the appearance of a paraprotein, leading to misleading diagnosis.

α_1 Antitrypsin Deficiency

α_1 (antitrypsin) deficiency associated with emphysema of the lung in adults, and juvenile cirrhosis.

3. PROTEIN FUNCTION
Serum proteins

A

Serum albumin is a transport protein (ligands in parenthesis are transported by albumin when primary carriers are filled).

A ligand is a compound bound to a larger molecule.

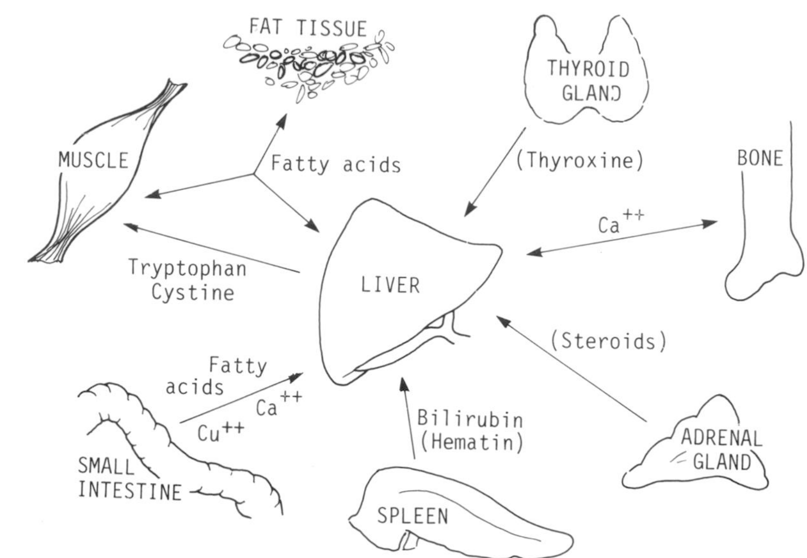

B

The globulins

α- and β-globulins contain a variety of proteins apart from α_1-antitrypsin (see below).
e.g. caeruloplasmin, carrier of copper; transferrin, carrier of iron; haptoglobins, bind haemoglobin

The γ-globulins are immune globulins.

C

α_1-antitrypsin and its role in emphysema

The predominant component of the α_1-globulin band consists of a protein named α_1-antitrypsin (AT). It is wrongly named because it is active against elastase rather than trypsin and is a member of a group of Serine Proteinase Inhibitors or Serpins.

In normal lung the destructive power of elastase released from the neutrophils is held in check by AT. Cigarette smoking increases the number and activity of lung neutrophils and consequently the amount of elastase. Moreover oxidation reduces the protection afforded by a given amount of circulating AT.

α_1-antitrypsin is less active after oxidation and elastase then causes tissue breakdown and loss of elasticity in the lungs i.e. emphesema.

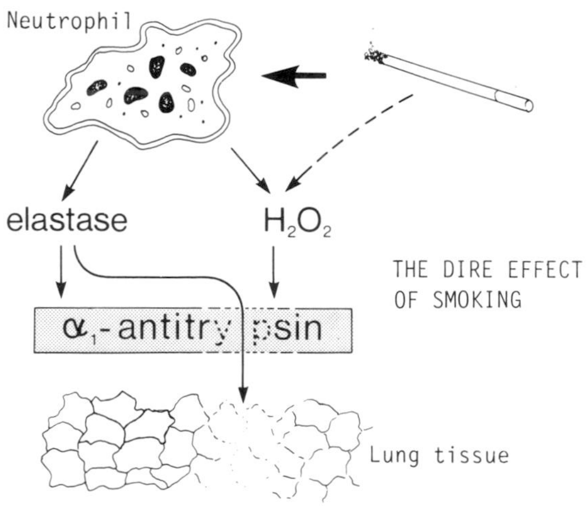

A

The unhappy result of mutants of α_1-antitrypsin (AT)

In addition to the normal protein there exist two variants, both confined to people of European descent: the S variant in 5–8% (genotype MS) and the Z variant in 4% (genotype MZ). The tissue specificity of AT depends on position 358 in the polypeptide chain of 394 amino acids being occupied by methionine (or valine). The methionine is susceptible to oxidation (see page 34C).

B

In non-smokers emphysema is unusual in people with either variant of AT.

oooooooo Glu oooooooo Met oooooooo
342 358

394 amino acids chain of normal AT

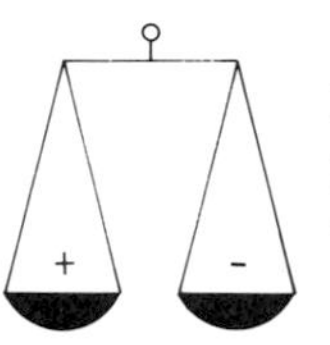

Elastase activity (+) properly controlled by circulating AT (−)

In the ZZ mutant Glu is replaced by Lys which causes a defect in secretion of AT and a drop in circulating AT

oooooooooo Lys oooooooooo Met oooooooooo

Mutant Z AT

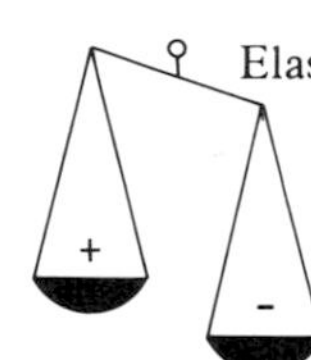

Elastase activity normal (+)
reduction in circulating AT (−)

C

Smoking leads to tendency to emphysema

Smoking not only increases neutrophil elastase but increases oxidation of Met.

oooooooooo Glu oooooooooo Met oooooooooo
oxid

Oxidation of AT involves Met 358

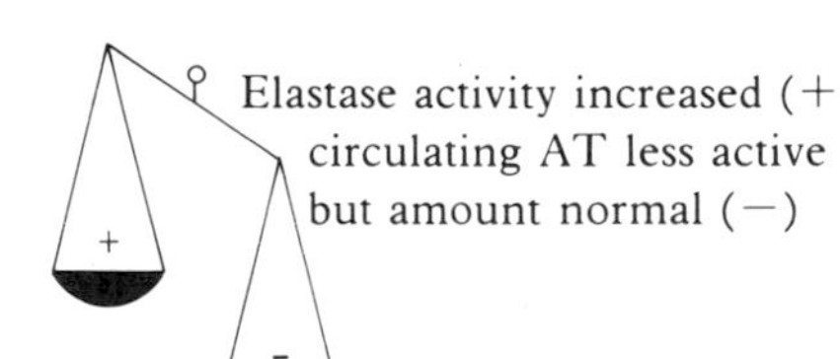

D

Smoking in a mutant makes emphysema probable.

If the ZZ mutant smokes there is a reduction in the amount of circulating AT and it is less active

oooooooooo Lys oooooooooo Met oooooooooo
oxid

Mutant Z AT in a smoker

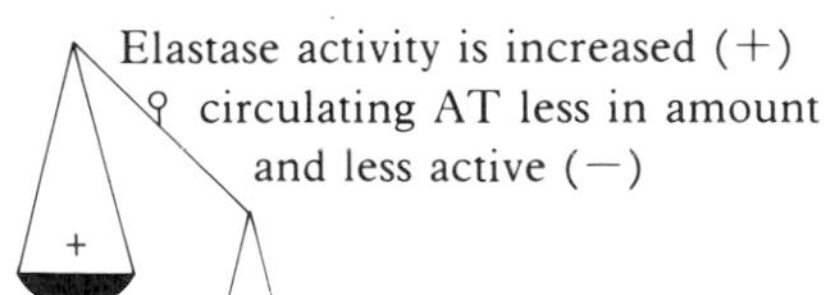

Proteins involved in the metabolism of iron.

Iron metabolism is regulated within tissues by the protein, *ferritin*, and transported in blood by the protein, *transferrin*. Apoferritin is the iron-free form of ferritin. Ferritin catalyses the oxidation of Fe^{2+} to Fe^{3+}.

Transferrin is synthesized in the liver and present in the β-globulin fraction of serum. It is a glycoprotein with two iron (ferric) binding sites. Many cells possess transferrin receptors on which the cells depend for their essential supply of iron (see page 261).

Ferritin, the iron store in liver and spleen, consists of 24 subunits and is a very large protein of M_r about 500 000.

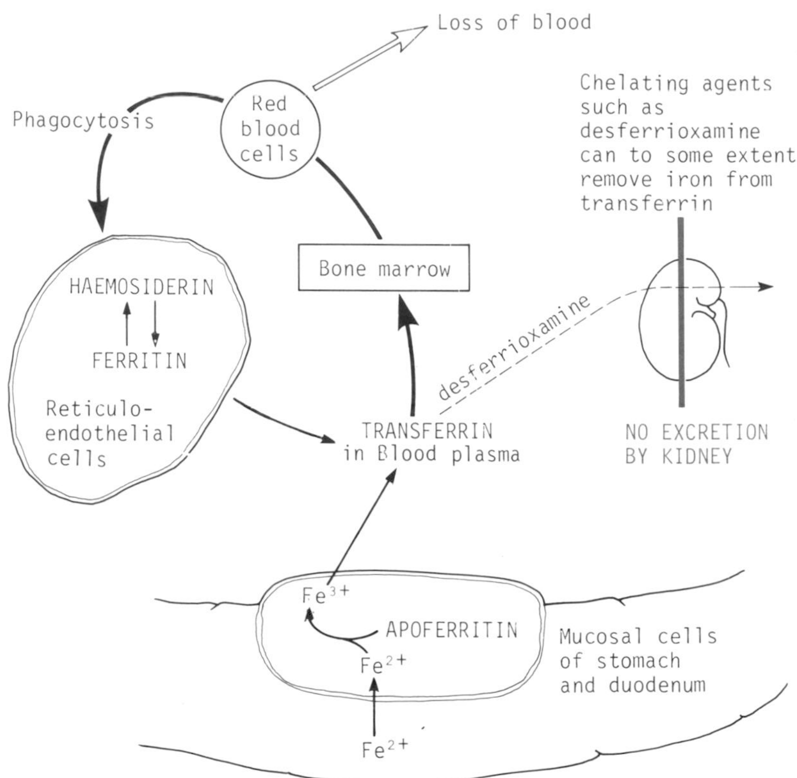

There is no excretory mechanism for iron, which slowly accumulates in the body throughout life, being absorbed in small quantity from the diet. Frequent loss of blood, such as in menstruation, may cause anaemia if dietary iron is inadequate, as iron is essential for haemoglobin synthesis, and thus red cell formation. When the red cells are destroyed, the iron is removed from the haemoglobin by reticulo-endothelial cells, where it remains until taken up by transferrin for utilization elsewhere. If excessive amounts of iron are absorbed it may accumulate in the reticulo-endothelial cells, such as the Kupffer cells of the liver, a condition known as iron overload or haemosiderosis due to the deposits of haemosiderin, a form of ferritin complexed with other proteins, and iron. The excretion of some iron through the kidneys may be induced by the chelating agent desferrioxamine, but loss of blood is the only really effective method of removing iron from the body. Ascorbic acid (vitamin C) strongly increases the absorption of iron.

Collagen

A connective tissue protein.

Collagen is the most abundant protein in mammals and is the major fibrous element of skin, bone, tendon, cartilage, blood vessels and teeth.

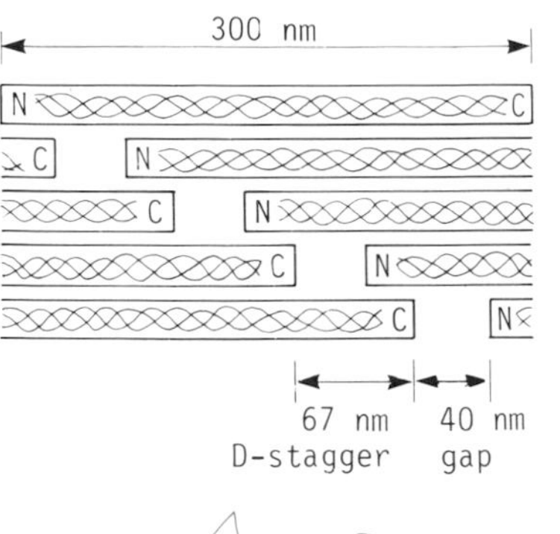

A microfibril is a staggered array of tropocollagen molecules. There is a gap between the ends. Microfibrils associate into a fibril and several fibrils form a collagen fibre.

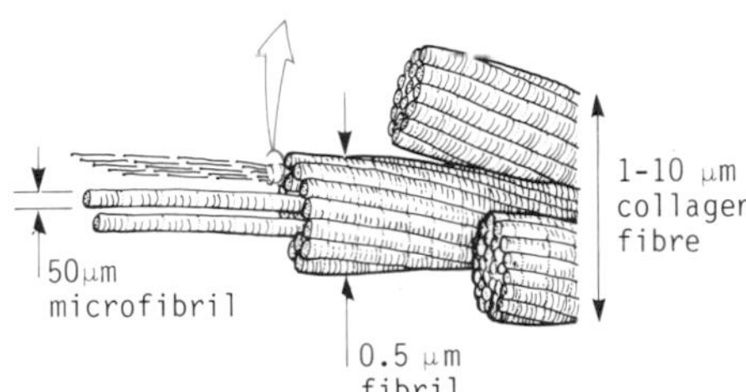

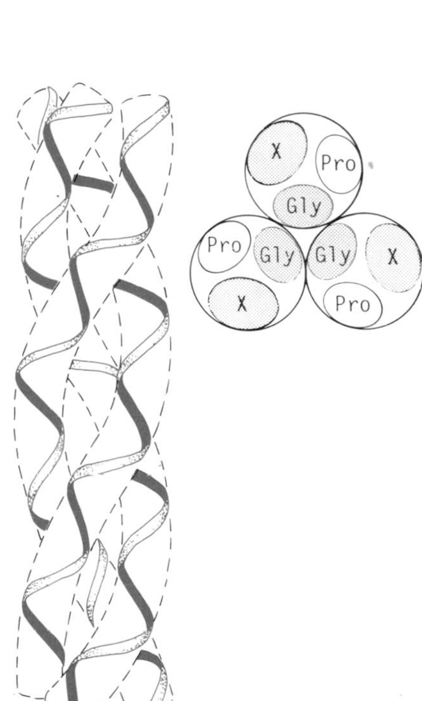

A model of the triple-stranded tropocollagen. Above is a cross-section showing outline of each strand and position of glycine.

In order to fulfill its many roles the basic structure of collagen is modified but always consists of three polypeptide chains of about 1000 amino acids each in left-handed helical conformation and each wound around each other to form a right-handed supercoil. There are in addition covalent cross-links involving the aldehydes derived from the lysine side chains; such links may be either within or between the chains. Collagen from young animals lacks these cross links and is called *tropocollagen*.

Depending on the amino acid composition of the three chains the collagens are classified into Types I to IV.

In contrast to globular proteins the amino acid sequence of tropocollagen is regular except for about the last 20 amino acids at each end, and every third amino acid is glycine usually followed by proline. Two unusual amino acids are also present, hydroxyproline and hydroxylysine. Hence a typical sequence might be:

-Gly-Pro-Met-Gly-Pro-Arg-Gly-Leu-Hyp-

Each of the three strands is hydrogen-bonded to the other two strands, there being no intrachain hydrogen bonds which are a feature of the α-helix. The H donors are the peptide NH groups of glycine and the acceptors are the peptide CO groups of residues on other chains. Thus the direction of the H bonds is transverse to the long axes of the tropocollagen rod.

The hydroxylation of the proline residues is effected by prolyl hydroxylase (for the role of ascorbic acid in this reaction see page 131).

The ACTH peptides

In the course of synthesis of many hormones the active peptide is split from a larger protein. An example of this is ACTH (adrenocorticotrophic hormone).

A precursor protein that gives rise to more than one mature protein is known as a *polyprotein*.

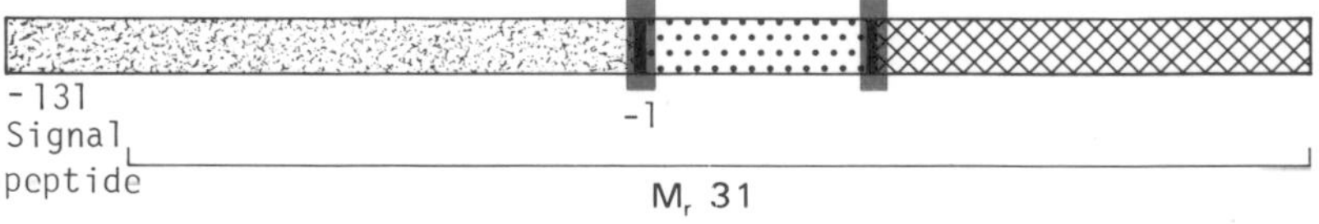

Proopiomelanocortin

The initial translation product in the pituitary is a large protein of M_r 31 000. (The significance of the signal peptide is explained on page 108). A protease splits the chain at points where there are two adjacent basic amino acids. In the figures the numbers following the M_r represent thousands.

One of the molecules formed is ACTH, and another is β-lipotropin (βLPH), a peptide once thought erroneously to possess the ability to mobilize fatty acids from adipose tissue.

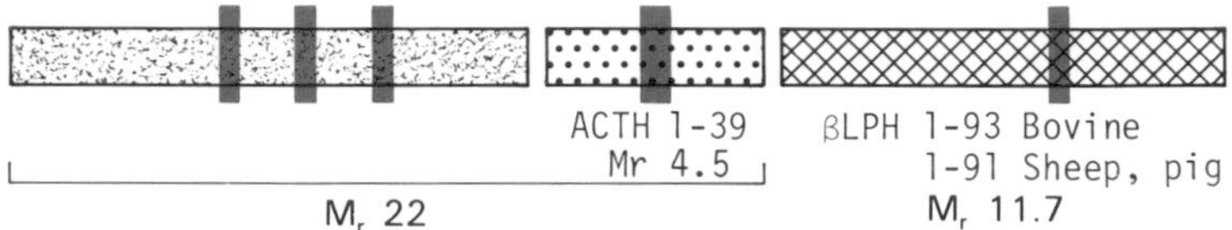

There is further cleavage to produce various peptides. In particular ACTH can be cleaved to melanocyte stimulating hormone (αMSH). βLPH can be cleaved to β-endorphin (βEND). The latter has morphine-like properties, in that it binds to opiate receptors in nervous tissue. γ-Lipotropin (γLPH) and 'corticotropin-like intermediate peptide' (CLIP) are also formed at this stage.

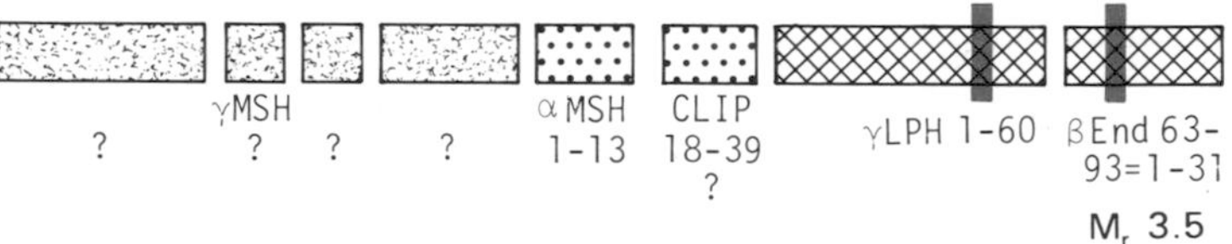

Subsequent protease activity yields 'β-melanocyte stimulating hormone, β-MSH' and enkephalin (Enk). The latter also binds to opiate receptors.

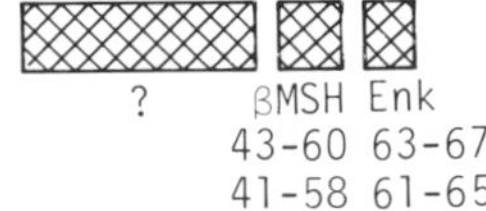

The numbers refer to the order of the amino acid residues starting at 1 for the amino terminus of ACTH or βLPH. Amino acid residues to the left of ACTH are indicated by a − sign.

A

Immunoglobulin structures

The structure and function of immunoglobulin.

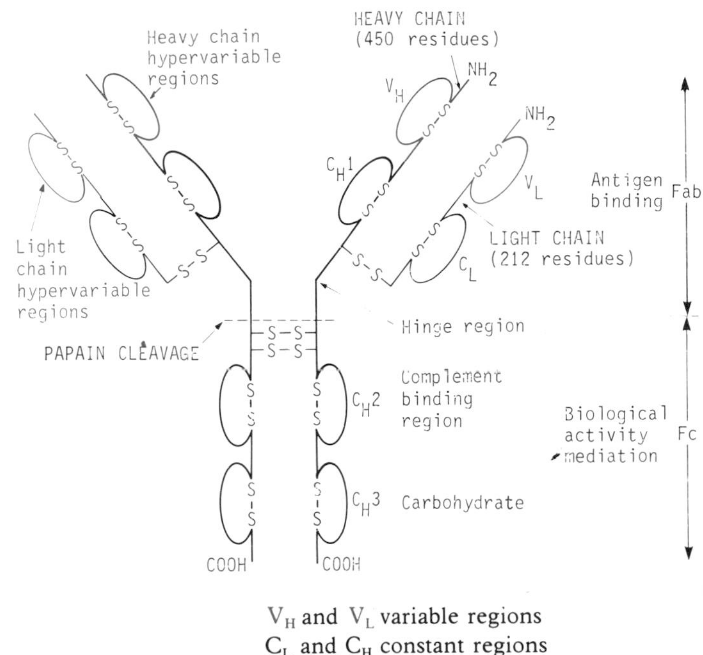

V_H and V_L variable regions
C_L and C_H constant regions

B

Three-dimensional structure of a human immunoglobulin

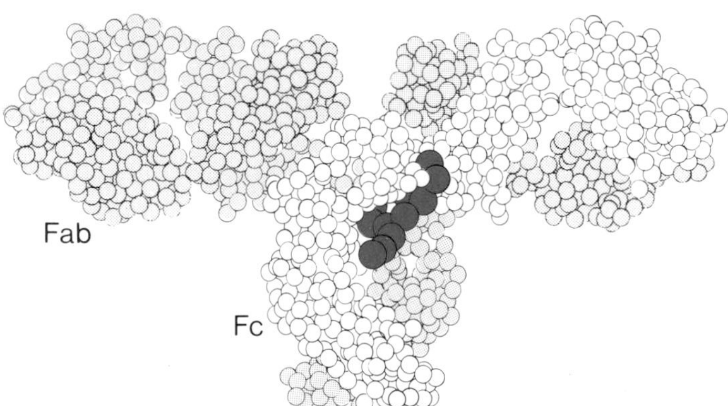

Space-filling representation of the three-dimensional structure of human IgG Dob. Dob has a 15 residue deletion in its hinge region as compared to normal IgG proteins. The two heavy chains are shown in *white* and in *pink*. Light chains are in *gray*. Carbohydrate is shown as *red* spheres.

(Dob is a human IgG1 (κ) cryoglobulin)

A

The various classes of immunoglobulins.
Physical, chemical and biological properties of human immunoglobulin classes.

Property	*IgG*	*IgA*	*IgM*	*IgD*	*IgE*
Usual molecular form	Monomer	Monomer, dimer etc	Pentamer	Monomer	Monomer
Molecular formula	$\kappa_2\gamma_2$ or $\lambda_2\gamma_2$	$(\kappa_2 a_2)\ \eta$ or $(\lambda_2 a_2)\ \eta$	$(\kappa_2\mu_2)_5$ or $(\lambda_2\mu_2)_5$	$\kappa_2\delta_2$ or $\lambda_2\delta_2$	$\kappa_2\varepsilon_2$ or $\lambda_2\varepsilon_2$
Other chains	—	J chain, S piece	J chain		
Subclasses	IgG1, IgG2, IgG3, IgG4	IgA1, IgA2	None established	None	None
Subclass heavy chains	$\gamma 1$, $\gamma 2$, $\gamma 3$, $\gamma 4$	$a1$, $a2$	-	-	-
M_r	150 000	160 000	950 000	175 000	190 000
Sedimentation constant (Sw20°)	6.6S	7S, 9S, 11S, 14S	19S	7S	8S
Carbohydrate content (%)	3	7	10	9	13
Serum level (mg/100 ml) (adult average)	1250 ± 300	210 ± 50	125 ± 50	4	0.03
Percentage of total serum immunoglobulin	75–85	7–15	5–10	0.3	0.003
Paraproteinaemia	Myeloma	Myeloma	Macroglobulinemia	Myeloma	Myeloma
Antibody valence	2	2	5 or 10	?	?
Binding to cells	Macrophages, Neutrophils	-	-	?	Mast cells
Other biological properties	Secondary Ab response; placental transfer	Characteristic Ab in mucous secretions	Primary Ab response rheumatoid factor	Main lymphocyte cell surface molecule	Homocytotropic Ab anaphylaxis; allergy

Each class of immunoglobulin has a characteristic type of heavy chain; γ for IgG, a for IgA etc. There are in addition subclasses $\gamma 1$, $\gamma 2$ etc. Each class may have either κ or λ light chains.

A

The formation of IgM polymer depends on the presence of J chain.

IgM differs from the other types of immunoglobulin in the possession of an extra domain, C_{H^4}. It also possesses J chain which is a polypeptide of M_r 15 000 rich in cysteine. J chain links two 7S monomers and this allows for the formation of S–S bridges between the C_{H^3} and C_{H^4} domains of other monomers so forming a pentamer. Similarly in IgA J chain is present. Each molecule of IgG has a valency of two, i.e. has the potential to combine with two molecules of antigen (in theory IgM has a valency of 10 but the effective valency is about 5).

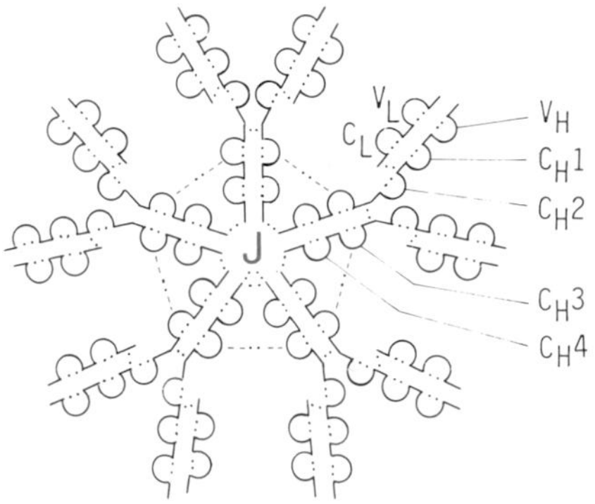

B

The valency of an antibody and an antigen may be defined as the number of sites of interaction. In most cases a given antigen will possess more than one antigenic site as shown in the figure.

Below is shown the interaction of a *hapten*, bis-N-dinitrophenyl (DNP)-octamethylene-diamine, with IgG. The two DNP groups are far enough apart not to interfere with each other's combination with antibody.

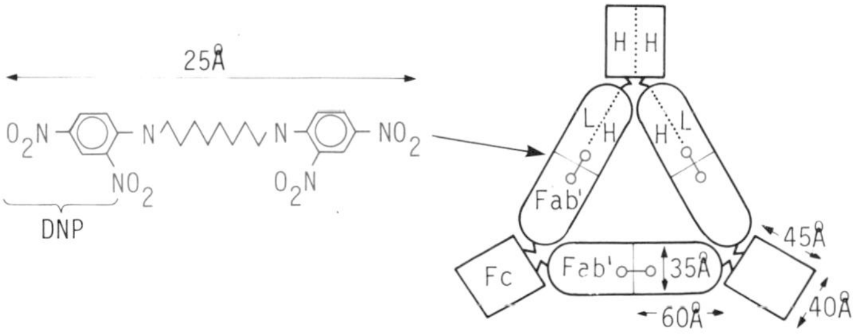

It is, of course, the interaction of antibody and antigen in this way that leads to precipitation.

A

The flexibility of the structure of IgG.

The shape of an IgG molecule can vary considerably so that the relative arrangement of the domains may change. The figure summarizes sites of movement. There are two peptides involved in these changes—the hinge peptide at the junction of C_{H1} and C_{H2} and the switch peptide at the junction of the V and C domains in Fab.

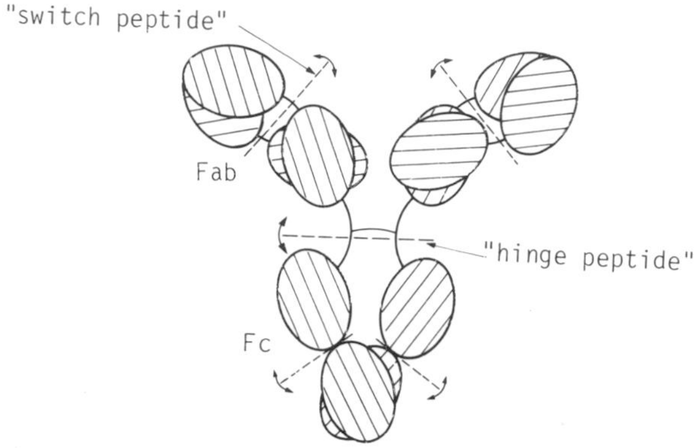

B

Antibody variants: Isotypes. Allotypes. Idiotypes.

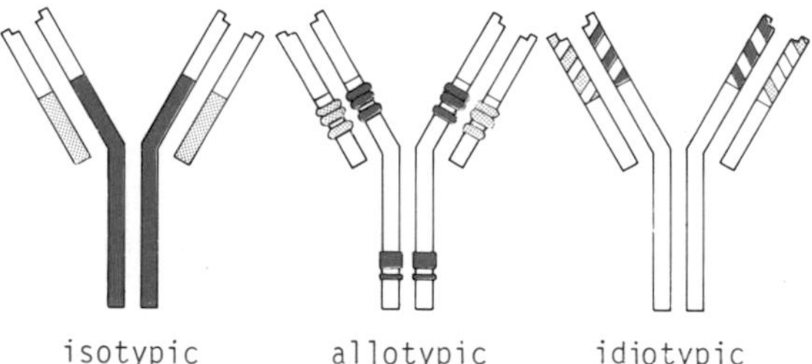

Antibody variants. Isotypic variation refers to the different heavy and light chain classes and subclasses (see page 41): the variants produced are present in *all* healthy members of a species. Allotypic variation occurs mostly in the constant region: *not all* variants are present in all healthy individuals. Idiotypic variation occurs in the variable region only and idiotypes are specific to each antibody molecule.

A

The structure of complement.
The structure of C1q is an example of a protein which combines the features of a globular and a fibrous protein.

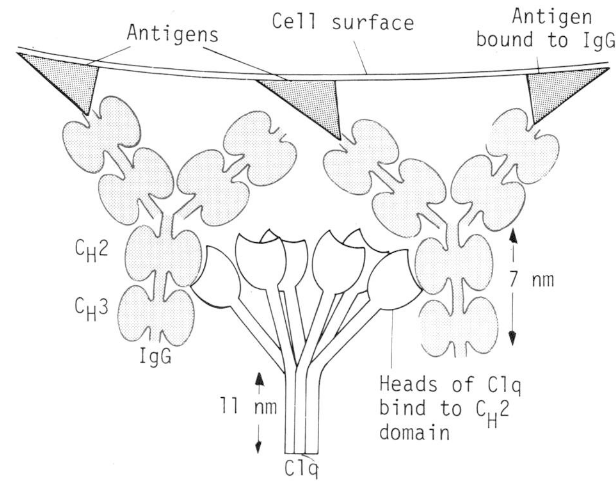

The Fc region of IgG is responsible for triggering pathways of the immune response that lead to the lysis of unwanted organisms. An example is the *Complement* system that consists of a series of proteases. The first stage is triggered by the binding of a molecule C1q (M_r 400 000) to the C_H2 domain of an IgG-antigen complex as shown.

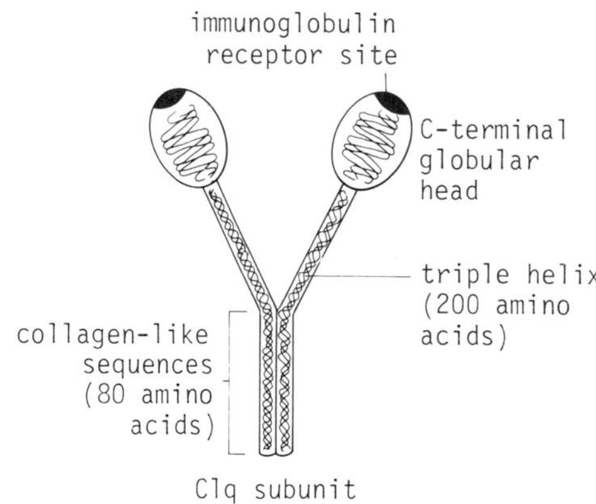

C1q structurally resembles a bunch of six tulips as shown. There are three chains in each tulip. The 'flower' is a globular structure while the 'stalk' is elongated and resembles the triple-helical structure of collagen (see p. 37). Indeed, it even contains the Gly-X-Y sequence with hydroxyproline.

A

The generation of antibody diversity.

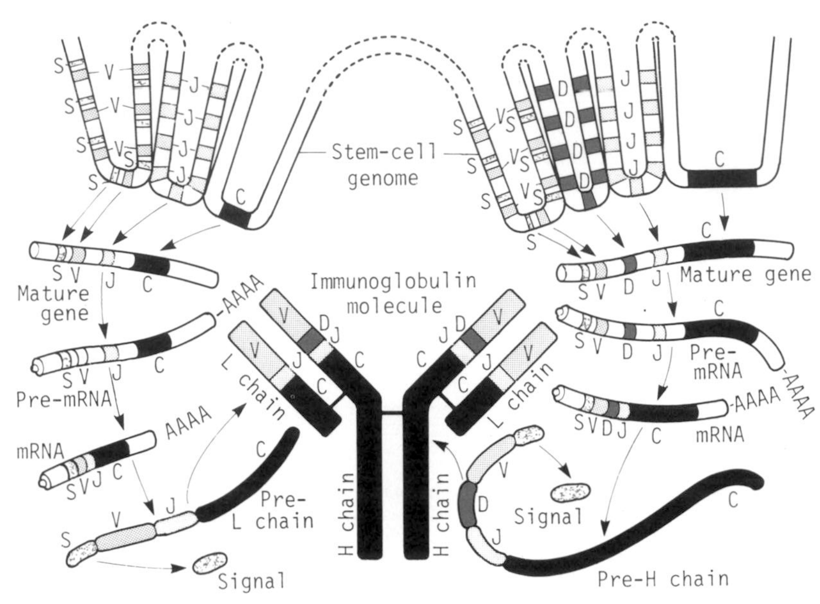

The stem-cell genome contains multiple variants of the L-chain V and J genes and of the H-chain V, J and D (Diversity) genes. The V genes are preceded by a small S segment coding for the Signal peptide (see p. 108). As lymphocytes mature, each differentiating cell constructs particular L and H genes of virtually unique structure by a recombination process that randomly selects one out of each set of gene segments and assembles them together with a C gene. The pre-mRNA is spliced (see p. 115) to give a mature mRNA which is translated and the Signal peptide removed. There is then oxidative formation of S-S bridges.

A

Immunoglobulin synthesis

IgM that is located on the plasma membrane differs in structure from that which is secreted.

IgM_{mem} (membrane)
IgM_{sec} (secreted)

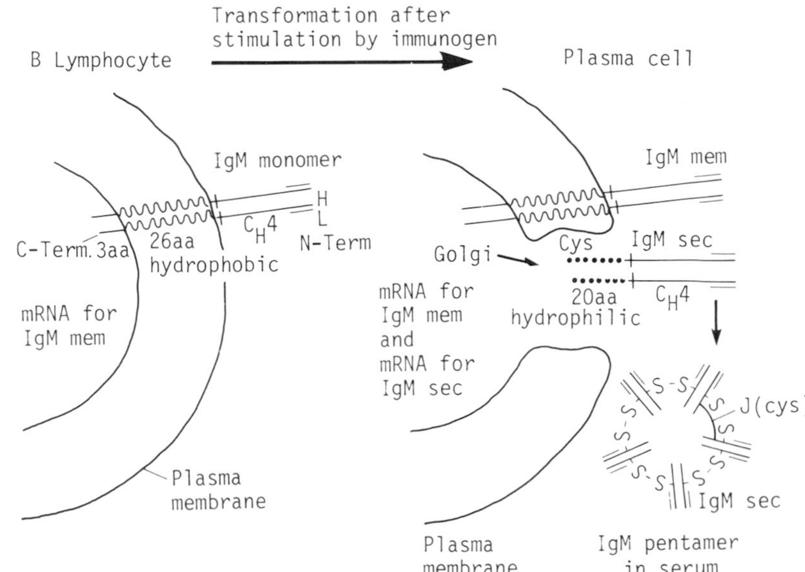

Comparison of the structure of the mouse IgM that is retained by the plasma membrane (left) and that which is secreted to form a pentamer (right)

The B lymphocyte bears a monomer of IgM (see p. 41A) on its surface. When such a lymphocyte is recognised by an immunogen it is stimulated to form a clone of plasma cells which synthesize and secrete the IgM pentamer. It is now known that the structure of the C-terminal region of the H chain accounts for the fact that the monomer IgM is an integral membrane protein (see p. 251) while the IgM pentamer is secreted. The monomer H chain possesses a hydrophobic peptide of 26 amino acids which has an affinity for the plasma membrane while the H chains of the pentamer have instead a hydrophilic peptide of 20 amino acids.

B

The immunoglobulin super family illustrated at a cell surface.

MHC antigens are the major histocompatibility antigens recognized in graft rejection and function as determinants that are recognized along with foreign antigen in T-lymphocyte responses. The V and C inside ○ shows whether the domains are more like IgV or C regions. N-linked carbohydrate is shown as —•.

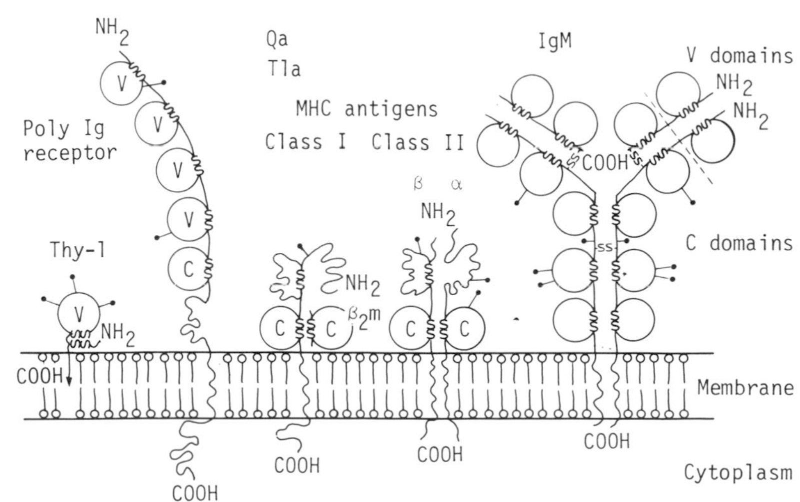

A

Immunoassay

The use of antibodies for the assay of substances of clinical significance.

Principles of radioimmunoassay

Principle of radioimmunoassay

1. A limiting and constant amount of antibody is incubated with a constant concentration of radioactively labelled analyte such that, for example, 40–50% of the labelled analyte is bound by the antibody.

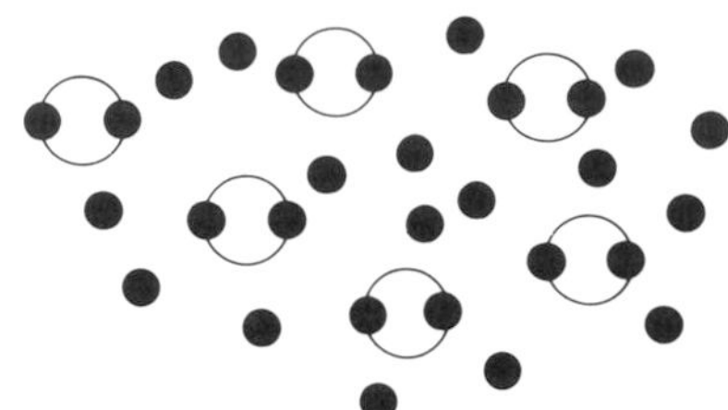

2. On the addition of unlabelled analyte, competition occurs between labelled and unlabelled antigen for the limiting concentration of binding sites in the antibody.

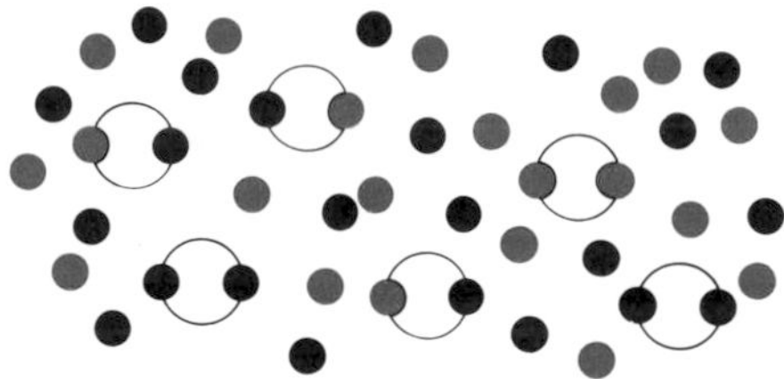

3. The amount of labelled analyte remaining bound to the antibody can be measured after separation of the antibody bound and unbound fractions. When no competing unlabelled analyte is present, the amount of labelled analyte bound will be high.
When competing analyte is present, less labelled analyte will be bound to antibody.

B

Enzyme-linked immunoassay.

In order to overcome the disadvantages of radioactivity, enzymes are used as markers

Antibody to Analyte

-Enz

Enzyme-analyte complex made by attaching enzyme to

Enzyme-analyte complex is inactive on enzyme substrate when bound to the analyte antibody

Procedure

Sample contains and -Enz

Free competes with -Enz for the binding sites on and displaces -Enz which can then react with its substrate.

Quantitative estimation of the enzyme reaction shows how much enzyme was displaced, which is related to the amount of analyte.
Enzyme activity is measured by adding substrate for the enzyme. Often a dehydrogenase such as glucose 6-phosphate dehydrogenase, which requires NAD^+, is used so that the reaction can be assayed by spectrophotometry. Another useful enzyme is xanthine peroxidase.

Enzyme-linked immunoassay contd.

An alternative strategy as applied in the Cambridge Life Sciences plc assay of progesterone in cow's milk.

1.

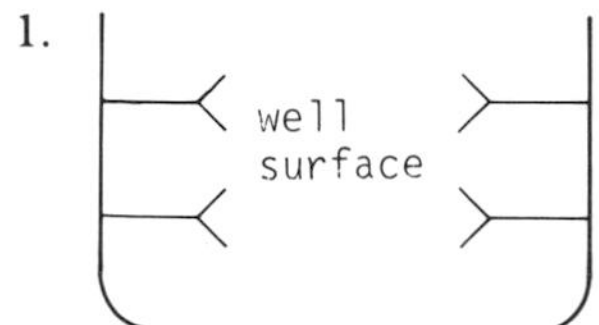

Progesterone antibody binding sites

Surfaces of wells of plate (solid phase) are pre-coated with progesterone antibody, which binds through the Fc domain.

2.

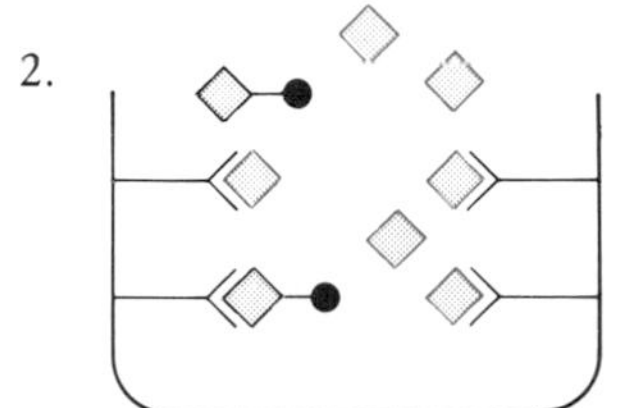

Progesterone in milk sample or standard

Enzyme labelled progesterone

Enzyme labelled progesterone competes for limited sites on antibody. Enzyme is alkaline phosphatase.

3.

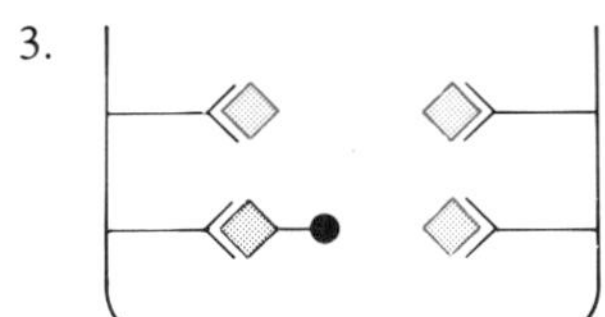

Unbound reactants removed by washing

4.

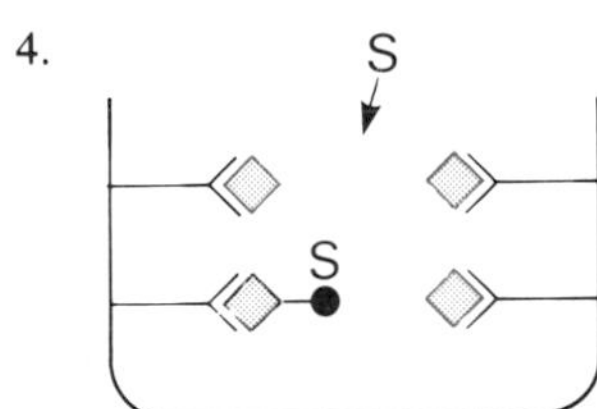

S = Substrate to enzyme, p-nitrophenyl phosphate

Incubation followed by direct measurement of optical density of solution or colour intensity. The amount of enzyme bound is inversely proportional to the concentration of progesterone in the original sample.

Oestrus detection and pregnancy diagnosis are interpreted directly from a standard curve

A glossary of some immunological terms that are useful to biochemists

Acquired Immune Deficiency Syndrome

AIDS is defined by the Centers for Disease Control in the USA as 'a reliably diagnosed disease that is at least moderately indicative of an underlying immune deficiency or any other cause of reduced resistance reported to be associated with that disease'. AIDS is caused by the human T cell lymphotrophic virus HIV I

Adjuvant and Freund's adjuvant.

A substance which non-specifically enhances the immune response to an antigen. A Freund's adjuvant is an emulsion of aqueous immunogen in oil: complete Freund's adjuvant also contains killed *Mycobacterium tuberculosis*, while incomplete Freund's does not.

Allotypes

The constant (C) regions of antibodies show some variation, that is genetically controlled, between one individual and another. The term 'allotype' is used to distinguish immunoglobulins that vary in this respect, between individuals.

Bence Jones Protein

Free immunoglobulin light chain dimers found in the serum and urine of patients and animals with multiple myeloma

β_2-microglobulin

A polypeptide which constitutes part of some membrane proteins including the Class 1 MHC molecules.

Chains

The heavy and two types of light (κ, λ) chains are coded for by genes on different chromosomes but sequence homologies suggest that all Ig domains originated from a common 'precursor' molecule about 110 amino acids long.

Cryoglobulin

An antibody or immune complex which forms a precipitate at 4°C.

Cyclosporin

An immunosuppressive drug which is particularly useful in suppression of graft rejection.

Fragments

Produced by chemical treatment: *H,L*: Heavy and light chains, which separate under reducing conditions.
Fab: Antigen Binding fragment (papain digestion).
Fc: Crystallizable (because relatively homogeneous) fragment (papain)
F(ab)$_2$: two Fab fragments united by disulphide bonds (pepsin). pFc^1: a dimer of $C_H{}^3$ domains (pepsin).
Facb: an Ig molecule lacking $C_H{}^3$ domains (plasmin)

Hinge region

Both flexibility and proteolytic digestion are facilitated by the repeated proline residues in this part of the molecule.

Hybridoma

Cell lines created in vitro by fusing of two different cell types, usually lymphocytes, one of which is a tumour cell to confer immortality on the product.

Idiotypes

The unique amino acid sequence, which forms an antibody combining site, can be recognized as a distinct antigenic site. This is known as an idiotype antigenic site and this specificity defines the idiotype.

Ig

Immunoglobulin; a term embracing all globulins with antibody activity, which has replaced 'gammaglobulin' because not all antibodies have gamma electrophoretic mobility.

Immunogen

A substance that when injected into an animal gives rise to an antibody

J chain

A glycopeptide molecule which aids polymerization of IgA and IgM monomers

Monoclonal

Derived from a single clone, for example monoclonal antibodies which are produced by a single clone and are homogeneous.

Myeloma

A lymphoma produced from cells of the B cell lineage

Polyclonal

A term which describes the products of a number of different cell types.

Tolerance

A state of specific immunological unresponsiveness.

A

Proteins of muscle

Muscle consists of fibres, which in turn are composed of fibrils. It is at the level of fibrils that the molecular level is approached.

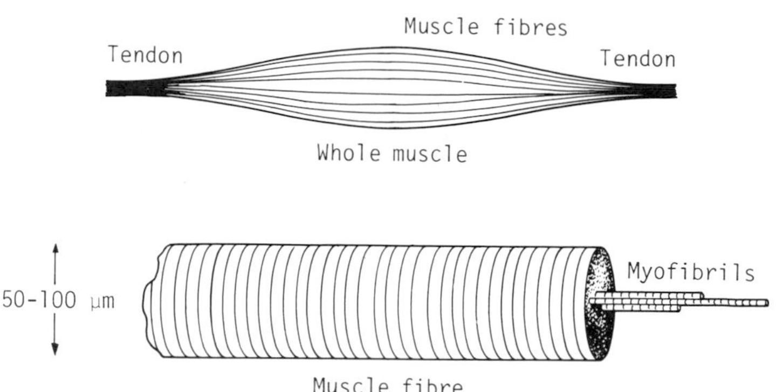

B

Within the muscle fibre, the fibrils are closely associated with the sarcoplasmic reticulum.

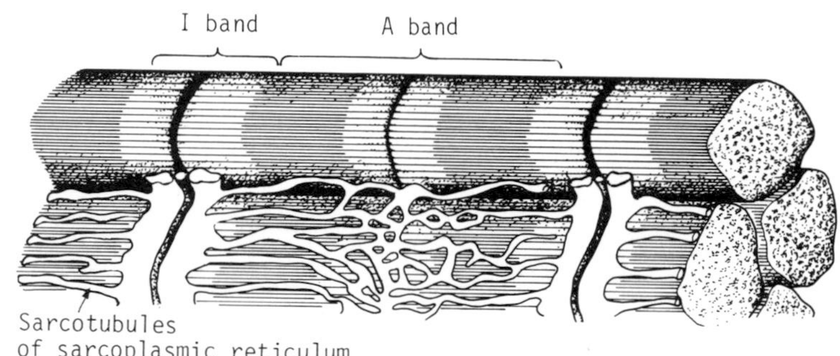

The sarcoplasmic reticulum contains one major protein, a Ca^{2+}-Stimulated ATPase. When the muscle is stimulated, the sarcoplasmic reticulum releases Ca^{2+}, which is needed for the contractile process. The Ca^{2+} is then pumped back into the sarcoplasmic reticulum by the Ca^{2+}-stimulated ATPase.

C

The myofibrils contain the protein myosin, and an assembly of the proteins actin, troponin and tropomyosin.

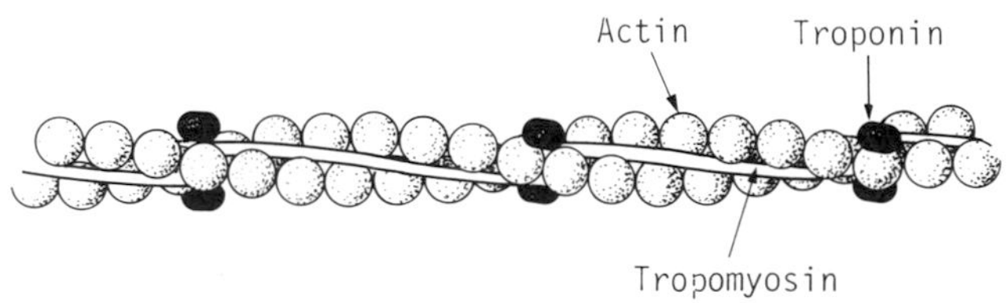

These assemblies of actin, troponin and tropomyosin are termed the *thin filaments*. These slide against the *thick filaments* of myosin during contraction.

D

The thick filaments are bundles of myosin molecules, with the myosin headpieces protruding from the side of the bundle.

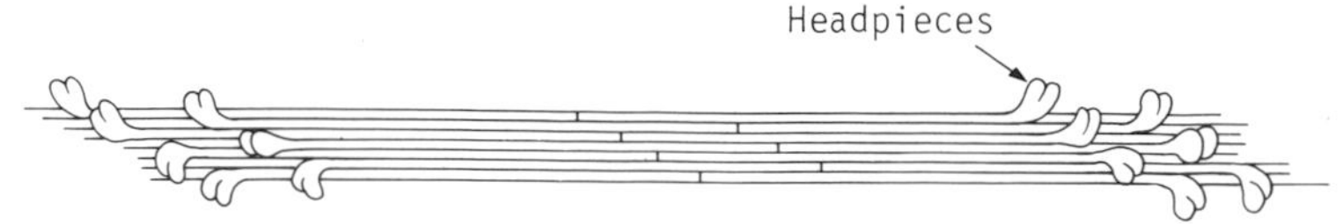

A

The three-dimensional structure of myosin conduces to the formation of complexes from which the myosin headpieces protrude.

HMM = Heavy meromyosin
HMM-S_1 = Heavy meromyosin headpiece
LMM = Light meromyosin

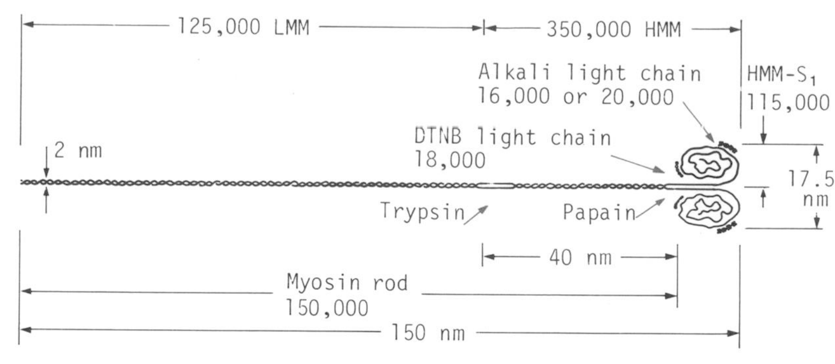

Red arrows indicate the points at which the myosin peptide chain can be cleaved by different agents. DTNB = Dithionitrobenzene

B

Under the action of ATP and Ca^{2+}, the myosin headpieces interact with the actin-troponin-tropomyosin complex to initiate the contraction process.

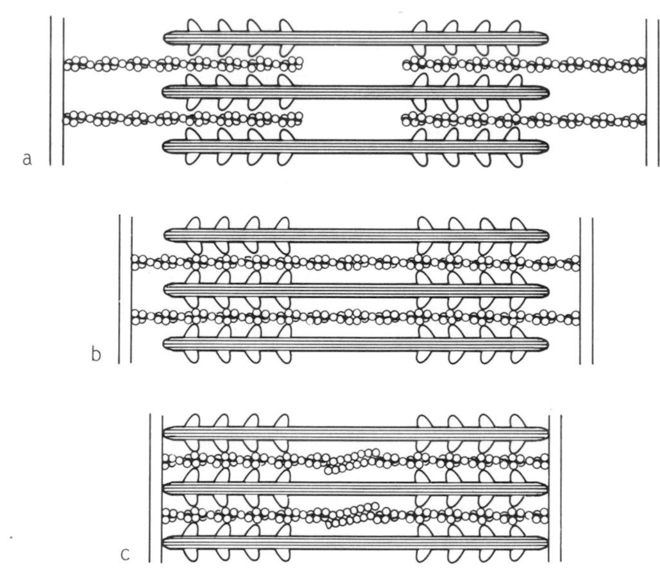

During contraction, the thin filaments (a) slide against the myosin headpieces until (b) they meet and (c) overlap. The headpieces of myosin contain ATPase activity that is associated with the mechanism by which the thick and thin filaments slide against each other.

A

The four classes of protein structures

X-ray crystallography of more than 100 different proteins has revealed common patterns.

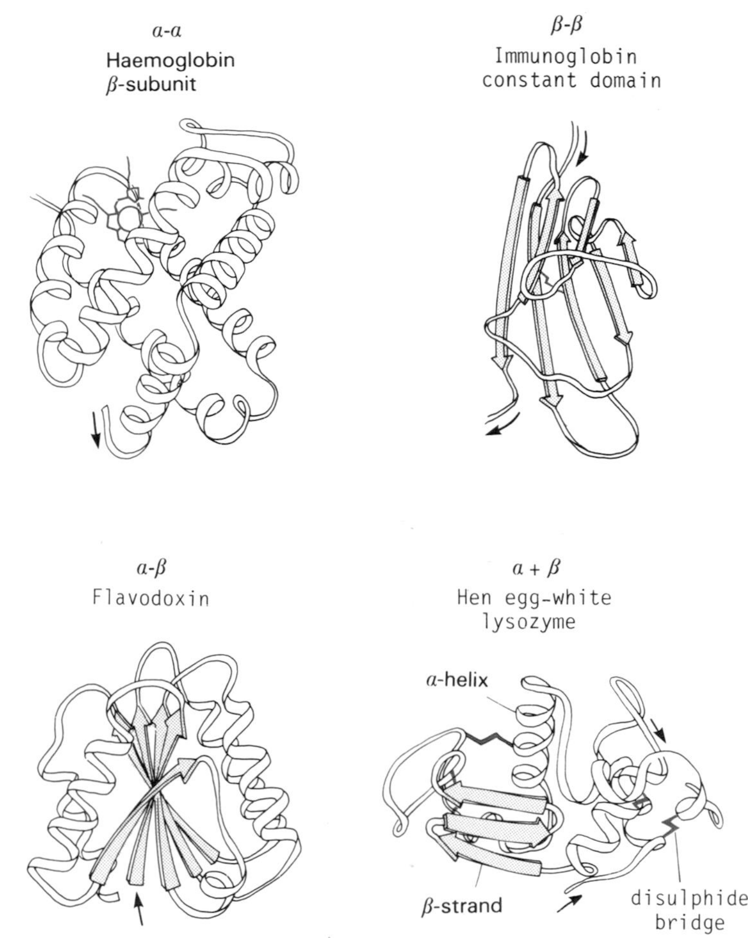

The classification is based on the relative order of α-helices and β-strands along the polypeptide chain. The α-helices and pleated (β) sheets are depicted as on page 18. In α/α structures there are mainly α-helices and little or no β-sheets. In β/β structures there are several β-strands but little or no α-helix. In IgG there are two β-sheets that pack together but they are twisted. In α/β structures α-helices and β-strands tend to alternate. Often the β-strands form a parallel sheet which is surrounded by α-helices. In α + β structures there are α-helices and β-strands that tend to segregate into different regions of the polypeptide chain. In the haemoglobin illustrated the haem is shown in red.

Summary of methods of investigation of structure and function of proteins

Modern techniques can be synchronized to unravel the relationship between structure and function of both soluble and membrane proteins (see for example, the section on membrane receptors page 267). The combination of techniques across the whole spectrum of biochemical knowledge (from molecular biology to physics) is one of the most significant advances in experimental capability in biochemical investigation.

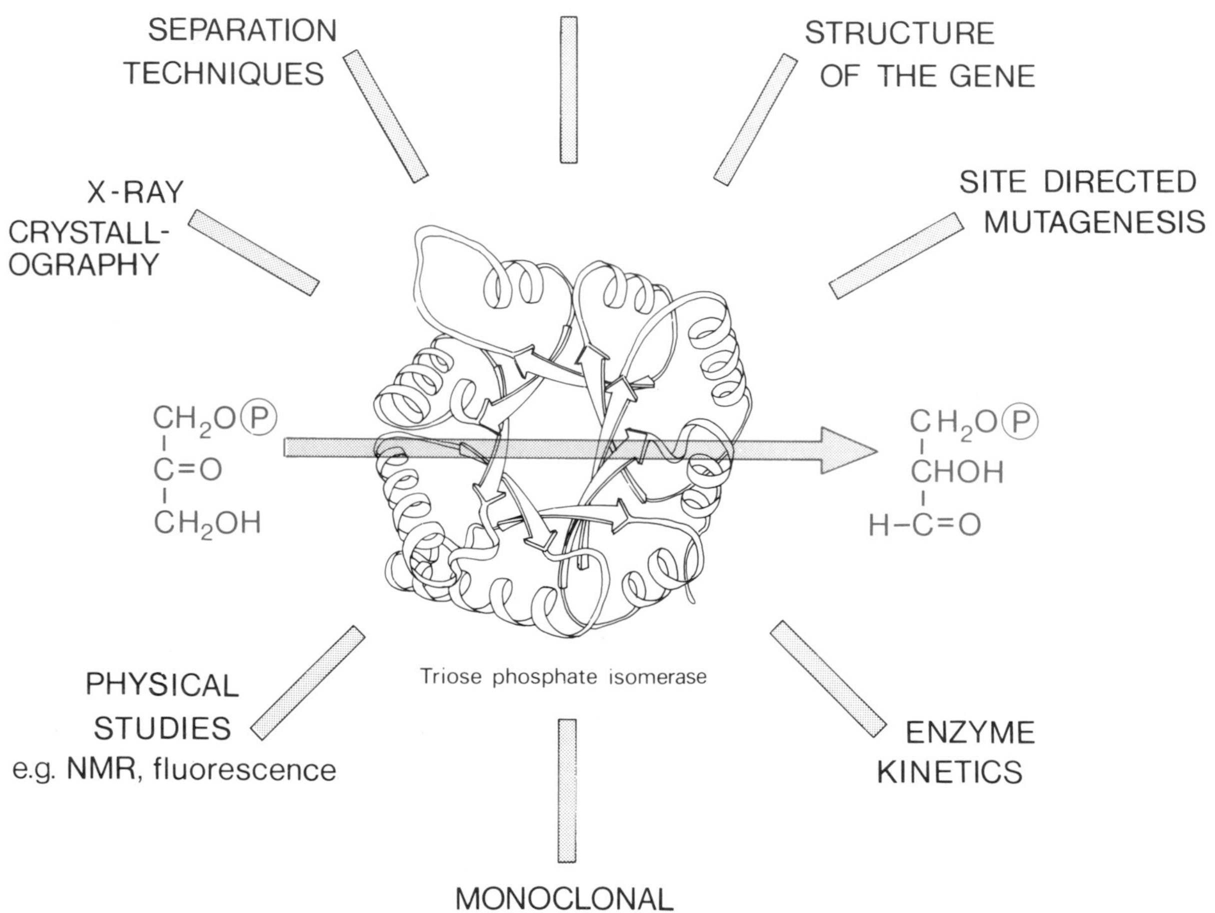

Triose phosphate isomerase

X-ray crystallography yields the 3-dimensional structure.

Separation techniques. During purification, the different separation techniques yield information regarding its size (gel permeation), charge (ion-exchange chromatography), iso-ionic point (iso-electric focussing).

Sequencing yields the primary structure, and allows predictions of regions that will be involved in hydrophobic interactions, such as in penetrating membrane bilayers.

Cloning techniques yield confirmation of protein sequence as a result of sequencing the gene. cDNA probes can be used to find variant sequences, and mutations of the clones yield information about the way that specific replacements affect function.

Enzyme kinetics yield information about the active site with which sequencing and three-dimensional structure must be compatible.

Monoclonal antibody studies probe different regions of the molecule.

Physical studies give data concerning dynamic properties of the molecule in solution (in particular nuclear magnetic resonance (NMR)).

3

THE STRUCTURE AND FUNCTION OF ENZYMES

A

Enzyme properties and kinetics

The 'initial rate' of an enzyme reaction is the rate at the earliest time that the reaction can be measured after mixing the reactants. It is used for all enzyme assays, and all calculations concerning enzyme kinetics. It is denoted by v_i, or more often just v.

Enzymes are proteins*. They act as catalysts to accelerate the rate at which a reaction proceeds towards equilibrium. (Enzymes cannot change the position of equilibrium.) Enzymes are specific with respect to the substrate with which they interact.

**Although recently it has been shown that RNA can also have enzymic properties, e.g. RNaseP from E. coli*

B

Important variables which affect the rate of an enzyme reaction are temperature, pH and amount of enzyme present.

A linear relationship between the amount of enzyme and the initial rate of reaction is often (but not always) found.

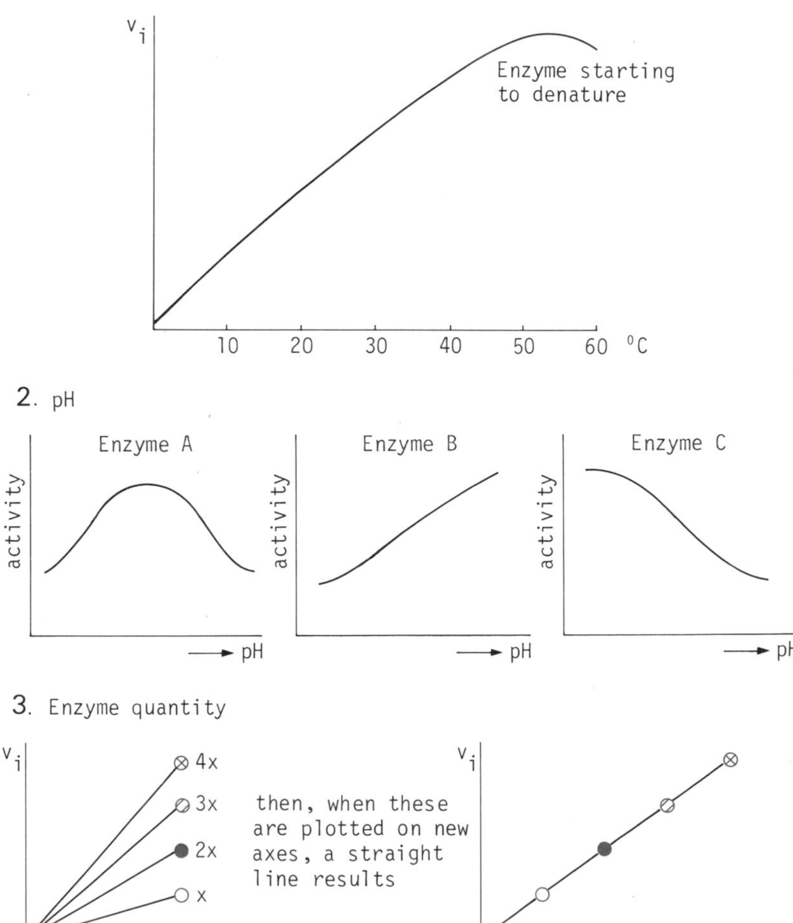

A

Maximum rate (V_{max}) is the maximum initial rate that can theoretically occur at a given concentration of enzyme and with infinitely high substrate concentration.

The Michaelis constant (K_m) corresponds to the substrate concentration at which the initial rate is half the maximum rate. K_m is independent of enzyme concentration.
(For a derivation of the Michaelis-Menten equation, see the following page).

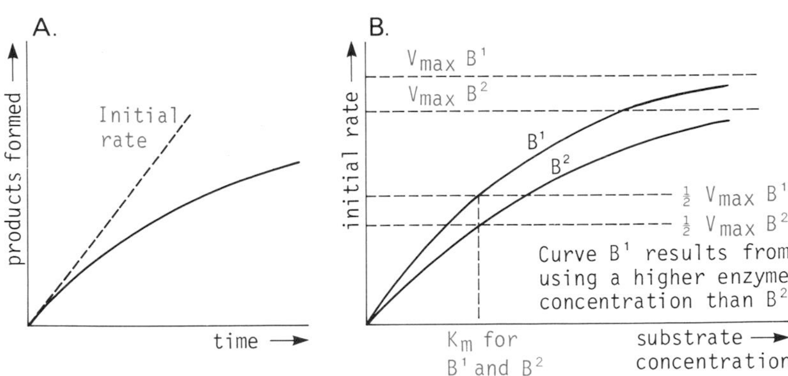

Although both of these curves look similar, B is derived from a series of measurements of initial rate as shown below.

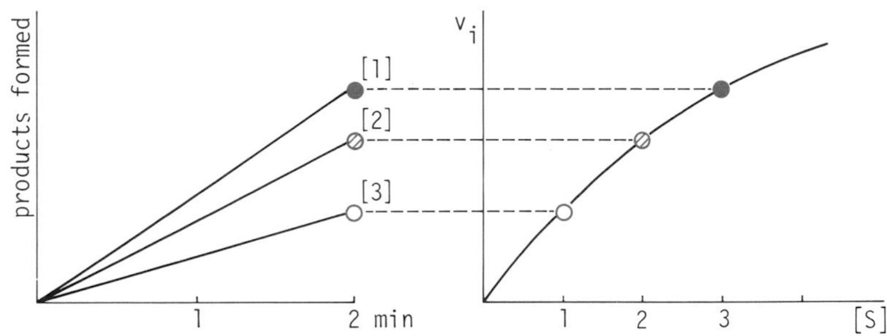

B

V_{max} and K_m are difficult to determine from plots such as those above. They are easier to determine from a plot of the reciprocal of the initial rate (1/v) against the reciprocal of the substrate concentration 1/[S], the *Lineweaver-Burk plot.*

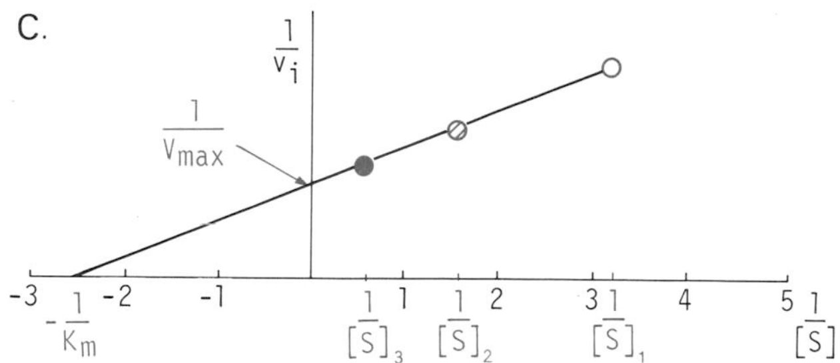

$1/v = 1/V_{max} + K_m/V_{max} \cdot 1/[S]$ is the equation for a straight line with intercept on the y axis $1/V_{max}$ and slope K_m/V_{max}. Hence the Lineweaver-Burk plot.

Derivation of the Michaelis-Menten equation

A mathematical analysis of the kinetics of an enzyme-catalysed reaction was proposed by Michaelis and Menten, whose names are always associated with the equation derived below. The derivation is based on a model of enzyme action in which the enzyme (E) binds to a single substrate (S) to form an enzyme-substrate complex (ES), which breaks down to form products (P) and to liberate free enzyme (E).

$$E + S \underset{k_{-1}}{\overset{k_1}{\rightleftharpoons}} E\,S \xrightarrow{k_2} E + P$$

k_1, k_{-1} and k_2 are rate constants for the various reactions as indicated above.

It is assumed that the reaction is in the steady state, with the concentration of ES constant, and the analysis applies only to the start of the reaction, at which point negligible amounts of products have been formed, so that the reverse reaction $E + P \longrightarrow ES$ is also negligible.
Then

$$\frac{d[ES]}{dt} = k_1([E_t] - [ES])\,[S] - k_{-1}\,[ES] - k_2\,[ES]$$

$$= 0 \quad (E_t = \text{the total amount of enzyme})$$

$$k_1\,[E_t]\,[S] - k_1\,[ES]\,[S] - k_{-1}\,[ES] - k_2[ES] = 0$$

$$k_1\,[E_t]\,[S] = [ES]\,(k_1\,[S] + k_{-1} + k_2)$$

$$\frac{k_1\,[E_t]\,[S]}{k_1[S] + k_{-1} + k_2} = [ES]$$

Divide by k_1

$$ES = \frac{[E_t]\,[S]}{[S] + \frac{k_{-1} + k_2}{k_1}}$$

Then initial rate (v) of formation of product $= k_2\,[ES]$
and when the enzyme is saturated with substrate maximum rate (V_{max}) $= k_2\,[E_t]$
and the Michaelis constant

$$K_m = \frac{k_{-1} + k_2}{k_1}$$

$$\text{So } v = k_2\,[ES] = \frac{k_2\,[E_t]\,[S]}{[S] + K_m}$$

$$v = \frac{V_{max}\,[S]}{[S] + K_m}$$

Strictly this derivation is restricted to a single substrate reaction. In multisubstrate reactions, if other factors are held constant and the concentration of one substrate is varied, the initial rate plotted against substrate concentration often produces a hyperbolic curve which can be analysed using the Michaelis-Menten equation.

Indeed, a hyperbolic relationship is observed when many biological events are examined kinetically. These, apart from enzyme reactions, include transport phenomena, the binding of ligands (eg. hormones) to receptors, and drug-related effects.

Enzyme Inhibition

Many substances inhibit enzymes and reduce the initial velocity (increase 1/v). The Lineweaver-Burk plot reveals the mechanism of the inhibition.

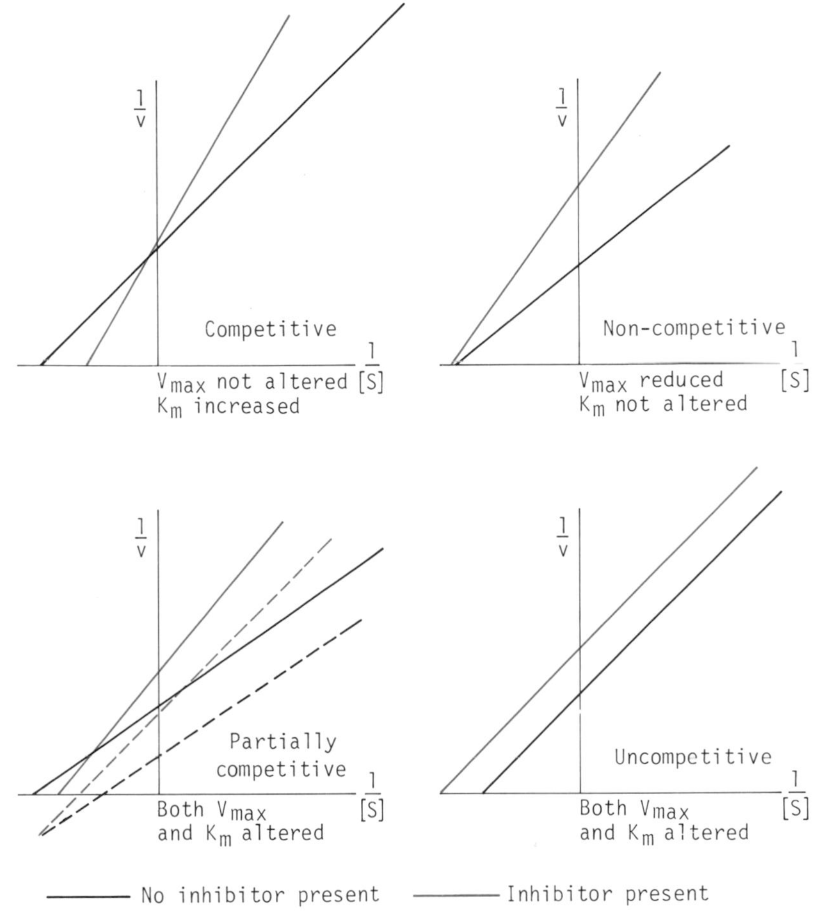

Competitive inhibition occurs when both substrate and inhibitor will fit the active site, and compete with each other to occupy it.

Non-competitive inhibition occurs when the inhibitor binds to a site on the enzyme other than the active site or binds irreversibly to the active site.

Partially competitive inhibition is a special instance of non-competitive inhibition, where the inhibitor binding constant is different for the free enzyme and the enzyme-substrate complex. Two alternatives are indicated by the solid and broken lines.

Uncompetitive inhibition occurs when the inhibitor binds after the substrate has bound to the enzyme, and then stops the reaction occurring.

A

International classification of enzymes (class names, code numbers and types of reactions catalysed)

1. Oxido-reductases (oxidation-reduction reactions)
   - 1.1 Acting on CH−OH
   - 1.2 Acting on C=O
   - 1.3 Acting on C=CH−
   - 1.4 Acting on $CH-NH_2$
   - 1.5 Acting on CH−NH−
   - 1.6 Acting on NADH; NADPH
2. Transferases (transfer of functional groups)
   - 2.1 One-carbon groups
   - 2.2 Aldehydic or ketonic groups
   - 2.3 Acyl groups
   - 2.4 Glycosyl groups
   - 2.7 Phosphate groups
   - 2.8 S-containing groups
3. Hydrolases (hydrolysis reactions)
   - 3.1 Esters
   - 3.2 Glycosidic bonds
   - 3.4 Peptide bonds
   - 3.5 Other C−N bonds
   - 3.6 Acid anhydrides
4. Lyases (addition to double bonds)
   - 4.1 C=C
   - 4.2 C=O
   - 4.3 C=N−
5. Isomerases (isomerization reactions)
   - 5.1 Racemases
6. Ligases (formation of bonds with ATP cleavage)
   - 6.1 C−O
   - 6.2 C−S
   - 6.3 C−N
   - 6.4 C−C

The term 'synthase' is used for enzymes in group 4 where emphasis on the synthetic activity of the enzyme is required.
The term "synthetase" is now not recommended and in any event should be confined to Group 6.

B

Nature of the active site and induced fit

The substrate binds to a region of the enzyme protein known as the active site.

The active site has regions responsible for binding the substrate and regions which catalyse the reaction. There is considerable overlap between these regions.

The term 'induced fit' implies that the shape of the binding site changes as a result of substrate binding, to be likened to the shape of a glove when a hand is inserted (see also p. 59).

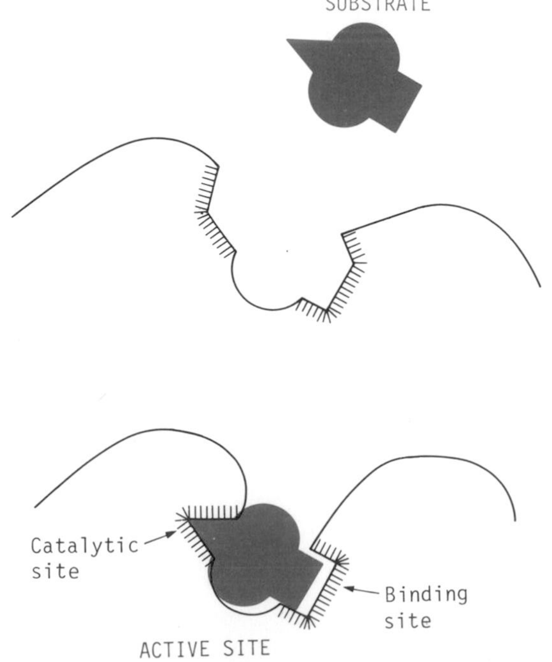

A

Protein folding is important in the formation of enzyme active sites.

Chymotrypsin is shown as an example.

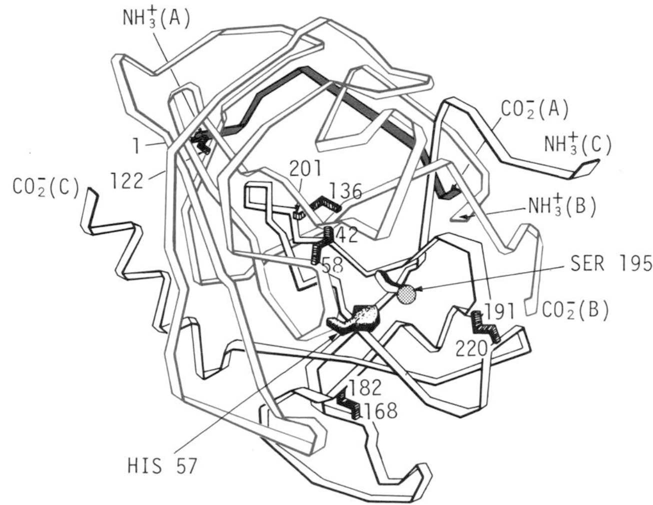

The folding of the peptide chains of the enzyme proteins is such that amino acid residues from different chains and from different regions of the same chain are brought into the correct position to form the active site, just as in haemoglobin the haem pocket is formed. Chymotrypsin is composed of three chains: A (dark red), B (light red) and C (black). His 57 is in the B chain whilst Ser 195 is in the C chain but both contribute to the active site. The amino and carboxyl terminal residues are on the outside as indicated.

B

The substrate may influence the conformation of the enzyme. This is known as *induced fit.*

The effect of glucose on hexokinase.

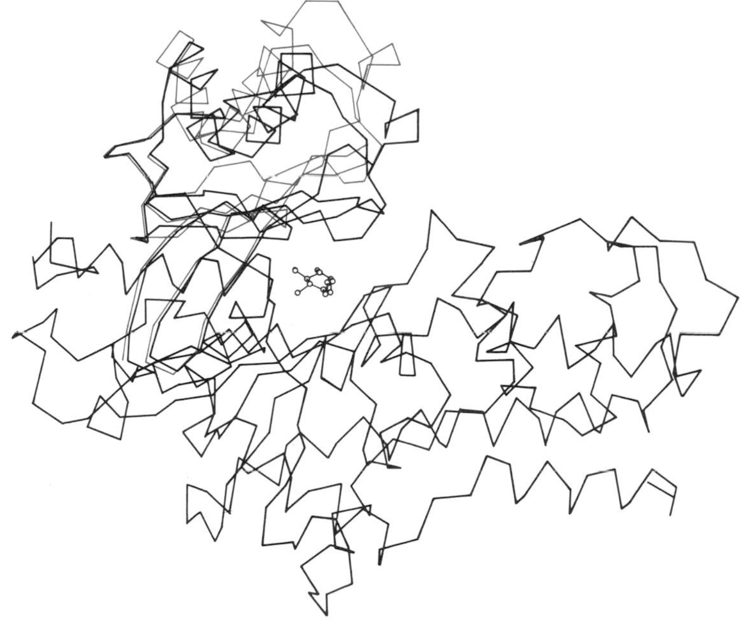

A conformational change in hexokinase is induced by glucose binding. The black lines show the α-carbon backbone of the enzyme (see p. 70) crystallized in the presence of glucose. The red lines show the backbone of that part of the enzyme that has a different structure when crystallized in the absence of glucose.

A

The active sites of two proteolytic enzymes, chymotrypsin and elastase, differ in that a bulky side chain intrudes into the pocket in elastase.

The peptide substrate is shown in red in both cases

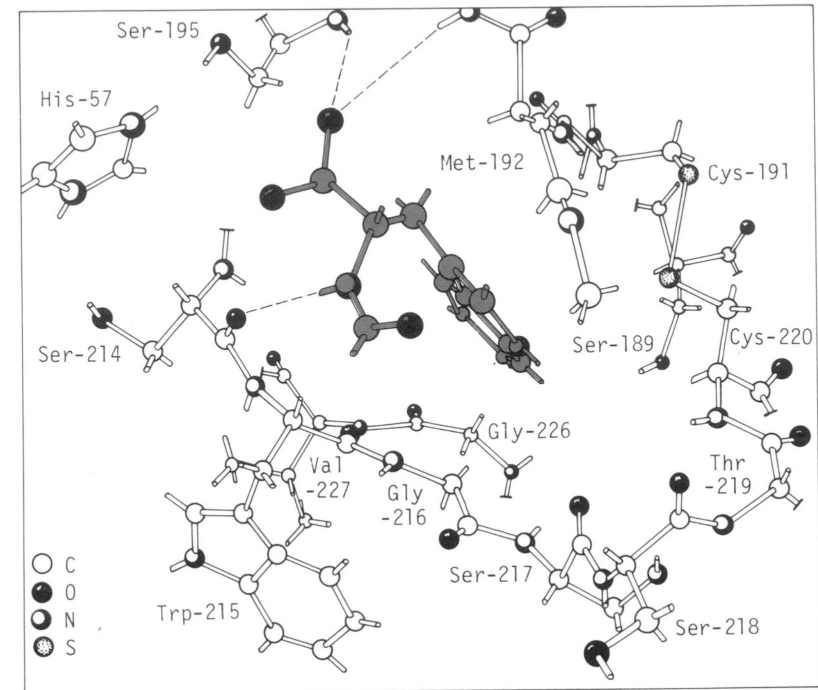

B

The substrate of elastase has a correspondingly smaller residue occupying the active site than is the case with chymotrypsin. The detailed structures reveal the precise 'tailoring' of the active site to the substrate.

Elastase, showing instrusion of the bulky side chain of valine and threonine (in pink) at positions 216 and 226 in contrast to the glycines at these positions in chymotrypsin above.

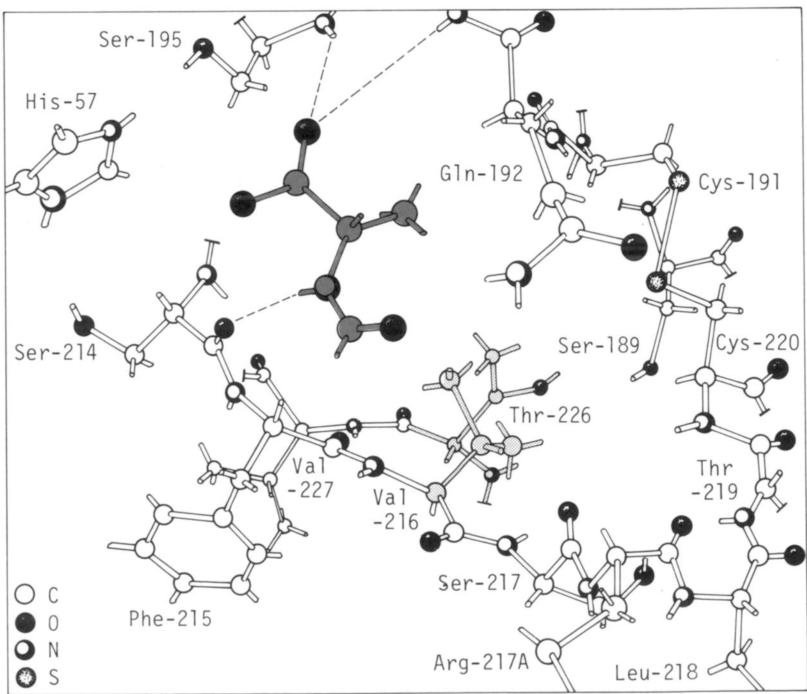

Comparison of the binding pockets in chymotrypsin (top, with N-formyl-L-tryptophan bound) and elastase (bottom, with N-formyl-L-alanine bound). The binding pocket in trypsin is very similar to that in chymotrypsin except that residue 189 is an asparate to bind positively charged side chains. Note the hydrogen bonds between the substrate and backbone of the enzyme.

A

Evolutionary aspects of protein structure and function

Enzymes in evolution. We may trace the change in structure of cytochrome c.

Analysis of the structure of *homologous proteins*, i.e. those having an identical function, from many species that have evolved over widely different evolutionary pathways, indicates to what extent a structure has been conserved during evolution.

An abbreviated phylogenetic tree of the evolution of cytochrome c is shown in B below. The figures in red indicate the number of amino acid residues (per 100) by which the cytochromes c differ between each evolutionary branch point.

The amino acid sequence of cytochrome c from more than 60 different species is known. These sequences vary considerably at certain positions. However every cytochrome c so far sequenced has histidine at a position corresponding to 18 in the human; that is to say it is invariant. Altogether there are 26 completely invariant residues.

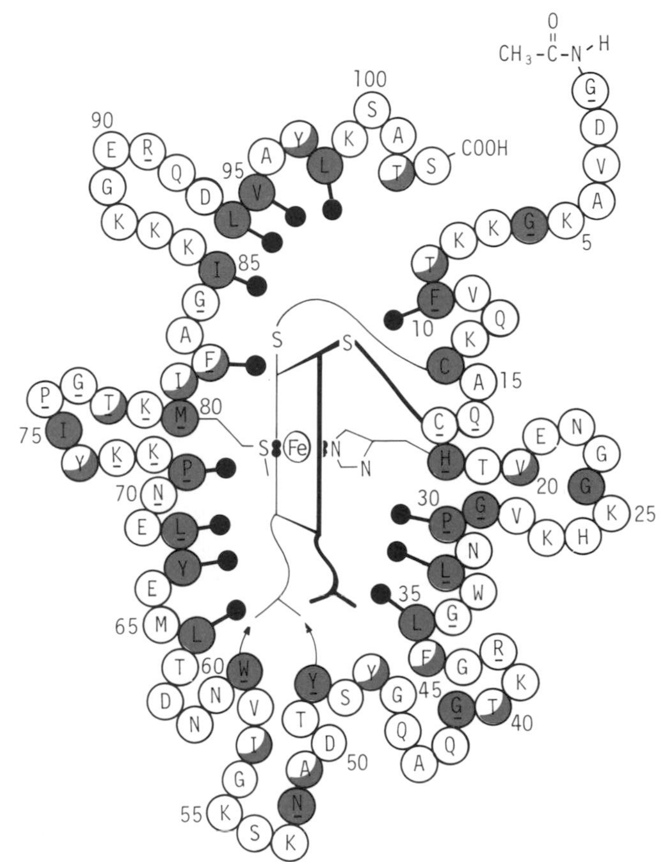

Chain-packing diagram of tuna-reduced cytochrome c, showing evolutionary invariant amino acids (underlined capitals), buried side chains (red circles), side chains packed against the haem (black dots) and side chains half buried (half-red circles).

B

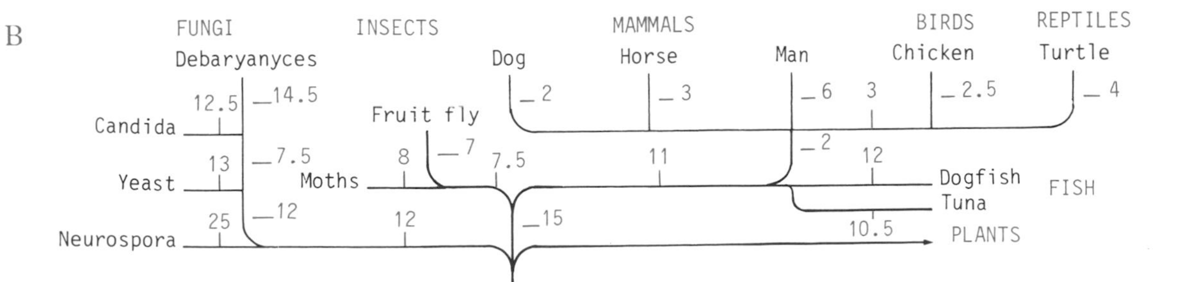

Cytochrome c phylogenetic tree

The *dehydrogenase structures* are an excellent example of the use of pleated sheet, helical and random coil regions in a globular protein.

Pleated sheets are represented by the broad arrows. The cylinders represent α-helical regions. The thin tubular portions indicate regions of random coil (i.e. no defined secondary structure). Numbers without prefixes refer to amino acid positions. Prefixes α and β label helical and pleated sheet regions respectively.

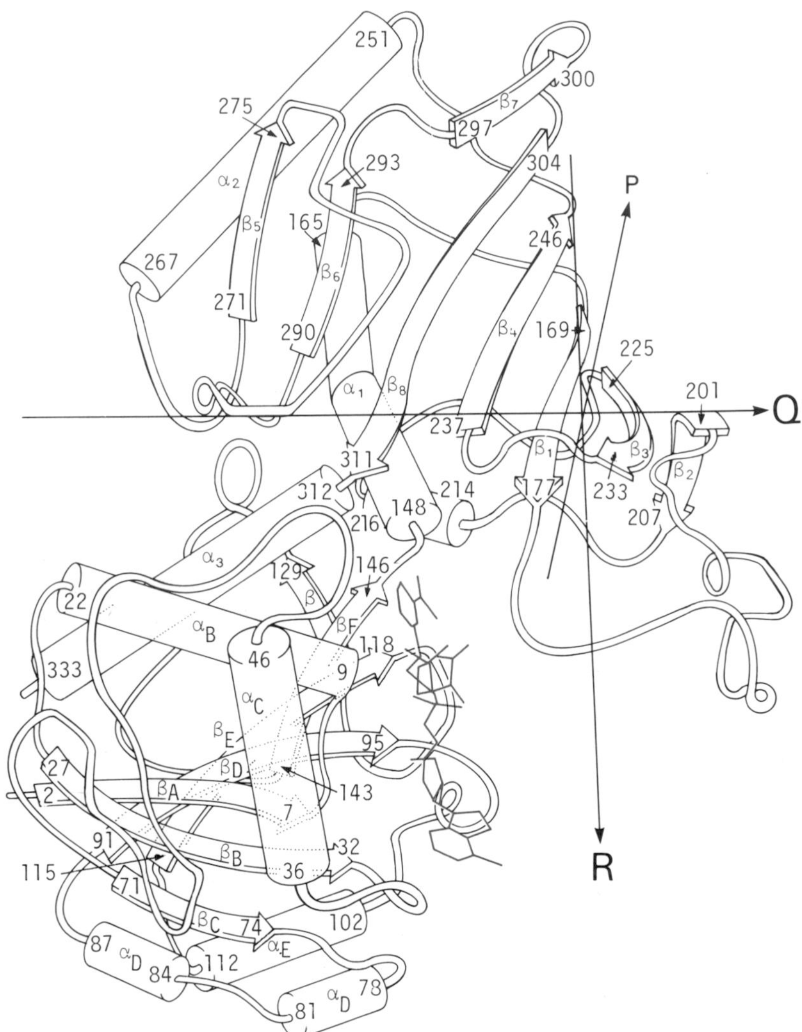

The structure of glyceraldehyde 3-phosphate dehydrogenase is shown with the NAD$^+$ molecule in red.

A

Convergent evolution may be shown by the binding sites in the dehydrogenase enzymes for NAD^+.

The binding of the coenzyme NAD^+ to the dehydrogenase enzymes involves a series of pleated sheets that seem to be common to many of these enzymes and are shown more clearly in this diagram than on the previous page.

If the structure of a group of enzymes having a similar but not identical catalytic role is studied, particularly with respect to their active sites, similarities are often found. This does not indicate the conservation of a structure but rather that beneficial mutations have been selected. Hence the term for this phenomenon—convergent evolution.

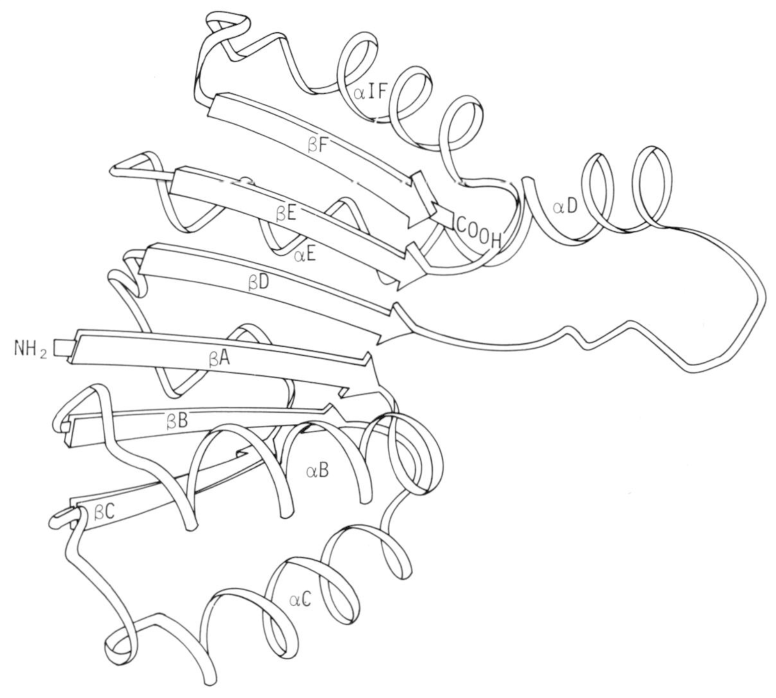

B

Allosteric properties of enzymes

The rate of an enzyme reaction can be affected by the presence of a substance that combines with the enzyme protein at a site other than the active site. Such a substance is known as a *modulator, modifier* or *effector*.

For the reason for the sigmoid curve *see* p. 64

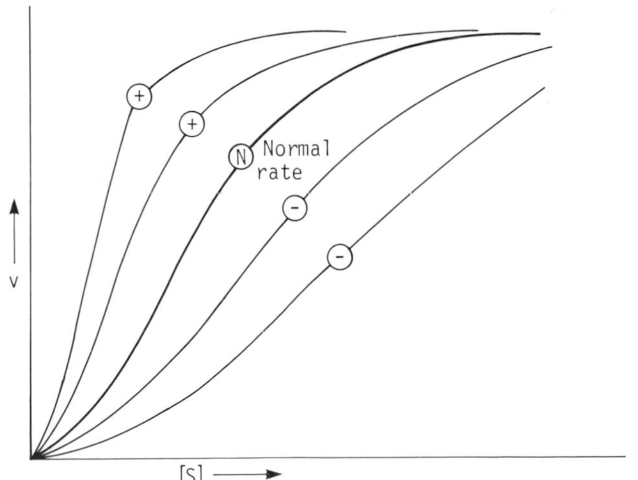

Influence of modulators on the substrate (S)—velocity (v) curves of a regulatory enzyme. (+) *Positive modulation* (−) *negative modulation.*

A

An enzyme can be inhibited by high concentrations of its own substrate. In this case some of the substrate binds to an allosteric site. One example of such an enzyme is *phosphofructokinase* (PFK). This enzyme is inhibited by its own substrate ATP.

The different sites indicated are for binding substrates and effectors as follows:
A F6P, C* ADP activator
A′ FBP, C′ ATP inhibitor
B ATP substrate
*not shown

Most allosteric enzymes are oligomeric proteins and it is for this reason that their kinetics are described by a sigmoid curve (compare haemoglobin). Interaction between subunits is important in *co-operative* effects. Phosphofructokinase is composed of subunits, each of which consists of two pear-shaped domains linked by a helical region.

Four of these subunits pack together in the mammalian enzyme. Such a tetramer carries four active sites, and a number of allosteric sites. The active site is in a cleft *on* each subunit as shown by B. The allosteric sites are thought to lie in clefts formed *between* subunits, as shown by C′ at the bottom and A and A′ in the clefts at the top.

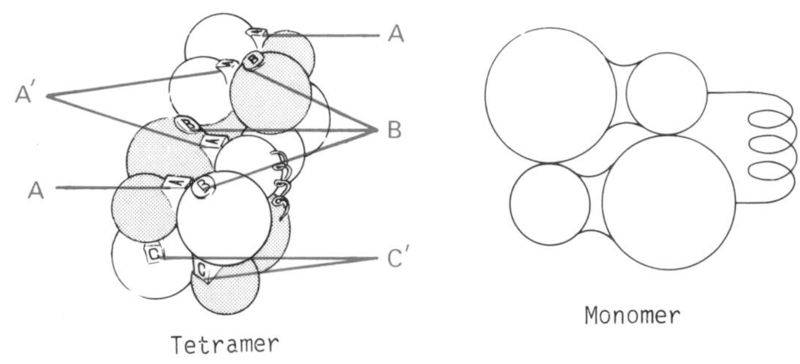

B

A more detailed diagram indicates more clearly the nature of the allosteric sites, in phosphofructokinase (PFK).

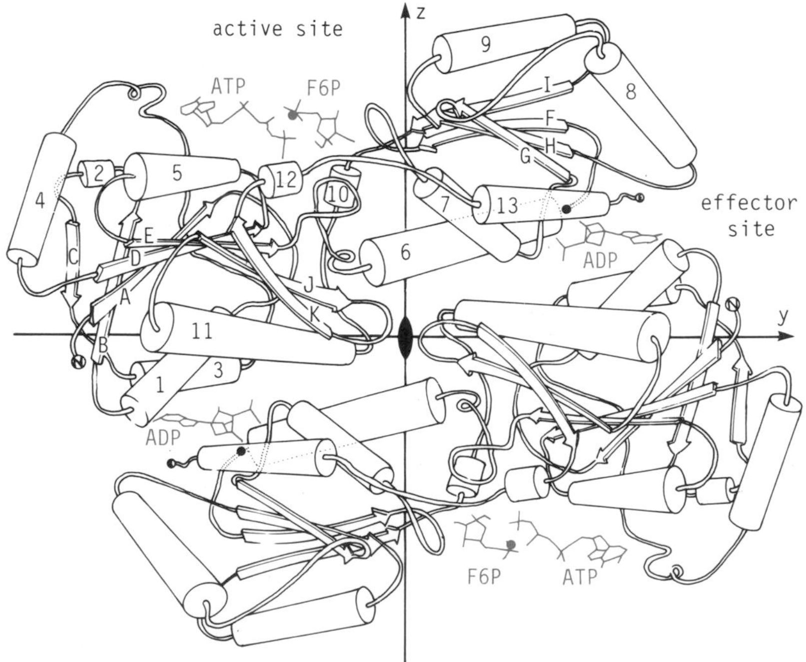

A schematic view of two subunits of the PFK tetramer, viewed along the x axis. The other two subunits lie behind the two shown. β-Sheet strands are represented by arrows (A–K) and α-helices by cylinders (1–13). Each subunit consists of two domains: domain 1 is on the left in the upper subunit. The substrates ATP and F6P are shown in the active site, and the activator ADP in the effector site.

A

Energy, Metabolic Control and Equilibria

Free energy

Associated with every compound is a quantity, G, known as its *free energy*

For a molecule in solution, free energy is concentration-dependent. The free energy of a particular component of a solution is equal to the free energy per mole times the number of moles of that component. *The standard free energy* (G^o) is equal to the free energy of the component at *unit activity* (in dilute solution this is approximately 1M). The molar free energy of a particular component of a solution is often referred to as the *chemical potential* of that component.

Thus for the *i* th component of a solution

$$G_i = G_i^o + RT\ln a_i \quad \text{———} (1)$$

where R = the gas constant (8.134 J $mole^{-1}$ K^{-1}), T = the absolute temperature, and a_i is the activity of the *i* th component (in dilute solution, activity approximates to concentration).

B

For a chemical reaction there is a quantity, $\triangle G$, known as the *free energy change* of the reaction, which is calculated as the difference between the free energies of the reactants and products at a particular instant in time.

For a reaction

$$A + B \rightleftharpoons C + D$$

$$\triangle G = G_C + G_D - G_A - G_B$$

Using equation (1), and observing that $G_C^o + G_D^o - G_A^o - G_B^o$ is termed the *standard free energy change* ($\triangle G^o$) of the reaction, then

$$\triangle G = \triangle G^o + RT\ln \frac{[C]_C\,[D]_D}{[A]_A\,[B]_B} \quad \text{———} (2)$$

Where $[A]_A$, $[B]_B$, $[C]_C$ and $[D]_D$ are the concentrations pertaining at the moment for which $\triangle G$ is calculated.

The magnitude and sign of $\triangle G$ determine the direction in which the reaction will proceed. It will always proceed in a direction that tends to bring the free energies of the components on one side of the equation equal to those on the other. When the free energies on both sides of the reaction are equal, equilibrium has been reached. It follows that at equilibrium the free energy change is zero. If $\triangle G$ is positive, $\triangle G$ for the reaction in the reverse direction will be negative, and thus the reaction will proceed in the reverse direction.

C

The standard free energy change is the quantity which is used to compare free energy change between one reaction and another, and is a constant at constant temperature.

The standard free energy change is the free energy change when [A], [B], [C] and [D] represent unit activity (for practical purposes 1M) since under these conditions, the second term on the right of equation (2) above becomes zero (ln 1 = 0) and $\triangle G = \triangle G^o$.

We now write an equation for the equilibrium position, where $\triangle G = 0$

$$0 = \triangle G^o + RT\ln \frac{[C]_E[D]_E}{[A]_E[B]_E}$$

$$\text{or } \triangle G^{o\prime} = -RT\ln K_{eq}$$

Where $[A]_E$, $[B]_E$, $[C]_E$ and $[D]_E$ are equilibrium concentrations of A,B,C and D.

If the equilibrium constant is known therefore, the standard free energy change can be readily found.

Note: $\triangle G^{o\prime}$ is used to denote that one component, H^+ is not at unit activity, when measurements are made at pH 7.0.

A

Free energy changes are involved in the reactions of biochemical pathways, which proceed in the direction of an overall negative free energy change.

It is often pointed out that the standard free energy change for the hydrolysis of ATP is negative and large compared with that for a phosphate ester, such as glycerol 3-phosphate. This in the past has given rise to the expression 'high-energy compound'.

$$ATP^{4-} + H_2O \longrightarrow ADP^{3-} + P_i + H^+ \qquad \Delta G^{0\prime} = -30.5 \text{ kJ/mol}$$
$$\text{Glycerol 3-phosphate} + H_2O \longrightarrow \text{glycerol} + P_i \qquad \Delta G^{0\prime} = -9.2 \text{ kJ/mol}$$

For this reason, reactions involving ATP often proceed in the direction that results in formation of ADP and P_i. However, in order that ATP stores may be replenished, the reverse reaction must also occur in biological systems. This will be the case when the reaction also involves other compounds of such a nature that the reaction results in a large negative free energy change, again dependent on the concentrations of reactants and products.

The most important example of this is when phosphorylation of ADP is coupled to the oxidation of substrates by the electron transport chain (p. 179).

$$NADH + H^+ + O_2 \longrightarrow NAD^+ + H_2O \qquad \Delta G^{0\prime} = -222 \text{ kJ/mol}$$

Another example is

$$\text{Phosphoenolpyruvate} + \text{ADP} \longrightarrow \text{Pyruvate} + \text{ATP}$$

(The standard free energy change for the hydrolysis of phosphoenolpyruvate is — 62 kJ/mol).

Metabolic Pathways

In a metabolic pathway, the rate at which each reaction in the pathway proceeds will depend on the activity of each enzyme and the concentrations of reactants and products. If the activity of the enzyme is high, and the thermodynamic considerations favourable, the reaction may be held near its equilibrium position. If the activity of the enzyme is low relative to that of other enzymes in the pathway, it will be saturated with substrate and will operate far from its equilibrium position. In either case, the enzyme may operate to regulate the rate of the pathway and may be the target for compounds that influence the rate of the pathway. However, only reactions operating far from equilibrium will direct the flow, or 'flux' of the pathway, and the reactions they catalyse are termed flux-generating reactions.

B *Equilibria*

A number of useful expressions can be derived from the equilibrium equation for any event that involves the association of two components. Such events include acid-base interactions, the binding of ligands (e.g. hormones) to receptors, membrane transport phenomena, and cooperativity as exemplified by the binding by haemoglobin of 0_2 and in some cases, the action of allosteric or non-allosteric enzymes. **The Henderson-Hasselbalch** equation is important for making calculations concerning the pH of a buffer solution.

In general, for an equilibrium $[AB] \rightleftharpoons [A] + [B]$.

$$\frac{[AB]}{[A][B]} = K_{association}, \quad \frac{[A][B]}{[AB]} = K_{dissociation}$$

For pH equilibria, HA is a conjugate acid and A^- a conjugate base

$$\frac{[H^+][A^-]}{[HA]} = K_a \quad \text{———} \ (1)$$

(K_a = acid dissociation constant)

Taking logs,

$$\log[H^+] + \log[A^-] - \log[HA] = \log K_a \quad \text{———} \ (2)$$

pH is defined as $-\log[H^+]$, pK_a as $-\log K_a$

Thus, multiplying (2) by -1 and rearranging

$$pH = pK_a + \log\frac{[A]}{[HA]} \quad \text{———} \ (3)$$

This is the Henderson-Hasselbalch equation.

A

The Hill equation is used to analyse binding in terms of the fraction of ligand bound e.g. for the binding of O_2 by myoglobin.

The concentration of oxygen $[O_2]$ is proportional to the partial pressure of oxygen, pO_2. Thus

$$K = \frac{[Mb].pO_2}{[MbO_2]} \text{ and } [Mb] = \frac{K[MbO_2]}{pO_2} \quad \text{———} \quad (4)$$

We can find an expression for the fraction (Y) of myoglobin that has bound oxygen.

$$Y = \frac{[MbO_2]}{[Mb] + [MbO_2]} \text{ but from (4) we can substitute for [Mb]}$$

$$Y = \frac{[MbO_2]}{\frac{K[MbO_2]}{pO_2} + [MbO_2]}$$

Rearranging,

$$Y = \frac{pO_2}{K + pO_2} \quad \text{———} \quad (5)$$

When exactly half the number of sites is occupied

$$Y = 0.5 \text{ and } K = pO_2$$

Thus K is designated P_{50}

Equation (5) can be rearranged to yield $\frac{Y}{1 - Y} = \frac{pO_2}{P_{50}}$ (the Hill equation)

Thus $\log \frac{Y}{1 - Y} = \log pO_2 - \log P_{50}$

A plot of $\log \frac{Y}{1 - Y}$ against $\log pO_2$ yields a straight line of slope 1. This is known as the Hill plot.

B

For haemoglobin, a plot of the fractional saturation against pO_2 is sigmoid. The sigmoidicity reflects cooperativity in the binding of the O_2. That is, binding of one oxygen promotes binding of the next.

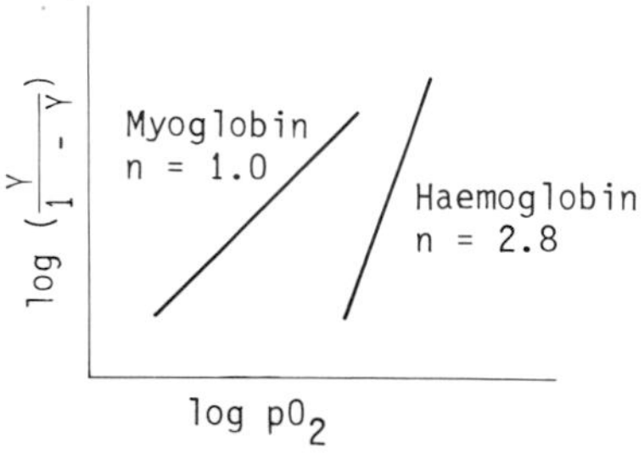

The degree of cooperativity can be determined as the Hill coefficient (n), obtained as the slope of the Hill plots. For Hill coefficients greater than 1, n must appear in the Hill equation.

$$\frac{Y}{1 - Y} = \frac{pO_2^{\,n}}{P_{50}}$$

and

$$\log \frac{Y}{1 - Y} = n\log pO_2 - \log P_{50}$$

This equation gives the Hill coefficient as the slope of a plot of $\log \frac{Y}{1 - Y}$ against $\log pO_2$.

The Hill plot is useful also for the analysis of cooperativity of allosteric enzymes.

A

It is possible to obtain useful derivations from the equilibrium equation for the binding of substrate to enzyme.

If [E] represents the concentration of free enzyme, [ES] the concentration of enzyme-substrate complex, and [S] the substrate concentration, then

$$\frac{[E][S]}{[ES]} = K_{diss} \text{ and } \frac{[E]}{[ES]} = \frac{K_{diss}}{[S]}$$

But $[E] = [E_t] - [ES]$ where $[E_t]$ = the total amount of enzyme.

Thus $$[E_t] - [ES] = \frac{K_{diss}[ES]}{[S]}$$

$$[E_t] = \frac{K_{diss}[ES] + [ES][S]}{[S]} \text{ and } \frac{[E_t]}{[ES]} = \frac{K_{diss} + [S]}{[S]}$$

In the most simple situation,

$$E + S \underset{k_{-1}}{\overset{k_1}{\rightleftharpoons}} ES \xrightarrow{k_2} \text{Products}$$

the velocity (v) of the reaction $= k_2[ES]$
and when the enzyme is fully saturated with substrate, the maximum velocity, $V_{max} = k_2[E_t]$.

Thus $$\frac{[E_t]}{[ES]} = \frac{V_{max}}{v} = \frac{K_{diss} + [S]}{[S]}$$

and $$v = \frac{V_{max}[S]}{K_{diss} + [S]}$$

If one substitutes the Michaelis constant (K_m) for K_{diss} in the above equation, the Michaelis-Menten equation results. This equation was derived in a more conventional way in a previous section (p. 56). It is by no means always true that the Michaelis constant represents the dissociation constant of the enzyme-substrate complex, but this is so in a number of cases where $k_2 \ll k_{-1}$. Such an equation can also be derived for the kinetics of membrane transport systems.

B

The binding of ligands to receptors can be analysed using the *Scatchard plot*.

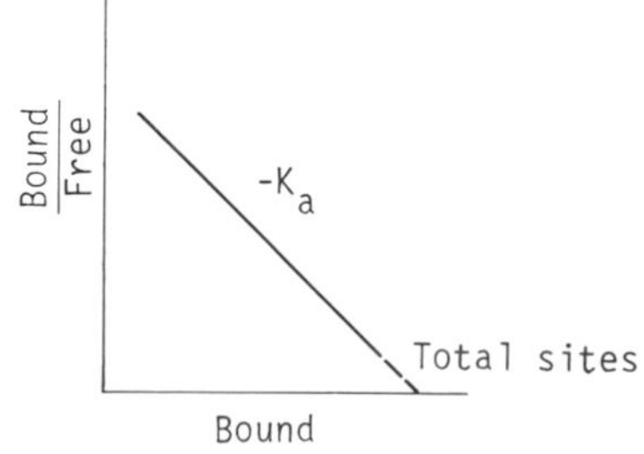

For ligand binding, if [L] is the concentration of free ligand (or F), [R] is the concentration of free receptors, and [LR] represents bound ligand (or B)

$$\frac{[LR]}{[L][R]} = K_{association}, \quad \frac{[LR]}{[L]} = K[R]$$

Thus,

$$\frac{\text{Bound}}{\text{Free}} = \frac{B}{F} = K[R] = K[T-LR]$$

Where T = the total number of sites (R + LR).

This is the equation of a straight line the slope of which yields the association constant K, and the total number of binding sites which can be determined from the intercept on the abscissa. This is known as a Scatchard plot.

Regulation of metabolic pathways—general principles

A

The enzyme reactions in a pathway may be operating near their equilibrium position, or far removed from it. The latter *non-equilibrium* reactions have an important function in pathway regulation.

The factor which determines whether an enzyme will be non-equilibrium is its activity. In the pathway below, if the activity of enzyme 2 is low compared with that of 1, it will remain saturated with substrate and will be a flux-generating reaction (see page 66)

1 2 3
A ⟹ B ⟹ C ⟹ D

B

Flux-generating reactions (red bars = A, D, E) are often found at the start of pathways or at branch points.

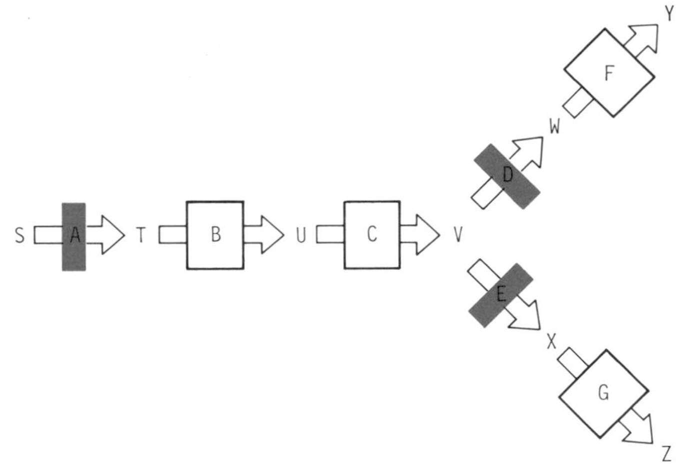

C

Feedback inhibition can occur within a pathway or between pathways if a metabolite (e.g. X in the diagram) acts as an allosteric inhibitor from one pathway acting on another.

Individual pathways can be regulated by allosteric inhibition (or activation) by certain metabolites of the same or other pathways.

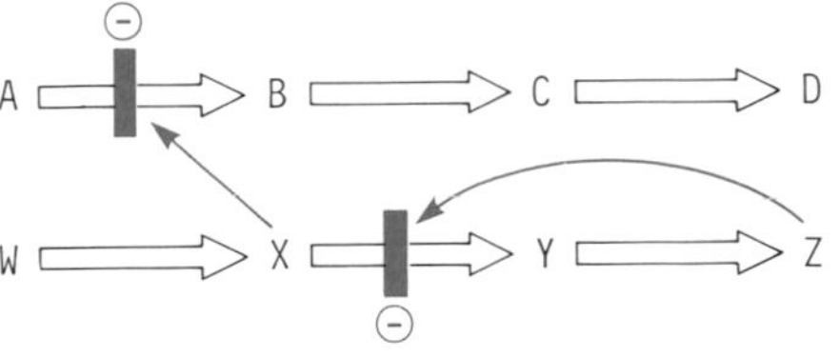

Regulation of this type operates in a specific way to regulate particular pathways.
More widespread and general regulation is exerted by changes in ATP/ADP, $NADH/NAD^+$, or $NADPH/NADP^+$ ratios. For example, changes in the $NADH/NAD^+$ ratio will affect all dehydrogenases utilizing this pair of coenzymes in a given cell compartment.

A

Multiple forms of enzymes, isoenzymes

Isoenzymes are proteins with different structures which catalyse identical reactions (also called isozymes).

Lactate dehydrogenases (LDH)

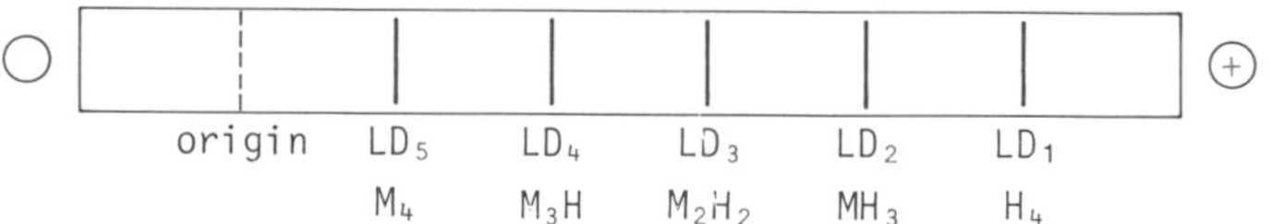

After electrophoresis of an extract containing LDH many protein bands possessing LDH activity may be detected. This arises because 2 different subunits of different charge are present in the cell. These are termed M (muscle) and H (heart). Permutations of these subunit types among the four subunits of the active enzyme give five possible combinations.

In the case of LDH M α-hydroxybutyrate is the preferred substrate so it is referred to as α-hydroxybutyrate dehydrogenase (αHBD) even though αHB does not occur in the body.

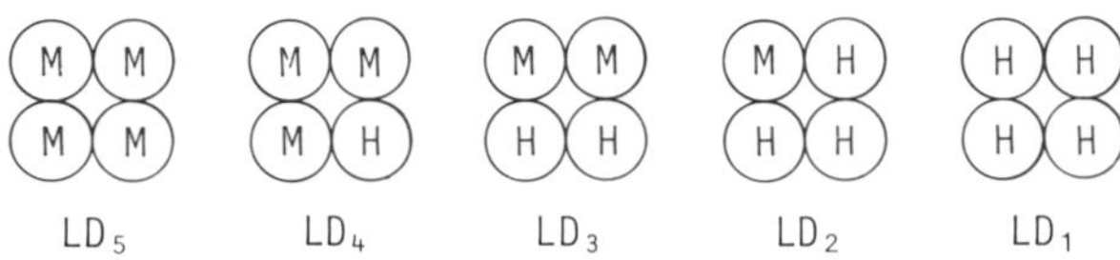

Although isoenzymes catalyse the same reaction the affinity of the enzyme for the substrate may be different, Thus hexokinase I and hexokinase IV (glucokinase) in liver have markedly different K_m values for glucose, but both catalyse the reaction:

Glucose + ATP ⟶ Glucose 6-phosphate + ADP

B

Multiple forms of enzymes can arise for a number of different reasons.

Group	*Reason of multiplicity*	*Example*
1.	Genetically independent proteins*	Malate dehydrogenase in mitochondria and cytosol
2.	Heteropolymers (hybrids) of two or more polypeptide chains, noncovalently bound*	Hybrid forms of lactate dehydrogenase
3.	Genetic variants (allelozymes)*	Glucose 6-phosphate dehydrogenases in man
4.	Conjugated or derived proteins	
	a. Proteins conjugated with other groups	Phosphorylase a, glycogen synthase D
	b. Proteins derived from single polypeptide chains	The family of chymotrypsins arising from chymotrypsinogen
5.	Polymers of a single subunit	Glutamate dehydrogenase of M_r 1 000 000 and 250 000
6.	Conformationally different forms	All allosteric modifications of enzymes

* These classes fall into the category of isoenzymes.

Enzyme assay

A

For an accurate estimation of the activity of an enzyme in an unknown solution, the amount of enzyme must be the only factor limiting the rate of reaction. The *amount* of enzyme protein can be determined by radioimmuno assay.

The concentration of substrate chosen for the assay of enzyme activity is well in excess of the K_m unless the enzyme is inhibited by excess substrate (e.g. enzyme 1 below).

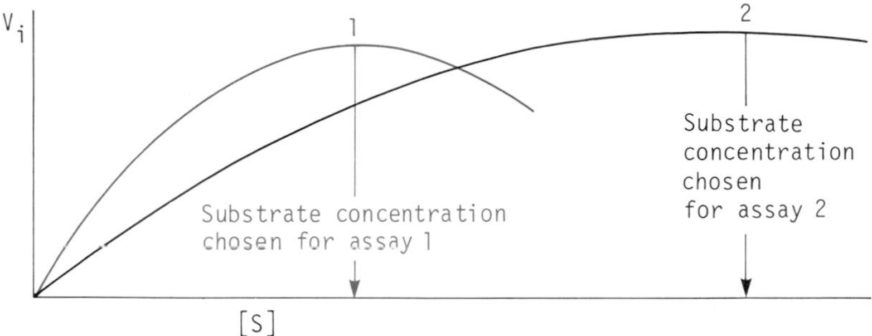

The sample to be assayed may contain activators or inhibitors in unknown concentration.

Note: The *amount* of enzyme *protein* is not revealed by determining the *activity* of an enzyme, because activators and inhibitors may be present in unknown quantities.

B

A coupled assay is one in which the enzyme reaction to be assayed is carried out in the presence of one or more other enzymes, to convert a product, which itself cannot easily be measured, into one that is easily measured (often this is NADH).

All cofactors are added in excess unless they inhibit at high concentrations.

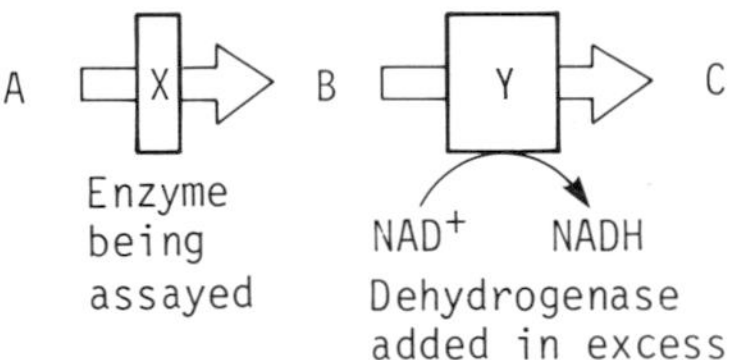

In measuring the formation of product B, NADH can be measured rather than B, provided that the NADH is formed at the same rate that B is formed. This requires that the dehydrogenase Y should be in excess, and must not limit the reaction rate in any way.

The K_m of Y with respect to B must be sufficiently low to allow Y to act on the small concentrations of B formed.

The equilibrium position of B $\longrightarrow$ C must be in the direction of C and NADH.

A

Enzymes in diagnosis

The assay of serum enzymes is used as an important aid to diagnosis. The level of activity of a number of enzymes is raised in different pathological conditions.

It is often useful to measure the level of activity at more than one time, as the time course over which the activity of each enzyme changes is characteristic of the condition.

Viral hepatitis.

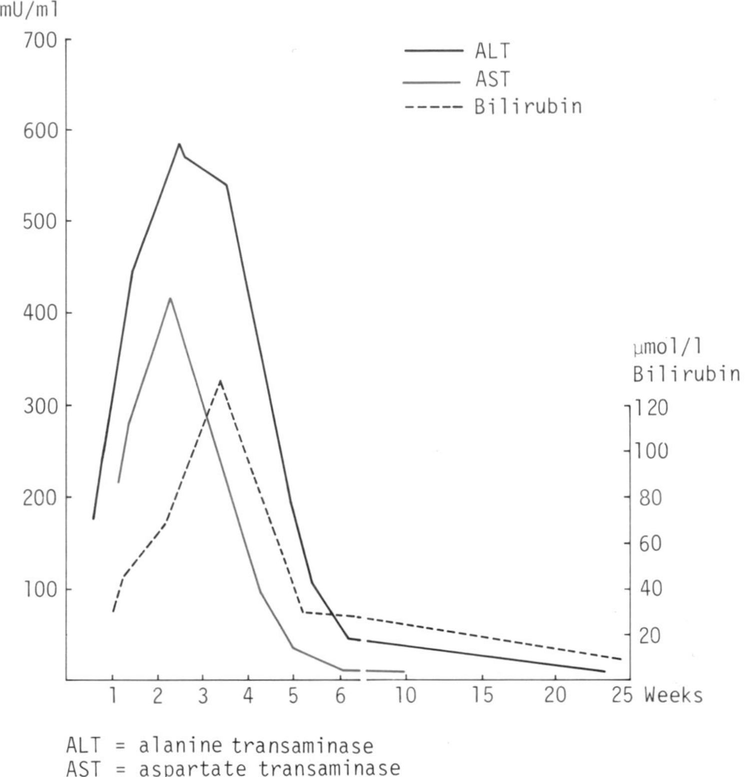

In cases of acute viral hepatitis there is a rapid rise in the level of serum transaminase activity and this occurs considerably before there is jaundice, shown by a rise in the serum bilirubin level.

B

The diagnosis of *myocardial infarction.*

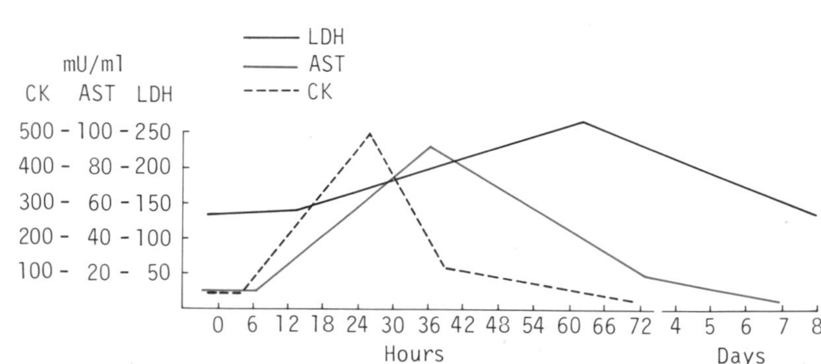

In myocardial infarction, the first enzyme to increase in activity is creatine kinase (CK), followed some hours later by aspartate transaminase (AST). Lactate dehydrogenase (LDH) rises more slowly.

Some enzymes of diagnostic interest

Enzyme	Principal conditions in which level of activity in serum is elevated
Amylase	Acute pancreatitis
Acid phosphatase (Optimum pH 5)	Prostatic carcinoma*
Alkaline phosphatase (Optimum pH 10)	Diagnosis of liver disease, especially biliary obstruction and detection of osteoblastic bone disease, e.g. rickets.
Aspartate transaminase (AST—previously GOT)	Myocardial infarction. Liver disease, especially with liver cell damage.
Alanine transaminase (ALT—previously GPT)	Liver disease especially with liver cell damage.
Lactate dehydrogenase (LDH)	Myocardial infarction, but also increased in many other diseases (liver disease, some blood diseases).
Creatine kinase (CK)	Myocardial infarction and skeletal muscle disease (muscular dystrophy, dermatomyositis).
γ-glutamyl transferase** (γGT)	Diagnosis of liver disease, particularly biliary obstruction, and alcoholism.

Urinary elevation

N-acetylglucosaminidase in the urine can be used to indicate renal transplant rejection.

* Assays based on conversion of substrate to product cannot reveal the amount of enzyme protein present, as possible variation in the amounts of activators and inhibitors cannot be ascertained. Radioimmunoassay is being increasingly used to determine the amounts of actual enzyme protein present in the serum, and this can even be specific for isoenzymes. For example, antibodies can be obtained that are specific for prostatic acid phosphatase, and distinguish it from acid phosphatase that has leaked from red cells and platelets, so that it is desirable that radioimmunoassay for acid phosphate be used in cases of suspected prostatic carcinoma. Radioimmunoassay can also be used to measure serum trypsin in suspected cases of cystic fibrosis.

** γGT is elevated in the plasma of alcoholics, and also of epileptics taking barbiturates. This is because consumption of alcohol or barbiturates causes considerable proliferation of the endoplasmic reticulum inducing high levels of the enzyme.

A

Metabolism of ethanol

Ethanol is metabolized in the human through the action of two enzymes.

Liver cytosolic alcohol dehydrogenase (ADH)

$$C_2H_5OH + NAD^+ \rightleftharpoons CH_3\,CHO + NADH$$

Acetaldehyde dehydrogenase (ALDH)

$$CH_3\,CHO + H_2O + NAD^+ \rightleftharpoons CH_3\,COOH + NADH + H^+$$

Formation of acetaldehyde and acetate.

Many of the unpleasant side effects of excess ethanol consumption are due to the formation of acetaldehyde.

B

Why you should be careful when you treat your oriental friend.

Human ADH is a dimer and is polymeric being formed from three different subunits, α, β and γ. The caucasian β_1 subunit is substituted in a very high proportion (90%) of orientals by β_2. On starch gel electrophoresis we find:

C

The role of the isoenzymes of the dehydrogenases of ethanol and acetaldehyde.

Kinetic properties of alcohol dehydrogenase isoenzymes

Composition of dimer	$\beta_1\beta_1$		$\beta_2\beta_2$	
Activity	pH 8.5	pH 10.0	pH 8.5	pH 10.0
Units/mg	0.39	0.80	32	5.0

It can be seen that at physiological pH the $\beta_2\beta_2$ enzyme is much more active than the $\beta_1\beta_1$ and hence much more acetaldehyde is formed. The structural difference between β_1 and β_2 resides in a change at the coenzyme binding site of his for arg.

Caucasians have two isoenzymes for the oxidation of acetaldehyde, ALDH-1 and ALDH-2 while about 50% of orientals have only ALDH-1. Since ALDH-2 has a high affinity for its substrate its absence will increase the concentration of acetaldehyde in the atypical subjects.

4

NUCLEIC ACIDS AND PROTEIN BIOSYNTHESIS

A

Dynamic state of body proteins

The metabolic state of the body proteins is an example of the dynamic state of the body constituents.

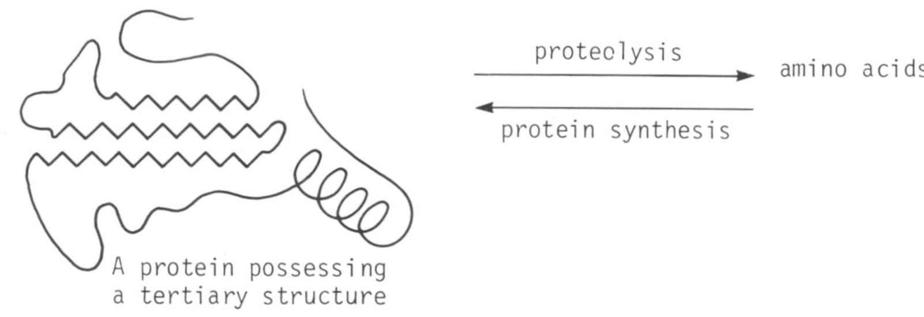

The body obviously needs to synthesize protein to provide a net increase for (a) growth, (b) the replacement of cells after their death and (c) for milk proteins during lactation. However even in an adult all the proteins, with the exception of collagen, are in a constant state of degradation and resynthesis so that the total amount of protein synthesized in the body is much greater than the net increase. An adult man synthesizes about 400 g of protein each day. The rate of turnover of a protein is measured as the *half life*, i.e., the time taken for half the molecules to be degraded to amino acids, usually measured in a few days.

B

Strategies for protein biosynthesis

The protein synthesizing activity of the body is subdivided among the various tissues.

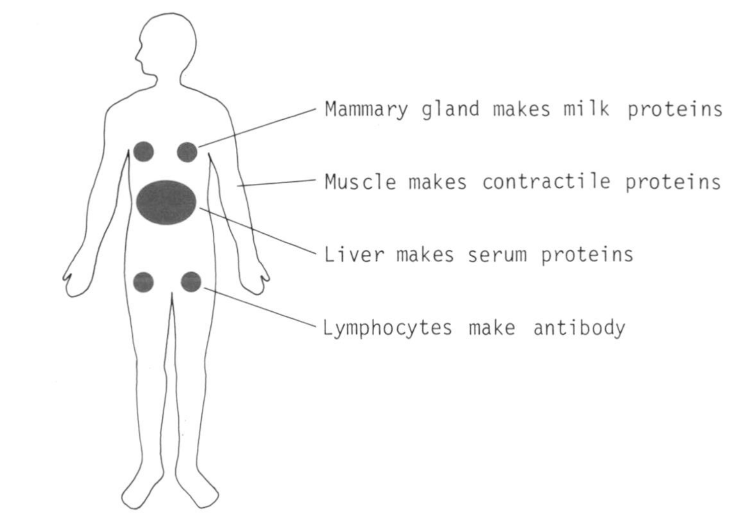

A

The polypeptide chain of proteins is assembled from 20 different amino acids.

The following amino acids participate in the formation of polypeptide chains:

Glycine	Proline	*Asparagine*
Valine	Tryptophan	Lysine
Alanine	Tyrosine	Arginine
Leucine	Phenylalanine	Histidine
Isoleucine	Glutamic acid	Methionine
Serine	*Glutamine*	Cysteine
Threonine	Aspartic acid	

Note the presence of *glutamine* and *asparagine.*
Note the absence of *hydroxyproline* and *hydroxylysine*, and of *phosphoserine* and *phosphothreonine*. These modified amino acids are important constituents of collagen and complement, and of the phosphoproteins, respectively. The parent amino acids are modified by hydroxylation or phosphorylation after incorporation into the polypeptide chains. Cystine is formed from two cysteine residues. More recently *phosphotyrosine* has been detected, particularly in proteins involved in the mediation of receptor action.

B

Multichain proteins may be assembled from individually synthesized chains or by limited proteolytic cleavage of larger proteins (e.g. insulin see p. 78A).

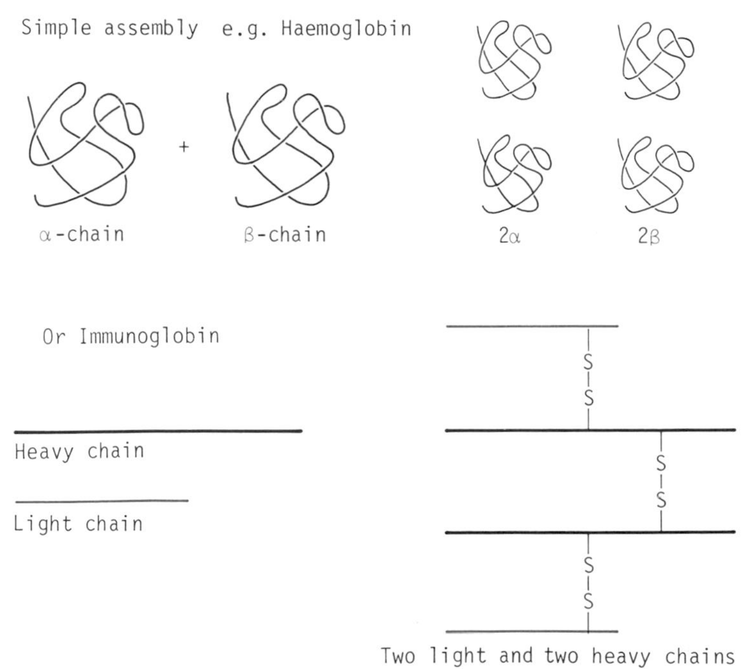

A

Insulin is synthesized as a precursor proinsulin (e.g. pig proinsulin).

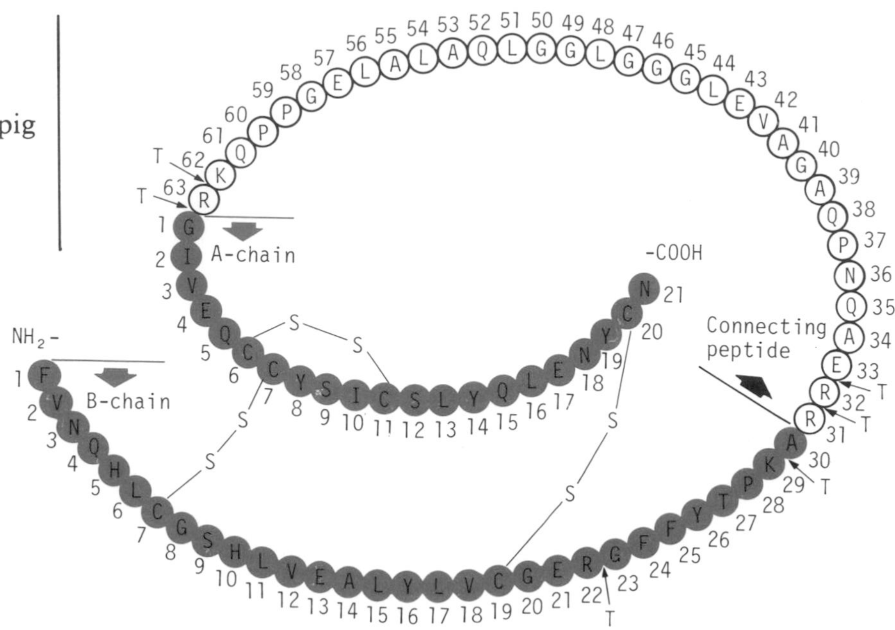

Removal of the *connecting peptide* (C-peptide) from proinsulin gives rise to insulin with the two chains A and B in the correct juxtaposition linked by S-S bonds. Proinsulin has only 5% of the biological activity of insulin and C-peptide is inactive. T indicates peptide bonds that would be cleaved by trypsin.

B

Various pro-proteins are known.

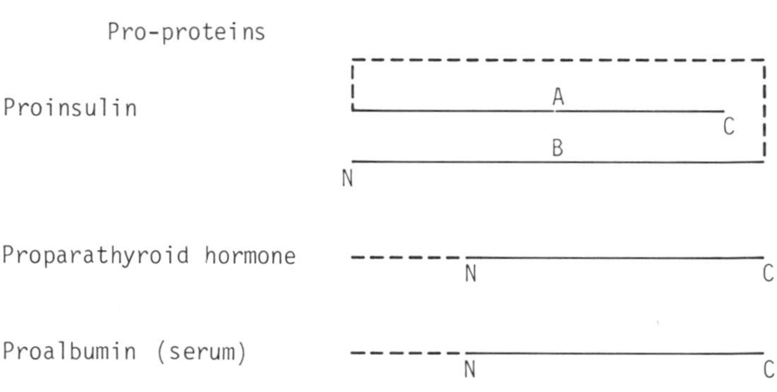

Three pro-proteins are shown. In two cases the additional peptide is at the N-terminus and contains six amino acids, four of which are basic. The link between the mature proteins and the pro-peptide always involves two basic amino acids.

A

Our knowledge of protein structure shows that the problem for the cell is to effect the linkage of amino acids by peptide bonds in the correct order.

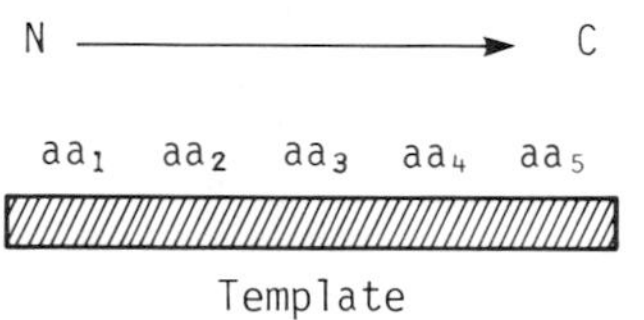

From our knowledge of metabolism it might have been expected that polypeptides would be synthesized by a series of enzymic reactions using enzymes which were specific for each amino acid and peptide. It soon became apparent that too many enzymes would be involved and moreover there was no evidence for the presence in the cell of peptide intermediates. It was concluded that the amino acids must be assembled on a *template*, probably of nucleic acid. The early experiments showed that the synthesis of the polypeptide chain was in the direction $NH_2 \longrightarrow COOH$.

1. NUCLEIC ACID STRUCTURE AND BIOSYNTHESIS

B

Replication, transcription and translation

The stages of nucleic acid and protein synthesis are usefully divided into four.

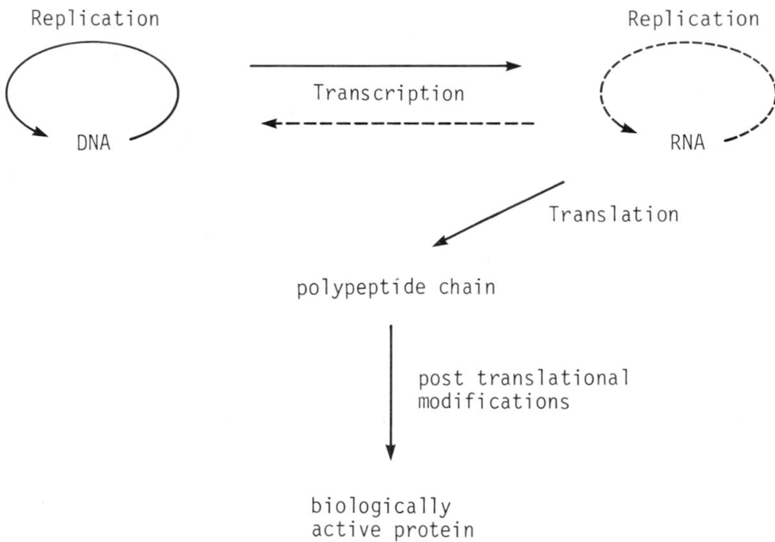

DNA is transcribed to RNA. One of the species of RNA is mRNA which is translated to give a polypeptide chain. Hence the template referred to in A is mRNA.

The dashed lines indicate activities that are essentially only concerned with RNA viruses.

Among the post-translational modifications are limited proteolysis, glycosylation, γ-carboxylation, phosphorylation and combination with prosthetic groups either by covalent or non-covalent bonds.

A

Nucleic acid and nucleotide structure

There are two kinds of nucleic acids:
deoxyribonucleic acid (DNA)
ribonucleic acid (RNA)
Nucleic acids are polymers, the monomeric unit being a *nucleotide* so that all nucleic acids are *polynucleotides.* A nucleotide has three components

phosphate $\frac{\text{ester}}{\text{link}}$ 5′ sugar 1′ —— N-base

In DNA the sugar is deoxyribose, in RNA it is ribose. The nucleotides are linked in the polynucleotides through the formation of a phosphodiester bond. The basic structure of a polynucleotide is therefore:

Polyribonucleotide ('RNA')

base | base
—phosphate—ribose—phosphate—ribose—
(deoxyribose in 'DNA')

B

The nucleotides are linked by a *phosphodiester bond* between the 3′ and 5′ positions of each pentose.

Since the polynucleotide chain has polarity (i.e. is not the same in each direction) by convention the order of the nucleotides containing the different bases is read from the 5′ end to the 3′ end. Chains of opposite polarity may be said to be *anti-parallel*

Illustrated is RNA.
In DNA the 2′ OH is replaced by H

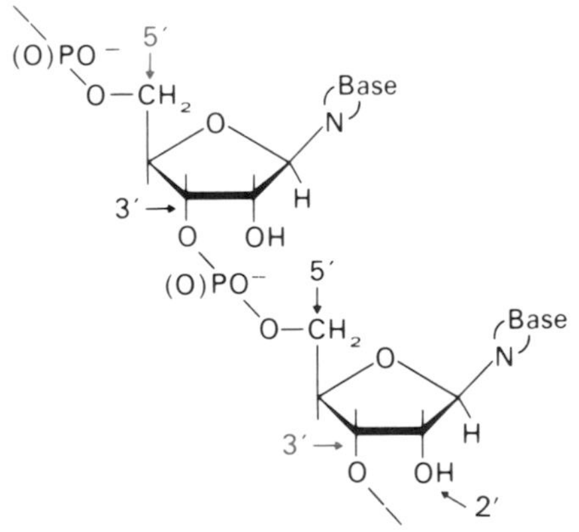

(O)PO⁻ 5′ O—CH_2 Base N O H 3′ O OH (O)PO⁻ 5′ O—CH_2 Base N O H 3′ O OH 2′

C

There are typically only four different bases in either DNA or RNA. The only difference between DNA and RNA in this respect is that DNA contains *thymine* and RNA *uracil.*

Pyrimidine bases

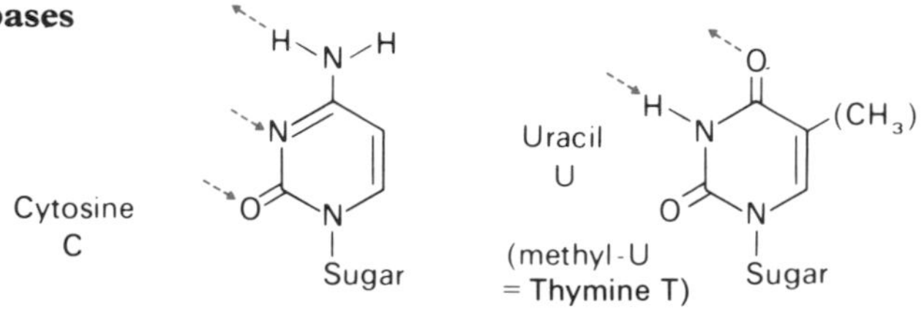

Cytosine C — Sugar

Uracil U — Sugar ((CH_3))
(methyl-U = Thymine T)

Purine bases

Guanine G — Sugar

Adenine A — Sugar

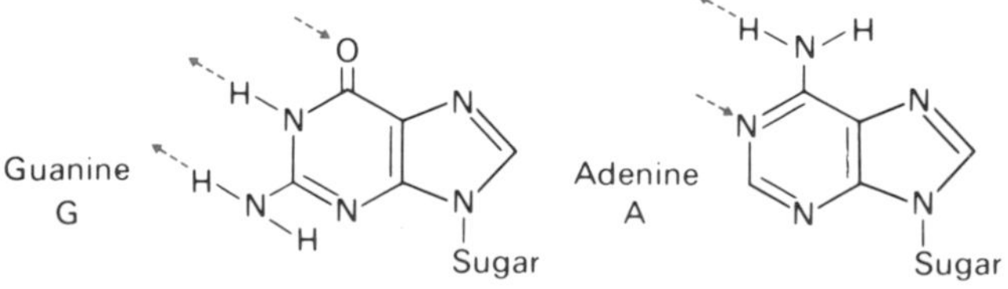

Sites and directions for hydrogen bonding shown by ------→

A

Nucleic acids differ one from another by the sequence of the bases.

The characteristics of a particular nucleic acid are determined by the order in which the four different bases occur in the polynucleotide chain. A polynucleotide is not symmetrical but has polarity, i.e., a 5′ and 3′ end. By convention the base sequence is specified 5′ ⟶ 3′. Hence pApCpGpApT specifies that the sequence in the polynucleotide starts with A at the 5′ and has T at the 3′ end. The p indicates the presence of a phosphate group and the capital letter the nucleotide.

B

DNA is usually *double-stranded*, the two chains being anti-parallel. The strands are coiled in the form of a *double helix* (see page 83).

Double-stranded DNA with antiparallel strands

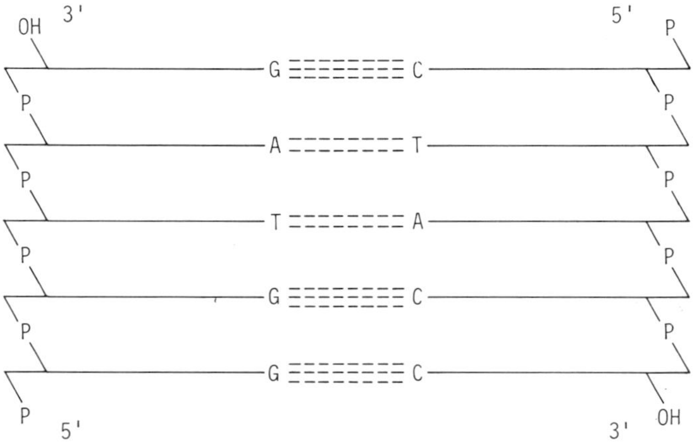

C

The proportion of the bases in DNA is such that A = T and G = C. This is in accord with the fact that A hydrogen bonds with T and G hydrogen bonds with C. This is known as *'base-pairing'*.

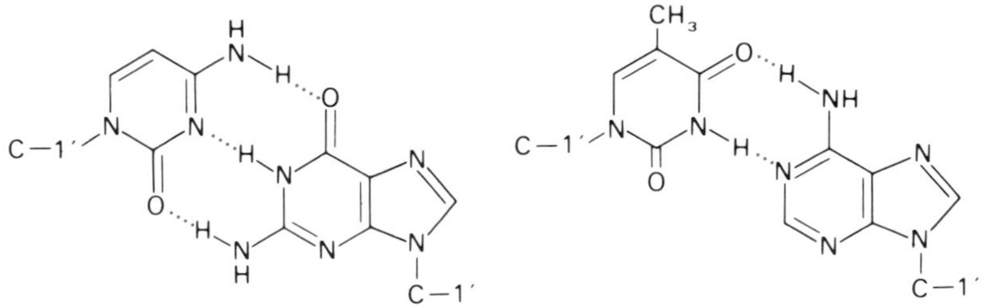

Structure of base pairs in DNA (C : G on the left, T : A on the right).

Note that the link between G : C is stronger than that between A : T. The antiparallel strands in DNA linked by base pairs are said to be *complementary*.

D

DNA comes in various sizes.

Viruses	**No. of base pairs**	**Length**
Polyoma (animal)	4 600	1.6 μm
Bacteriophage	53 000	16 μm
Bacteria		
E. coli	3.4×10^6	1.2 mm
Human	1.2×10^{10}	200 cm

A *bacteriophage* is a virus which infects bacteria and is commonly known as a *phage*.

A

Nucleosides and nucleotides.

While a *nucleotide* has three components a *nucleoside* consists only of a sugar and a base

5′————1′————N-base
sugar

It follows that a nucleotide with 3 PO_4 groups attached to a nucleoside at the 5′ position of the sugar is called a nucleoside triphosphate, so that ATP stands for adenosine triphosphate.

P—P—P—5′ ribose 1′-adenine.

Similarly one obtains the terms adenosine diphosphate (ADP) and adenosine monophosphate (AMP) and related terms for other bases.

B

Nomenclature in nucleic acids.

Base	*Nucleoside*[*†]	*Nucleotide*[†]
Purines		
adenine	adenosine (A)	adenylic acid (AMP)
guanine	guanosine (G)	guanylic acid (GMP)
hypoxanthine	inosine (I)	inosinic acid (IMP)
Pyrimidines		
cytosine	cytidine (C)	cytidylic acid (CMP)
uracil (in RNA)	uridine (U)	uridylic acid (UMP)
thymine (= 5-methyluracil in DNA)	thymidine (T)	thymidylic acid (TMP)

*In polymers with repeating units the letters indicate nucleotides e.g. poly (A) or poly (dT).
†When the sugar is deoxyribose the nucleoside or nucleotide is abbreviated dT, dAMP etc. dNTP signifies unspecified deoxynucleotide triphosphate.

C

Hybridization of DNA

Double-stranded DNA can be denatured by heat and the strands can be re-annealed.

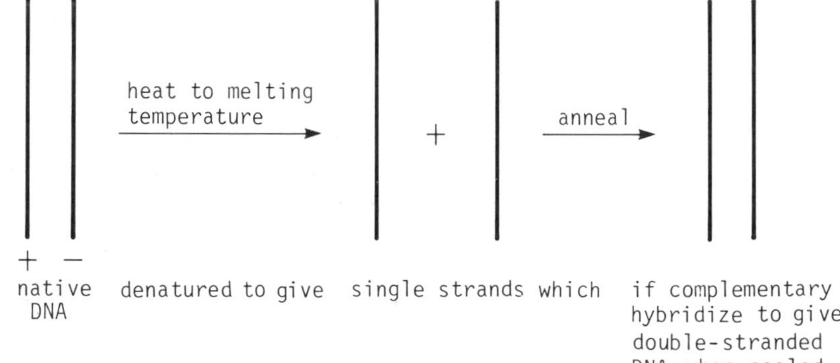

Complete denaturation of a molecule of DNA leads to separation of the two *complementary* strands. If a solution of *denatured* DNA is cooled quickly the denatured strands remain separated. If the temperature is held for some time just below T_m (the *melting temperature*), a process known as *annealing*, the native double stranded structure can be reformed. This is termed *hybridization* when the nucleic acid is from different sources.

The DNA double helix

DNA exists as a *double helix* but may vary in conformation.

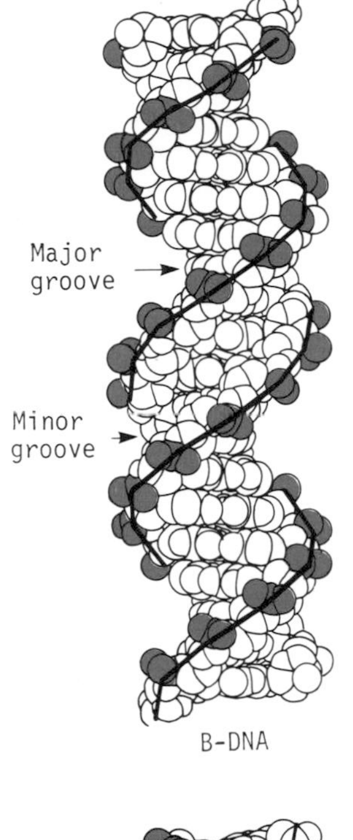

B-DNA

DNA exists in solution as a right-handed helix known as the B-form. (The chains turn to the right as they move upward)
Note that this is characterized by a regular repeat of a Minor and Major groove

The phosphate groups are red.

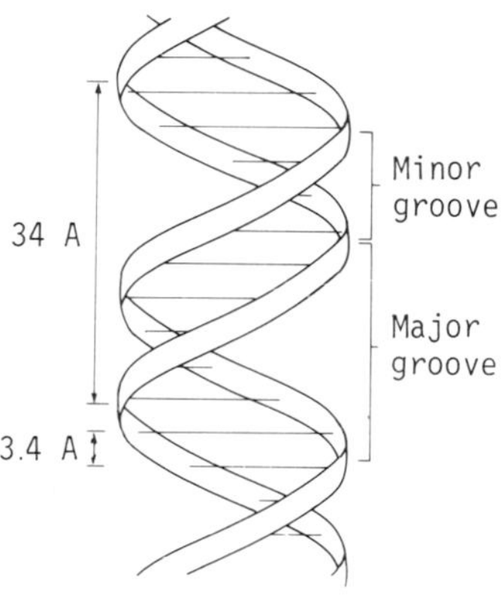

There is increasing speculation that DNA may change *in vivo* in conformation and that this may be important in gene expression.

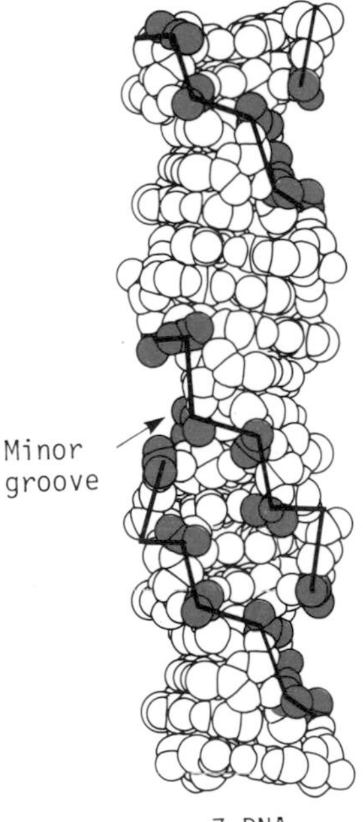

Z-DNA

Another quite different form of DNA has been found, Z-DNA which has a left-handed helix. The irregularity of the Z-DNA backbone is illustrated by the heavy line that connects the phosphate groups.

A

The use of hybridization techniques

Hybridization can be used to compare the *homology* (degree of complementarity) of nucleic acids from different species. A DNA is broken into pieces of moderate length and is denatured. The denatured fragments from the two species are mixed and those that have similar nucleotide sequences tend to hybridize, whereas those where the sequences are very different do not. The reassociation of DNA fragments obeys second order kinetics. A plot of the fraction of molecules reassociated versus the log of C_0t (where C_0 is the initial concentration of denatured DNA and t is time) is a convenient way of displaying data. The value of C_0t increases in direct proportion to the length of the DNA chain in the genome but it is very much decreased if the sequence of bases is highly repetitive [poly(U) and poly(A)]. The slope at the midpoint gives an indication of the heterogeneity of the DNA fragments in solution.

B

Measurement of the rate of hybridization.

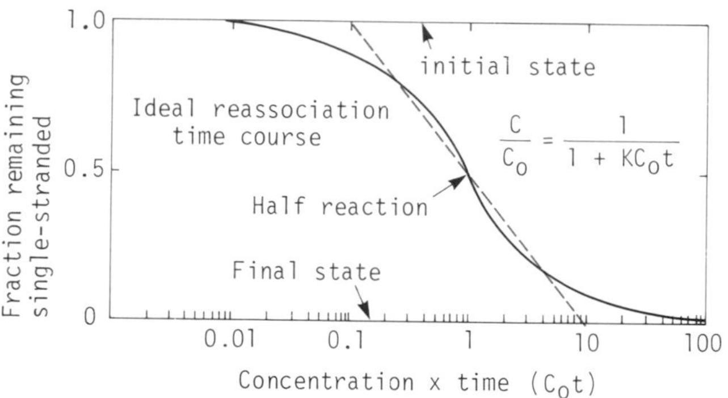

Time course of an ideal, second-order reaction to illustrate the features of the log C_0t plot. The equation represents the fraction of DNA which remains single-stranded at any time after the initiation of the reaction. For this example, K is taken to be 1.0, and the fraction remaining single-stranded is plotted against the product of total concentration and time on a logarithmic scale.

C

An application:
As the size of the DNA increases so the rate of reassociation is reduced.

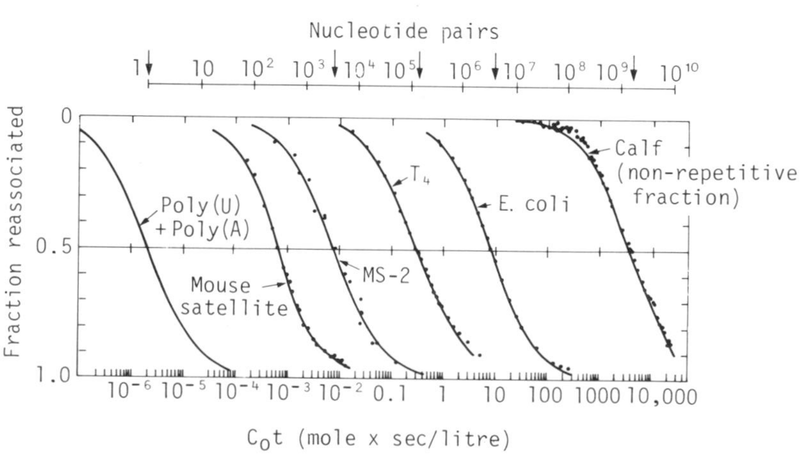

Reassociation of double-stranded nucleic acids from various sources. MS-2 and T_4 are from bacteriophages.

A

The cell cycle

The amount of DNA in an animal cell is vastly greater than that in a bacterium. This may in part be because animal cells participate in a cell cycle, the extra DNA being involved in the control of the cycle. Other explanations are the presence of many repetitive sequences of nucleotides and the presence of *introns* (split genes), see page 115.

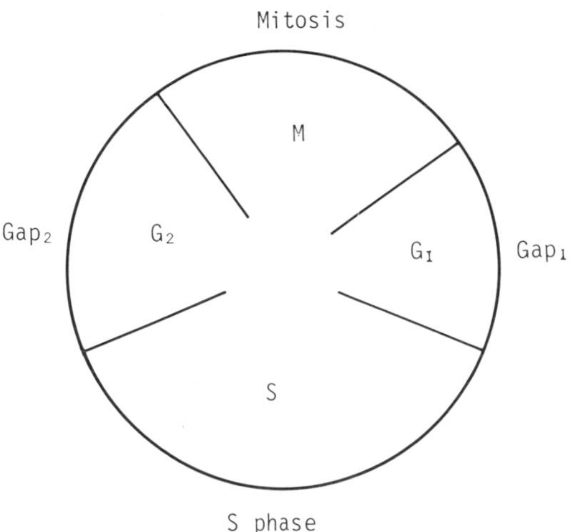

After mitosis and cell division the cell enters a quiescent phase known as Gap_1 (G_1). The *chromosomes* are uncoiled and consist of *euchromatin.* DNA is synthesized when the cell is in S phase. The cell then enters Gap_2 (G_2) when DNA synthesis ceases and the chromatin becomes coiled, *heterochromatin.*

B

Chromatin structure

In chromatin the histones are associated with the DNA in the form of nucleosomes.

A characteristic feature of the nucleus of eukaryotic cells, unlike prokaryotes, is the association of small basic proteins with the DNA. These are called *histones.* Four of the five different histones group in pairs to give a particle around which the DNA is wrapped. The product is known as a *nucleosome.* The nucleosomes are themselves coiled so our concept of the structure of chromatin is as shown on the next page.

Types of histones

Type	*Lys/Arg ratio*	*Number of amino acid residues*	M_r ($\times 10^{-3}$)	
H1	20.0	215	21.0	
H2A	1.25	129	14.5	Components of Nucleosomes
H2B	2.5	125	13.8	
H3	0.72	135	15.3	
H4	0.79	102	11.3	

A schematic view of the structure of *chromatin*.

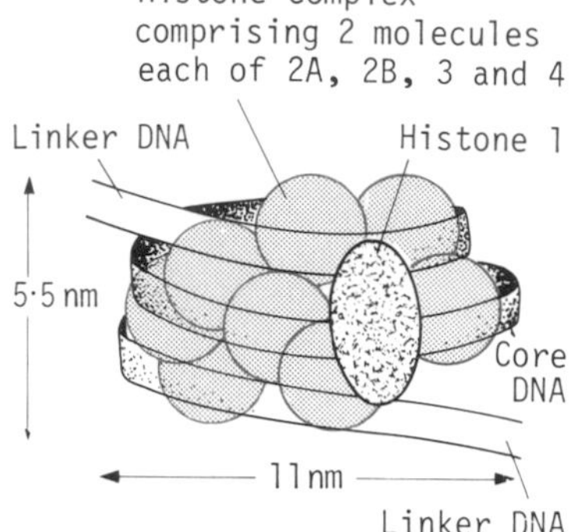

The nucleosome showing the association of the 5 different types of histones.

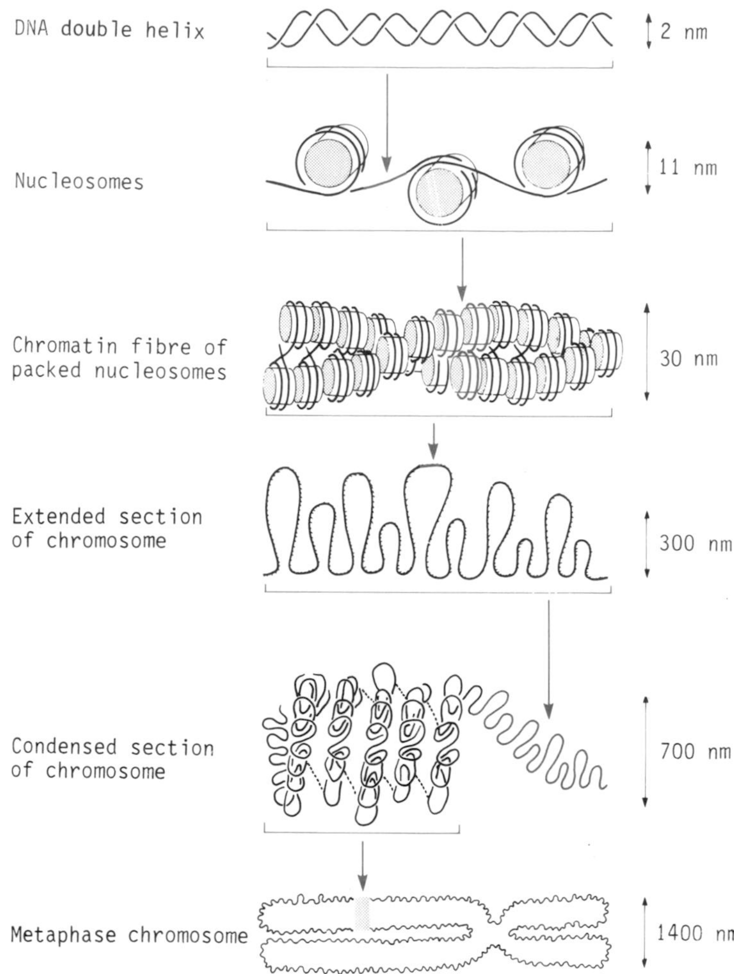

The DNA molecule of a chromosome of a higher organism would stretch over many centimetres if were laid out straight (see p. 81C) and so must be highly folded to form the compact structure found in the nucleus. The DNA must also be organized into functional units which permit the reading of genes and their replication in a regulated manner.

DNA is organized in regular repeating units, nucleosomes, which contain successive segments of the double helix, about 200 base pairs long, associated with an octamer of four histone proteins, H3, H4 H2A and H2B. The DNA is wrapped in two turns around the octamer. At the point of entry and leaving, the DNA is sealed by another histone, H1. These repeating structural elements along the DNA form a filament which can coil into a helical or 'solenoidal' structure to form the second level of folding.

Chromosomes can be digested with a DNAase to produce a particle containing 14 helical turns of DNA and a histone protein core; this nucleosome core has been crystallized.

A

Biological activity and replication of DNA

The biological activity of DNA was first demonstrated by injecting mice with extracts of pneumococci.

R strain: non-pathogenic;
S strain: pathogenic.

Two strains of pneumococci were used. The wild strain is pathogenic and is known as smooth or S because of the kind of colony it forms and the fact that it contains capsular polysaccharide. A mutant Rough or R strain lacks capsular polysaccharide and is non-pathogenic. The crucial experiment is illustrated. The effect of injection into the mice was:

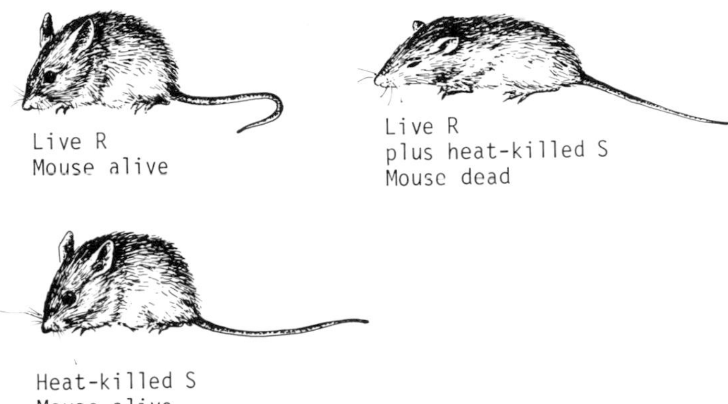

The heat-killed S was extracted and the active constituent shown by Avery and his colleagues in 1944 to be DNA. Since the blood of the dead mice contained live S strain pneumococci, the DNA of the S strain had *transformed* the live R strain into live S strain.

B

The infectivity of viral nucleic acid.

While the experiment illustrated represents the 'classical' way in which the biological activity of DNA was first demonstrated, with the passage of time simpler methods have been devised. In their most sophisticated form these are described under the section on 'Recombinant DNA'. Other methods involve the use of bacterial (bacteriophage) and plant viruses. The DNA from a bacteriophage can be shown to "infect" a suitable host cell and cause it to 'lyse' as a result of phage replication. In some types of virus the genome is RNA rather than DNA. The RNA from a plant virus (e.g. tobacco mosaic virus) similarly can be shown to be infective. The primary structure of the coat protein of the resulting virus depends crucially on the structure of the infecting RNA.

DNA is replicated in the cell by a process that ensures that one of the parent strands is present in the daughter molecules.

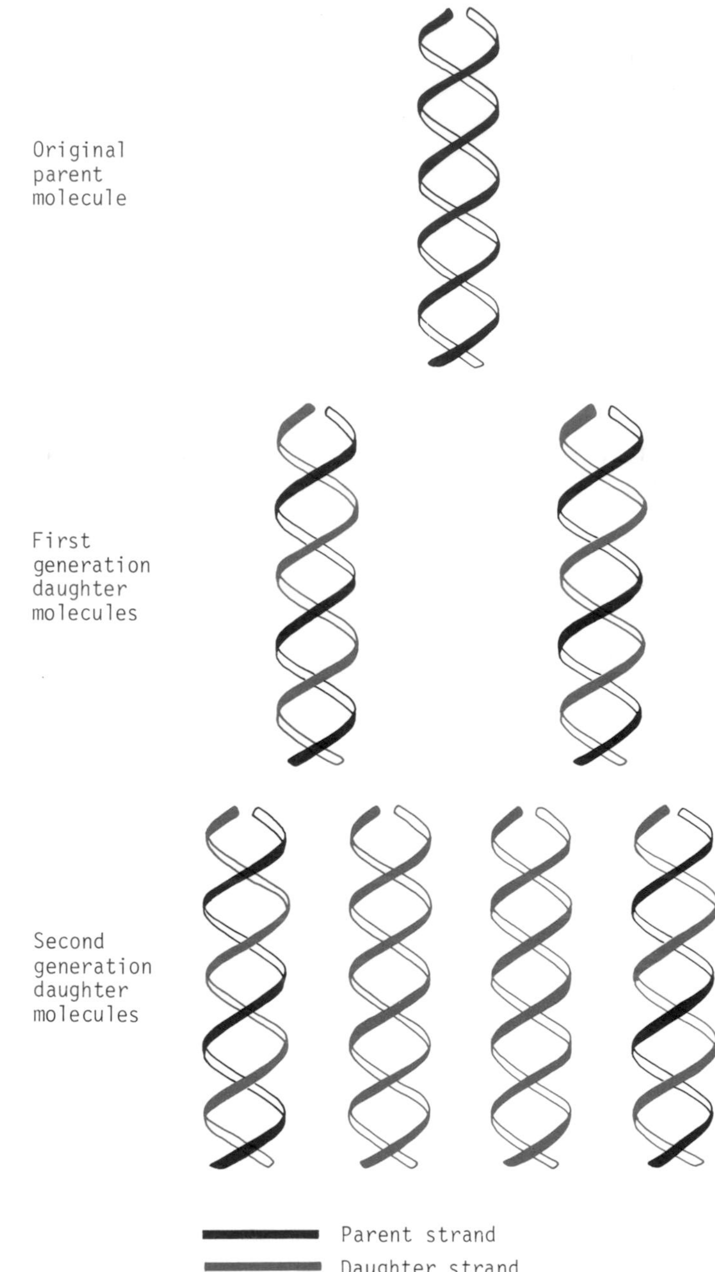

The experiments that proved that DNA was synthesized by a process of *semi-conservative* replication in which one of the parental strands is retained in the daughter molecules were done by Meselson and Stahl. They grew E. coli in the presence of the heavy isotope of N, ^{15}N. They determined the density of the DNA and showed that after one generation it was intermediate between the density of light DNA (containing ^{14}N) and heavy DNA (containing ^{15}N).

A

Replication and transcription of DNA and RNA

It is possible to isolate enzymes from cells which effect the synthesis of polynucleotides in the presence of deoxynucleoside triphosphates and a DNA template.

DNA is synthesized by enzymes called DNA polymerases, all of which utilize deoxynucleoside triphosphates as substrates and the polynucleotide is formed in the direction 5′ ⟶ 3′

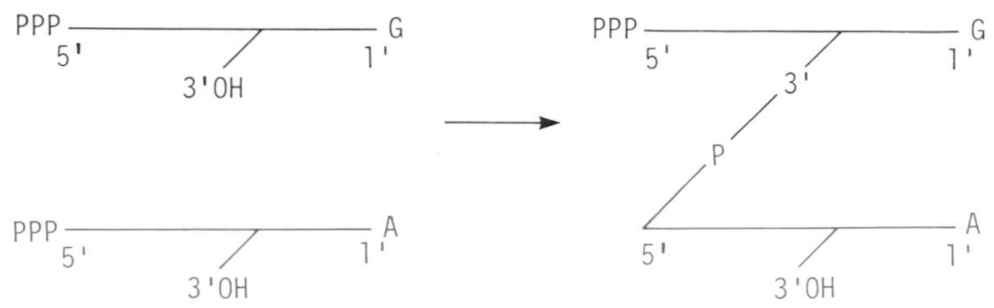

The horizontal line indicates the presence of deoxyribose and the numbers the identity of the carbon atoms of the sugar.

B

The role of parental DNA.

A DNA template is used to direct the order of bases in the newly synthesized polynucleotide such that it is *complementary* to the template DNA.

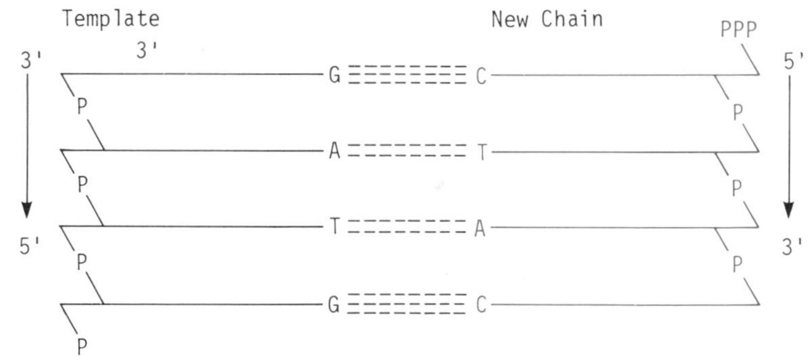

C

DNA ligase. An important enzyme for sealing nicks.

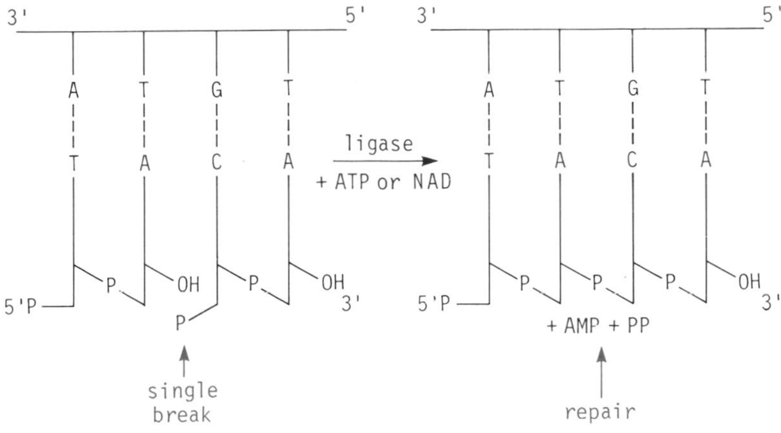

DNA *ligase* plays a role in the repair of a single break in a double-stranded DNA. Although ligase was discovered by studying DNA repair it is believed to play a key role in DNA synthesis in a normal cell.

A

(DNA Replication contd.) In the living cell the most common mechanism of DNA replication is for it to proceed in both directions along the two stands of DNA simultaneously. In the simplest case of *E.Coli* unwinding and replication of the double-stranded circular DNA leads to the production of a structure that resembles a circle with an inner loop. In this case replication is from a single initiation point.

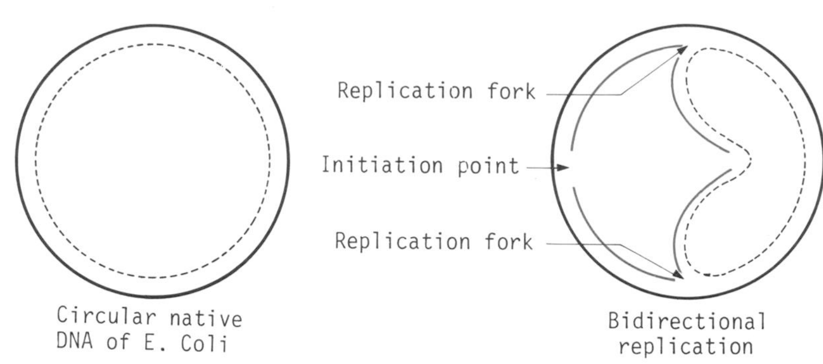

The problem posed is that all known DNA polymerases cause the synthesis of new polynucleotides in the direction 5′ ⟶ 3′ None have been found which synthesize strands in the opposite direction.

B

In order to explain the apparent growth of both strands of DNA in the same direction, it is postulated that at least one of the strands is first synthesized in fragments.

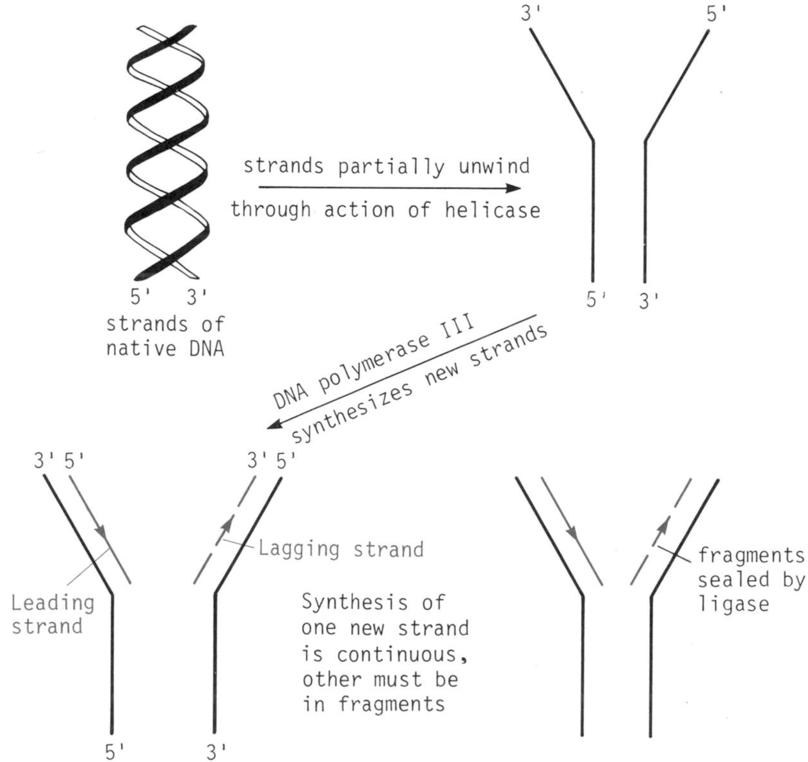

The experiments that led to this concept involved the examination of DNA shortly after its synthesis when a number of small fragments were found. The work was done by Okazaki and the fragments are therefore named after him.

A

Transcription

Transcription is said to be *asymmetric* in that only one strand of DNA is *transcribed*, and *conservative* in that the double stranded DNA template is *conserved.*

Three kinds of RNA are produced:
Ribosomal RNA (rRNA)
Transfer RNA (tRNA)
Messenger RNA (mRNA)

The base sequence of the synthesized mRNA is complementary to that of the DNA strand used as a template.

In *E. Coli* the relative amount of the three types of RNA are rRNA 80%, tRNA 15%, mRNA 5%.

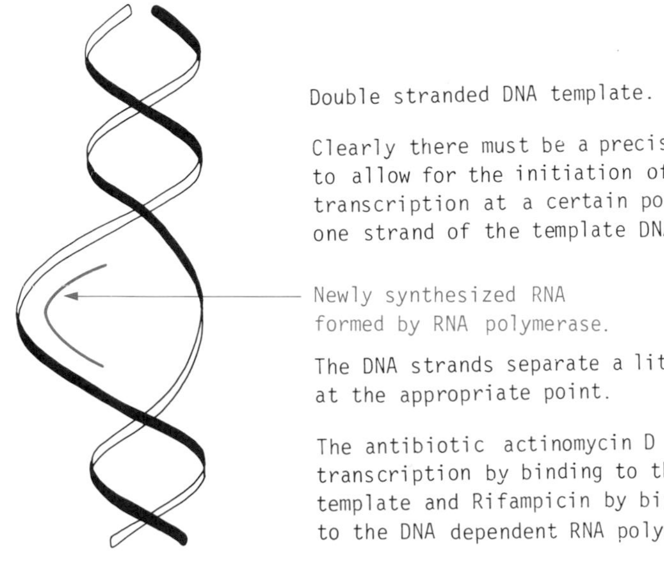

B

Viral replication

THE RNA of RNA virus is replicated by RNA-dependent RNA polymerase.

An RNA virus such as poliomyelitis must arrange for the replication of many (+) strands. It does this by forming a (−) strand from which many (+) strands are formed by asymmetric synthesis.

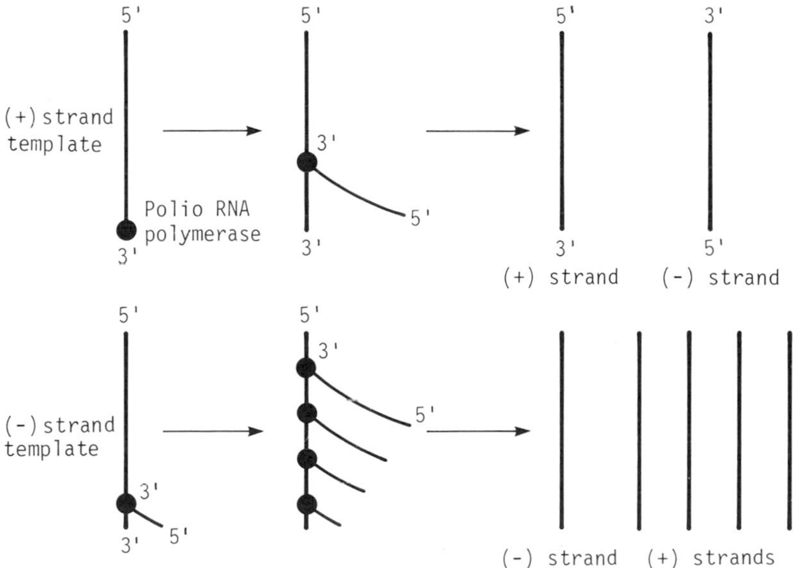

The polymerase is a multisubunit protein (4) in which only 1 of the protein subunits is virus-specific. The other subunits are provided by the host cell. RNA virus other than polio is replicated by a variety of different methods.

A

The role of RNA in DNA biosynthesis.

It came as a surprise that DNA synthesis also required the cell to be synthesizing RNA. A specific enzyme, 'primase', is responsible.

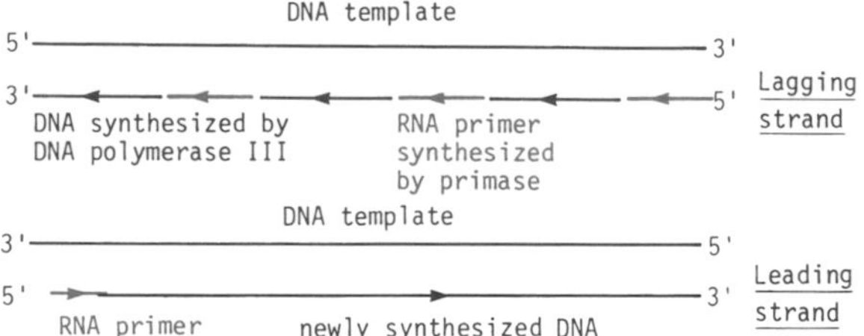

On the *lagging strand* the RNA primer is removed by DNA polymerase I, the gaps filled by the same enzyme and all fragments sealed by ligase. On the *leading strand* the RNA primer is synthesized by RNA primase and removed by DNA polymerase I, the space filled by DNA polymerase I and the two DNA fragments joined by ligase. Both DNA strands appear to grow in the same direction i.e. from left to right.

B

DNA biosynthesis is a multienzyme process.

Three DNA polymerases have been isolated from *E. coli:* Polymerase III is for synthesis, Polymerase I mainly for editing and has exonuclease activity, Polymerase II is membrane bound.

At least 15 replication proteins (comprising over 30 polypeptides) have been characterized. The duplex is opened in advance of replication by the action of helicases such as rep protein acting on the 3′ ⟶ 5′ strand and the helicase II on the 5′ ⟶ 3′ strand. The latter action is augmented or replaced by helicase activity of the primosome. The bared single strands are promptly covered by binding protein. [P] indicates the activation of the 3′ OH required for chain initiation.

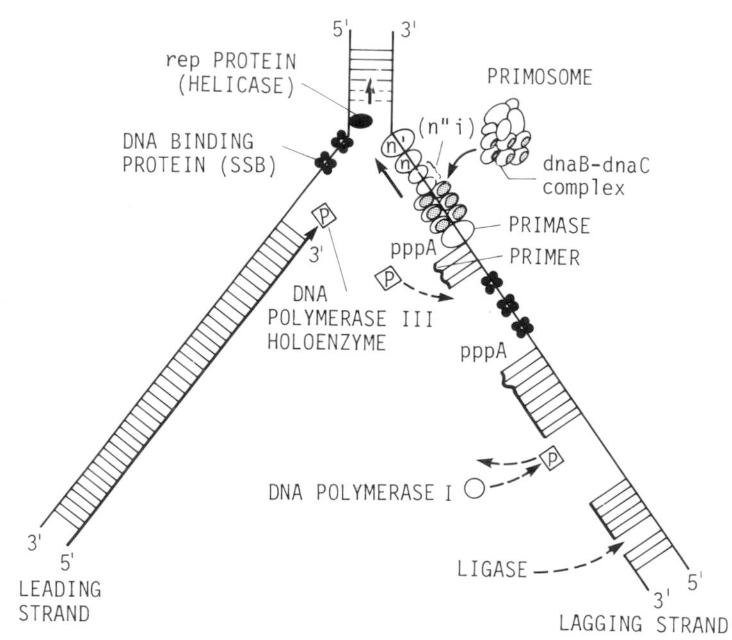

Scheme for enzymes operating at one of the forks in the bidirectional replication of an *E.coli* chromosome.

C

Reverse transcriptase

RNA oncogenic virus may contain a RNA directed DNA polymerase.

The formation of double-stranded DNA complementary to viral RNA by *reverse transcriptase.*

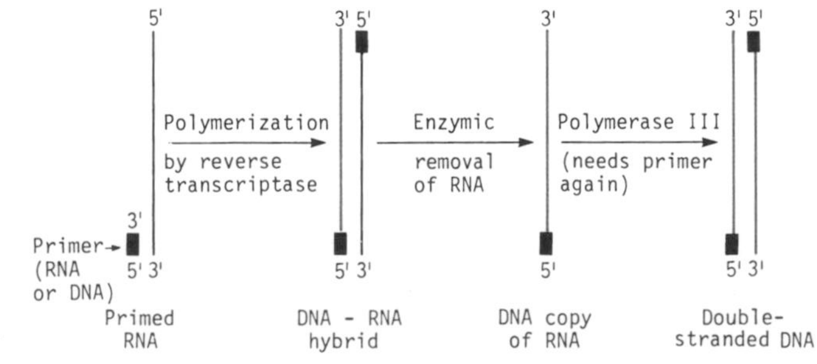

A

The copy of the viral RNA is integrated into the host chromosome becoming a *provirus*.

Such a virus is known as a retrovirus

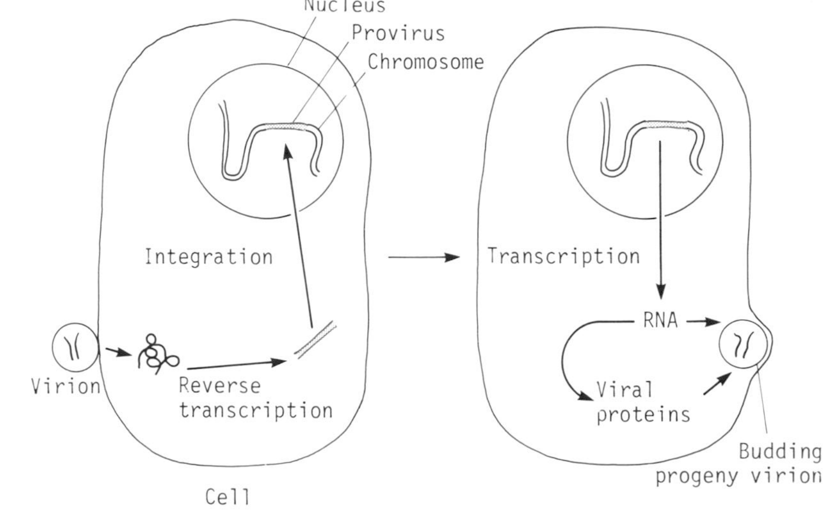

B

Action of antibiotics and antimetabolites

The site of action of various antibiotics and other substances that inhibit DNA polymerase, RNA polymerase and the synthesis of nucleotides. For further explanation, see pages 94, 105.

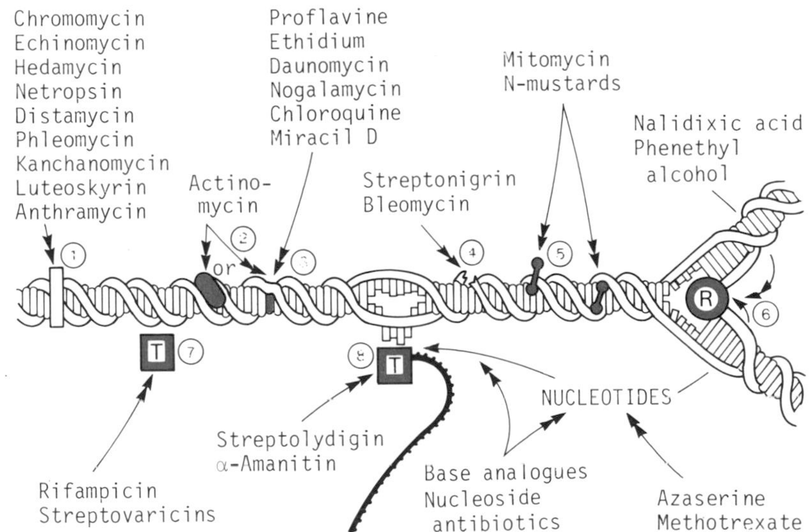

1. Non-covalent binding to DNA by unknown mechanism.
2. Intercalation without distortion of helix structure.
3. Intercalation between bases with distortion of helix structure.
4. Breaking of strands.
5. Alkylation and crosslinking.
6. Inhibition of DNA polymerase (replication).
7. Inhibition of initiation of RNA polymerase (transcription), not effective if added after initiation.
8. Inhibition of RNA polymerase at any time.

T = Transcription;
R = Replication.

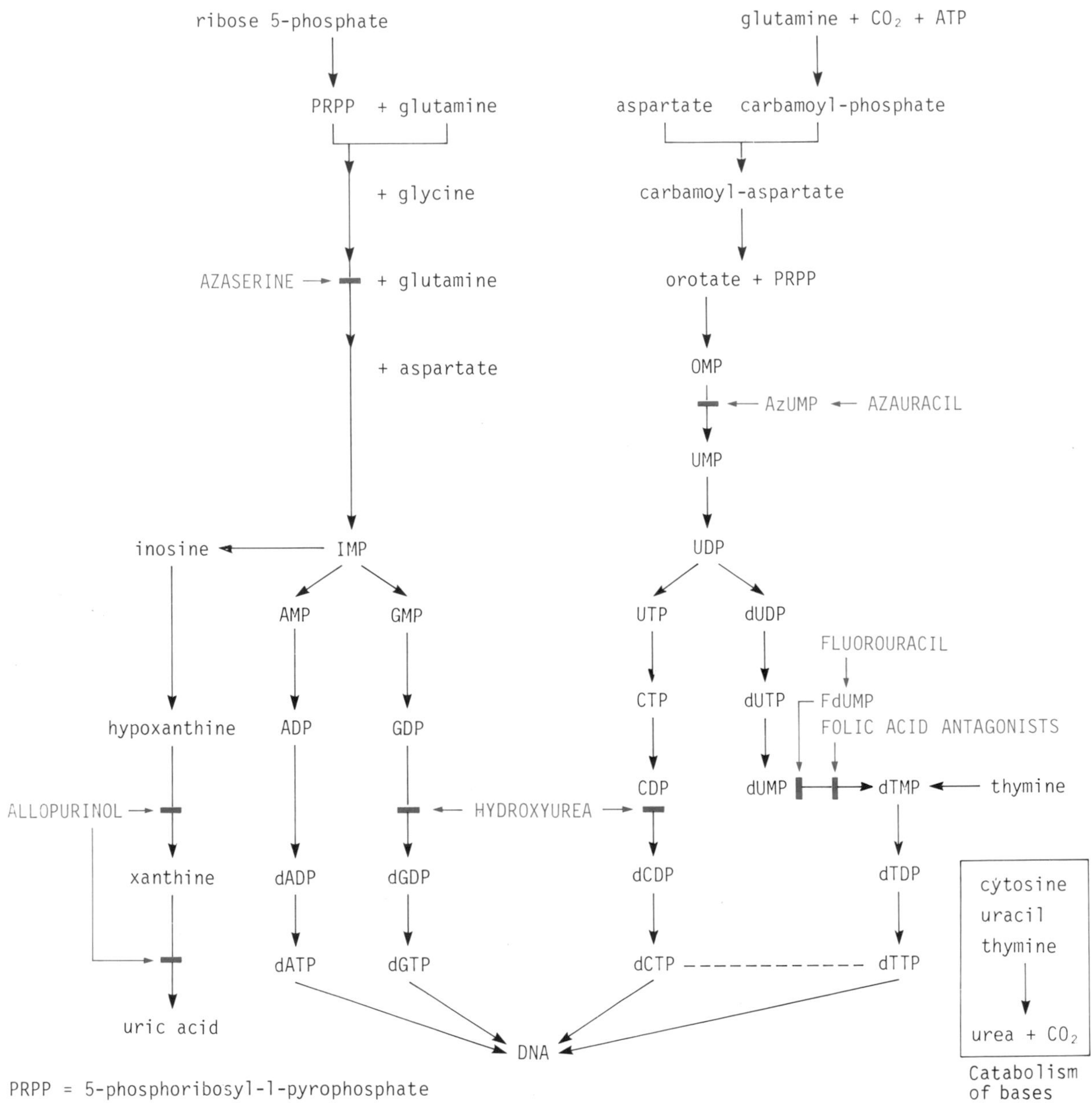

Pathways involved in the synthesis and degradation of deoxyribonucleoside triphosphates. The blocking action of some antimetabolites is shown in red.

In cancer chemotherapy, antimetabolites of the pathways of nucleotide synthesis are utilized as drugs.

2. PROTEIN BIOSYNTHESIS

A

Role of endoplasmic reticulum

The microsome fraction from disrupted liver cells contains the most active structures for protein synthesis in vivo.

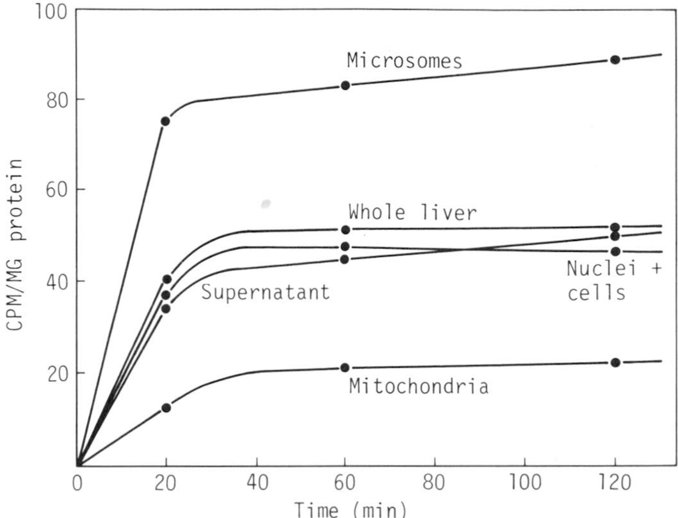

It was Zamecnik and his colleagues who showed that if [^{14}C]leucine was injected into a rat and pieces of liver were removed at the times shown, the microsome fraction was the site of protein synthesis. This fraction is rich with respect to fragments of the rough-surfaced endoplasmic reticulum. If this isolated microsome fraction is incubated with [^{14}C]leucine, radioactive protein is formed.

B

Biosynthesis of protein by the isolated microsome fraction.

Constituents of incubation mix

1. Particles rough endoplasmic reticulum	The electron micrograph shows that the liver microsome fraction contains fragments of various morphological constituents but is enriched with respect to the rough-surfaced endoplasmic reticulum.
2. Supernatant	containing enzymes
3. Energy	ATP. GTP is also required.
4. Amino acid donor	[^{14}C]leucine

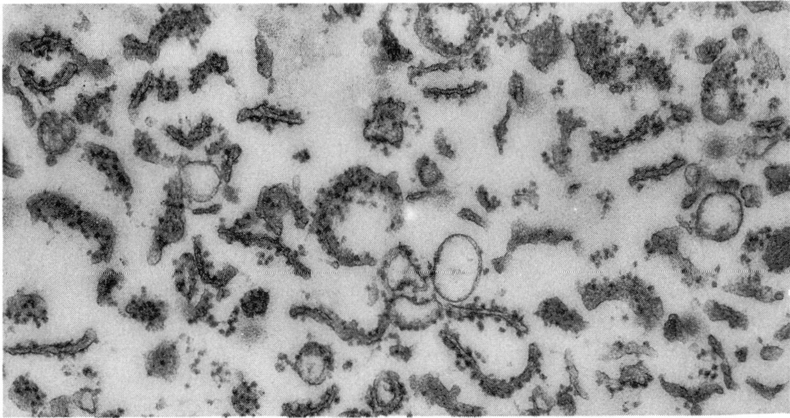

A

Amino acid adenylates, transfer RNA

The first step in protein synthesis involves the activation of the amino acids.

Since polypeptides are polymers, energy must be utilized in their formation. Hence an 'activated' amino acid was sought. There is a specific enzyme for the activation of each of the 20 different amino acids.

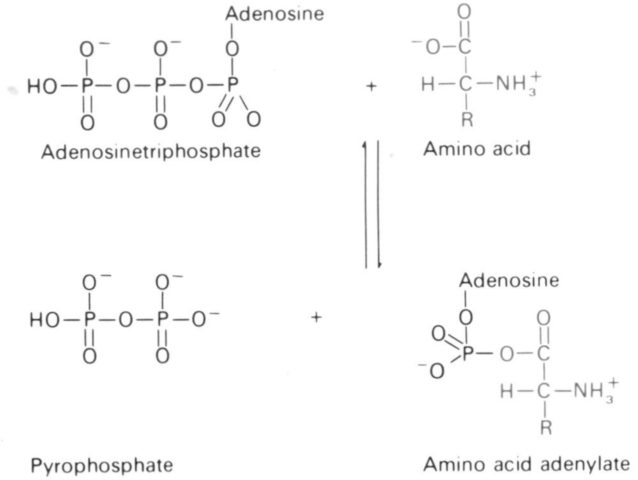

Adenosinetriphosphate + Amino acid ⇌ Pyrophosphate + Amino acid adenylate

The "activation" of an amino acid by reaction with ATP yields an amino acid adenylate and pyrophosphate.

B

The second step involves the transfer of the activated amino acid to *transfer RNA*.

The same enzyme is involved as in step 1. There is at least one specific tRNA for each of the 20 amino acids but all have CCA at the 3′ end.

The amino acid residue is attached to tRNA by an ester bond which facilitates its subsequent utilization in the formation of a peptide bond.

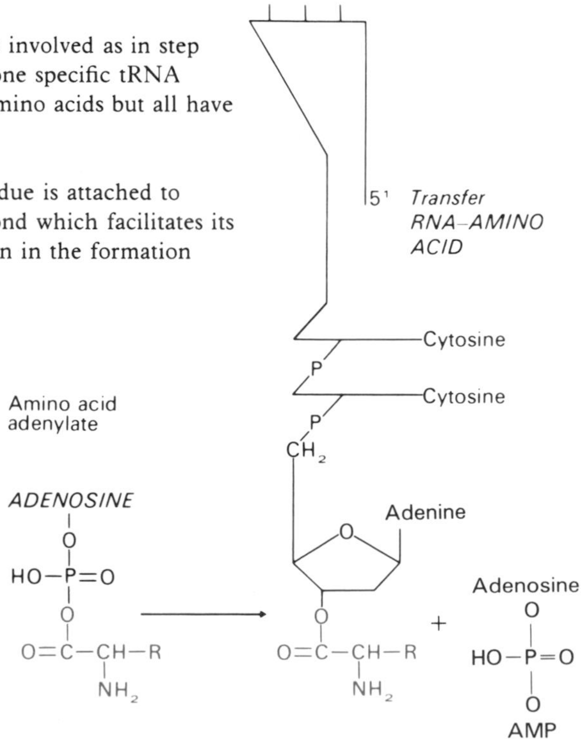

5′ *Transfer RNA–AMINO ACID*

Cytosine

P

Cytosine

P

CH_2

Adenine

Amino acid adenylate

ADENOSINE

HO—P=O

O=C—CH—R

NH_2

→

O=C—CH—R

NH_2

\+

Adenosine

HO—P=O

AMP

A

Properties of ribosomes

Ribosomes may be characterized by their sedimentation constant.

The site of protein synthesis in the cell is the cytoplasmic ribosome. This contains Mg^{2+} ions, proteins and RNA. Ribosomes consist of two subunits one about twice the size of the other.

Sedimentation constants of ribosomes and their subunits

	Intact ribosome	**Large subunit**	**Small subunit**
Prokaryotes	70S	50S	30S
Eukaryotes	80S	60S	40S

The ribosomes present in mitochondria are smaller than those in the cytoplasm. They vary in size according to source but are more like 70S than 80S ribosomes.

B

Characteristics of rRNA.

Subunit	**S of RNA**	**M_r**	**Subunit**	**S of RNA**	**M_r**
50S	23S	0.98×10^6	60S	28S	1.7×10^6
	5S	40 000		5.8S	51 000
30S	16S	0.9×10^6		5S	39 000
			40S	18S	0.7×10^6

C

The genetic code

Codons which code for an amino acid are called *sense* codons. The three codons that do not code for an amino acid are called *nonsense* codons.

First base of codon:	Second base of codon: U	C	A	G	Third base of codon:
U	UUU Phe	UCU Ser	UAU Tyr	UGU Cys	U
	UUC Phe	UCC Ser	UAC Tyr	UGC Cys	C
	UUA Leu	UCA Ser	UAA Term.*	UGA Term.*	A
	UUG Leu	UCG Ser	UAG Term.*	UGG Trp	G
C	CUU Leu	CUU Pro	CAU His	CGU Arg	U
	CUC Leu	CCC Pro	CAC His	CGC Arg	C
	CUA Leu	CCA Pro	CAA Gln	CGA Arg	A
	CUG Leu	CCG Pro	CAG Gln	CAG Arg	G
A	AUU Ile	ACU Thr	AAU Asn	AGU Ser	U
	AUC Ile	AAC Thr	AAC Asn	AGC Ser	C
	AUA Ile	ACA Thr	AAA Lys	AGA Arg	A
	AUG Met+Init.†	ACG Thr	AAG Lys	AGG Arg	G
G	GUU Val	GCU Ala	GAU Asp	GGU Gly	U
	GUC Val	GCC Ala	GAC Asp	GGC Gly	C
	GUA Val	GCA Ala	GAA Glu	GGA Gly	A
	GUG Val+Init.†	GCG Ala	GAG Glu	GGG Gly	G

Theoretical considerations led to the prediction that the code, for the translation of the base sequence in mRNA into the order of amino acids in a polypeptide, would be a 'triplet' of bases called a codon. The code shows that 61 triplets code for an amino acid. Three triplets (marked *) indicate the termination of a polypeptide chain, two other triplets, are used for initiation as well as for Met and Val respectively (marked †).
The genetic code has generally been considered to be universal in that it appears to be valid for all living cells whether eukaryote or prokaryote. (GUG seems seldom to be used for initiation).
The genetic code for mitochondria differs from that indicated above. Thus in humans UGA is used for Trp rather than Term. and AUA for Met rather than Ile.

A

The reality of the code *in vivo*.

We may check the 'code' in two ways. First to see whether it accounts for single amino acid changes in a haemoglobin chain, such as occurs in sickle-cell disease, on the basis of a *single point mutation*. It does. Secondly, we can now determine the primary structure of mRNA and relate it to the structure of a protein as shown in a hypothetical peptide. The polypeptide and mRNA are found to be *collinear*. The mRNA is read 5′ ⟶ 3′

B

The gene for *Haemoglobin S* is derived from that for normal adult Haemoglobin A by a single point mutation.

HbS differs from HbA by the change of glutamic acid at position 6 in the β chain of HbA for valine in HbS.

	HbA	**HbS**
	Glu	Val
possible codons	GAA	GUA
	GAG	GUG

In each case A has been changed to U, i.e. only one base has been changed. In fact the base change is GAA to GUA.

C

Collinearity of polypeptide and mRNA.

Recognition of the initiator codon determines the *reading frame* i.e. which bases will form the triplets for successive codons.

NH_2– term COOH term

Met Pro Ser Ala Gly Gly
AUGCCGUCGGCGGGCGGUUAG
5′ 3′

The sequence of bases starts with AUG which is not only the codon for Met but also is used as the initiator codon (see later) and ends with UAG which is one of the three codons used for termination. (No amino acid is specified as it is a *nonsense* codon.) In the mRNA for any particular protein it is found that several codons for a particular amino acid such as Gly may be used as indicated. Thus there is *degeneracy* within a particular mRNA.

D

Structure of eukaryotic mRNA.

A typical mRNA contains a region at the 5′ end which is not translated and is hence designated as the 5′ non-coding region. (In bacteria there is evidence that the base sequence is complementary to the 3′ end of the RNA of the smaller of the two ribosomal subunits (16S RNA) and hence it is assumed that the mRNA non-coding region has a role in the association of the mRNA with the ribosomal subunit). There is also a non-coding region at the 3′ end of the mRNA. As described on page 115 this end usually bears a poly (A) tail. Many eukaryote mRNA's also have a 'Cap' of 7-methylguanosine at the 5′ terminus. This is attached by the triphosphate bridge to the 5′-terminal nucleotide.

A

The role of transfer RNA (tRNA)

In the early studies there was much speculation as to the way any particular amino acid would become associated with its codon on the mRNA. Crick predicted that it would be through the medium of a small RNA which would bear an *anticodon*. This hypothetical RNA was called *adaptor RNA*. tRNA fulfils the role of the adaptor RNA. Thus it was shown that tRNA bearing an amino acid has the unique property of locating itself at the correct codon. tRNA must therefore bear an anticodon. The structure of the anticodon XYZ (3′⟶ 5′) can be predicted from our knowledge of base pairing (A-U, G-C).

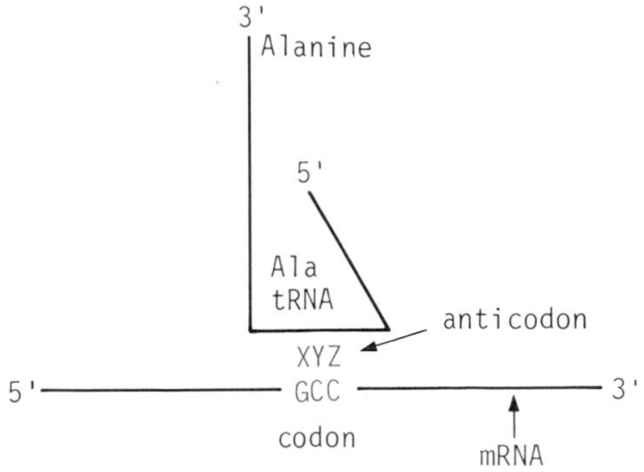

B

Structural studies show that transfer RNA has the expected anticodon. tRNA contains many intramolecular base pairs.

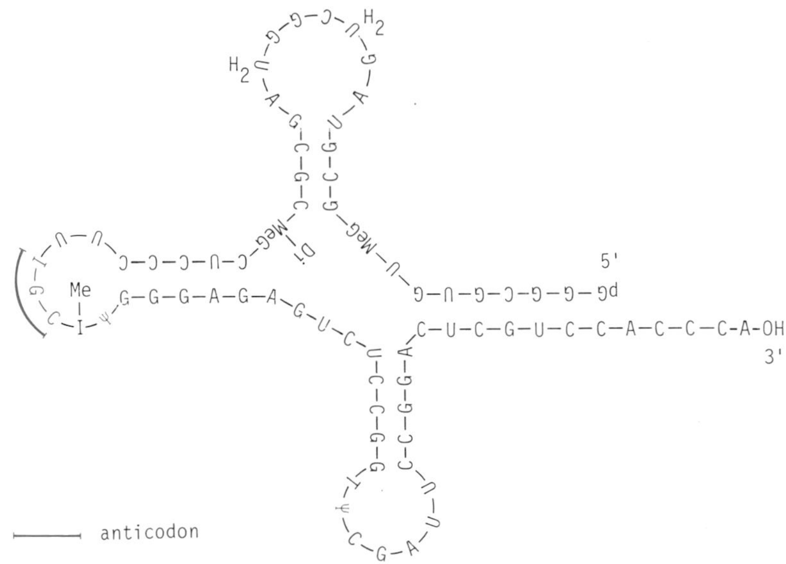

From the genetic code one might have expected there to exist 61 different tRNA's (one for each sense codon) but there are probably only about 40 different tRNA's since one anticodon may pair with more than one codon (The so-called wobble hypothesis).

Having determined the primary structure of tRNA, Holley studied the way in which the maximum number of intramolecular base pairs could be formed. All the bases except those shown in the loops could be so base-paired and the predicted structure was a clover leaf. This structure has been confirmed from the X-ray crystallography of some tRNA molecules. As predicted one finds a triplet of bases which could serve as an anticodon in a loop, (the anticodon loop). I stands for inosine which is deaminated guanosine (G). Hence we have:

Codon for Alanine	5′	⟵ GCC	3′
Anticodon found	3′	CGI ⟶	5′

A

The Role of Organelles

Many of the various organelles in the cell are involved in the biosynthesis of proteins on the rough-surfaced endoplasmic reticulum.

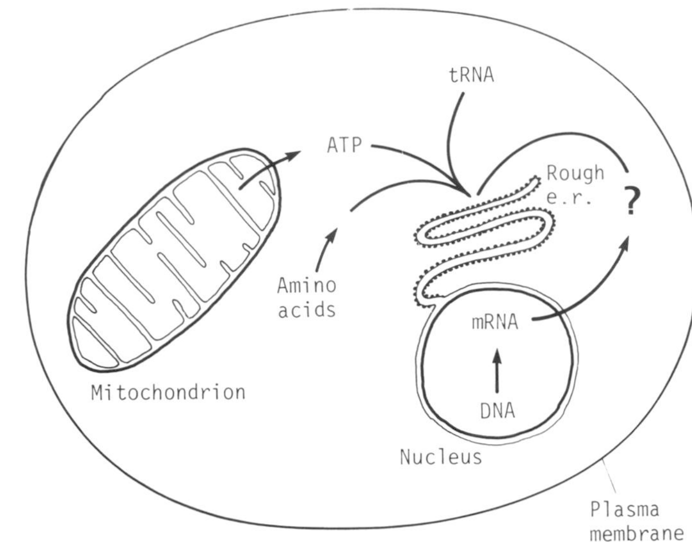

We are now in the position to realise that many of the cell organelles are involved in protein synthesis. Hence if the delicate internal structure of the cell is affected by, for example, the ingestion by the body of barbitone then the protein synthesizing activity of the cell will be affected. The ? merely indicates that we are not yet certain of the form in which the mRNA travels from the nucleus to the cytoplasm.

B

Polyribosomes

Ribosomes combine to give polyribosomes which are the active units in protein biosynthesis.

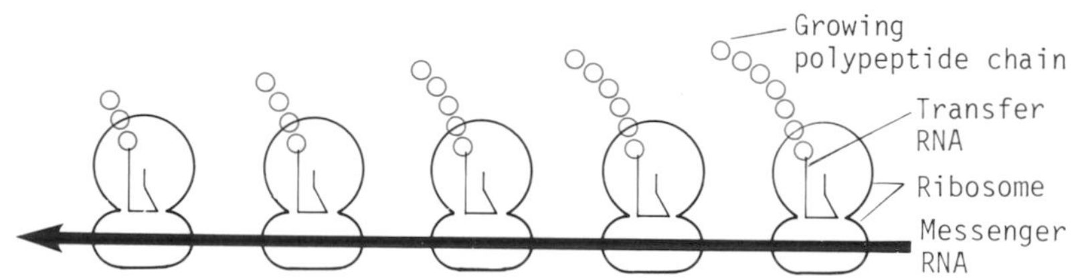

The mRNA is associated with the smaller of the two ribosome subunits, while the tRNA and growing polypeptide chain are associated with the larger of the two subunits. mRNA moves with respect to the ribosomes as shown. As is to be shown the large ribosomal subunit has two sites for tRNA.

A

Chain elongation, initiation and termination

Chain elongation involves a shuttle movement of tRNA between sites P and A.

When the two RNA's are assembled on the ribosome a peptide bond is formed by a nucleophilic attack of the α-amino group attached to the incoming tRNA on the carboxyl group of the tRNA to which the nascent chain is attached. The tRNA is then released. The aminoacyl tRNA normally enters site A.

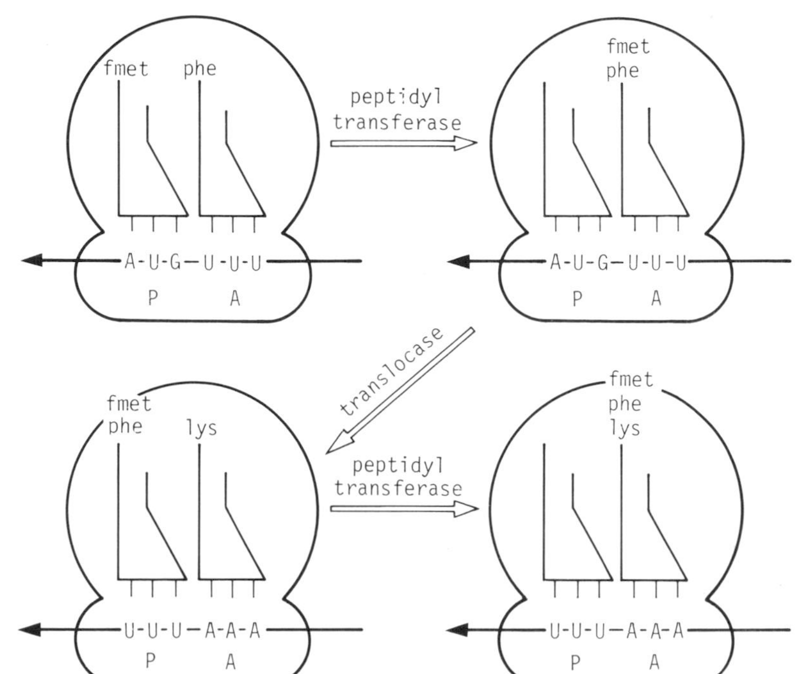

The enzyme peptidyl transferase catalyses the transfer of the incoming amino acid to the COOH end of the nascent peptide chain. Translocase is the enzyme that catalyses the movement of the new tRNA with its nascent peptide from site A to site P, known as translocation.

B

The problem of chain initiation.

The problem of chain initiation is to get the tRNA bearing the amino acid which will serve as the NH_2-terminal amino acid of the nascent peptide, into Site P. In eukaryotes Met and in prokaryotes, Formyl Met, always function as the NH_2-terminal amino acid. There are two different tRNA molecules for Met even though there is only one codon. The tRNA which has the unique property of entering site P is called initiator tRNA. A Met bound to this tRNA may be formylated in the presence of an enzyme, as it is in prokaryotes, but in any case is designated $tRNA^{fMet}$. Hence in principle the mechanism of chain initiaiton in prokaryotes and eukaryotes is similar, the eukaryotes lacking only the formylating enzyme. (The mechanism in mitochondria is more like that in prokaryotes as explained on p. 102B). After completion of the chain in prokaryotes the formyl group is always removed and the N-terminal methionine also if it is not the terminal amino acid residue of the mature protein. In eukaryotes the methionine is removed in appropriate cases when about 30 amino acid residues have been added to the chain.

A

Termination of the nascent polypeptide chain.

Release factors are involved.

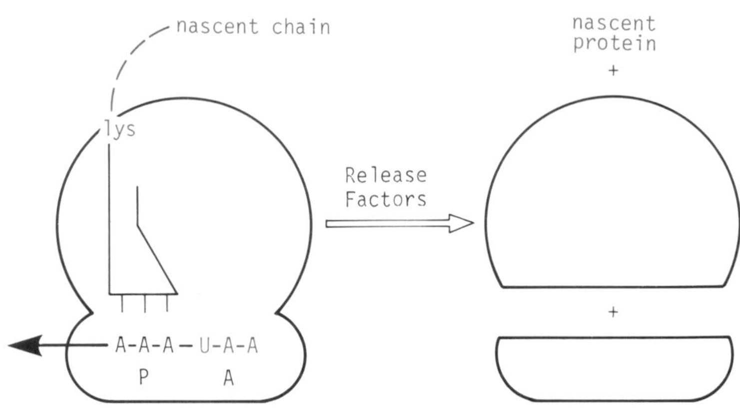

When UAA or one of the other nonsense termination codons (UAG, UGA) occupies site A, a protein factor (RF) binds to the A site in the presence of GTP and catalyses the termination reaction. The reaction involves the hydrolysis of the peptidyl tRNA ester bond, the hydroylsis of GTP and the release of the completed polypeptide chain, the deacylated tRNA and the ribosome from mRNA. In prokaryotes there are three release factors, but only one is required in eukaryotes. Following the release of the polypeptide chain the ribosome dissociates into its two subunits ready for another round of protein synthesis.

B

Mitochondrial protein synthesis

Mitochondria have their own semi-independent existence with respect to the synthesis of mRNA and protein.

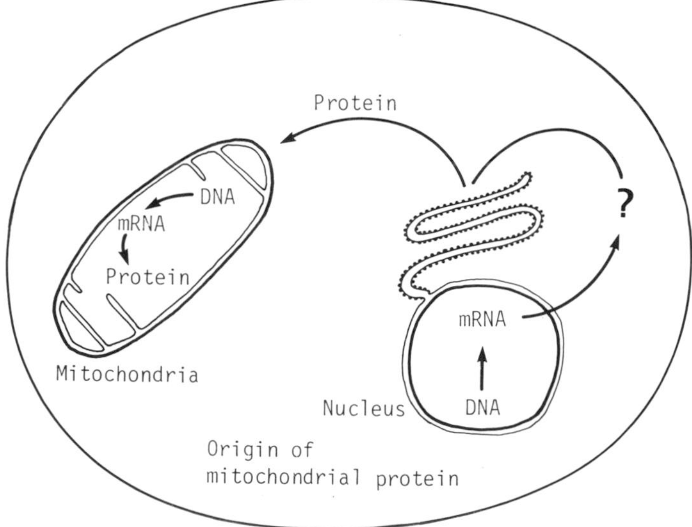

Only a few of the proteins required for the metabolic activity of mitochondria are synthesized within the organelle; the other proteins required are derived from the cytoplasm. In many respects the mechanism of protein synthesis in mitochondria more closely resembles that of 70S ribosomes than 80S ribosomes.

A

The site of action of antibiotics

Puromycin interrupts protein synthesis by virtue of the similarity of its structure to the 3′ end of amino acyl tRNA.

Many antibiotics are effective because they inhibit protein synthesis and are specific for either eukaryotes or prokaryotes. *Puromycin* has been particularly important in research on the mechanism of protein synthesis but is not specific with respect to cell type.

PUROMYCIN

Transfer RNA-AMINO ACID

B

Puromycin enters the A site on the ribosome.

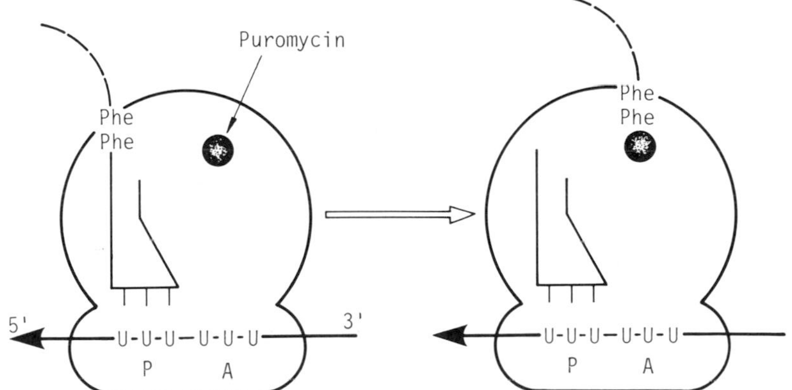

Chain elongation is prematurely terminated and C-terminal puromycinyl polypeptides are released from the ribosome.

A

Various antibiotics inhibit protein synthesis by interfering with either peptidyl transferase or translocase.

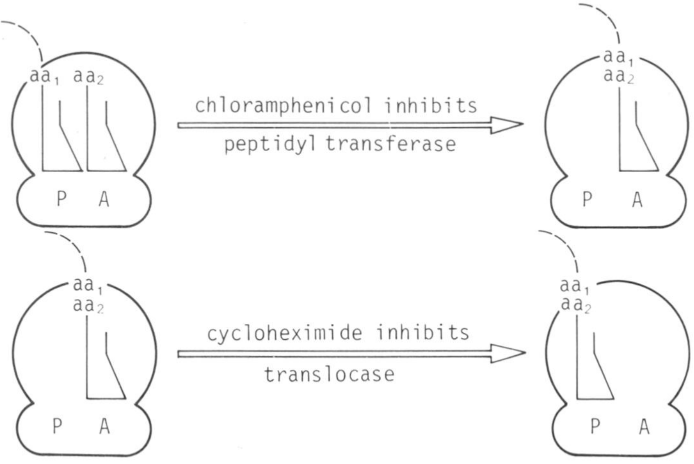

Chloramphenicol is specific for prokaryotes and mitochondrial protein synthesis.

Cycloheximide is specific for eukaryotes.

B

The site of action of various antibiotics that inhibit translation.

G factor and T factor are proteins involved in elongation.

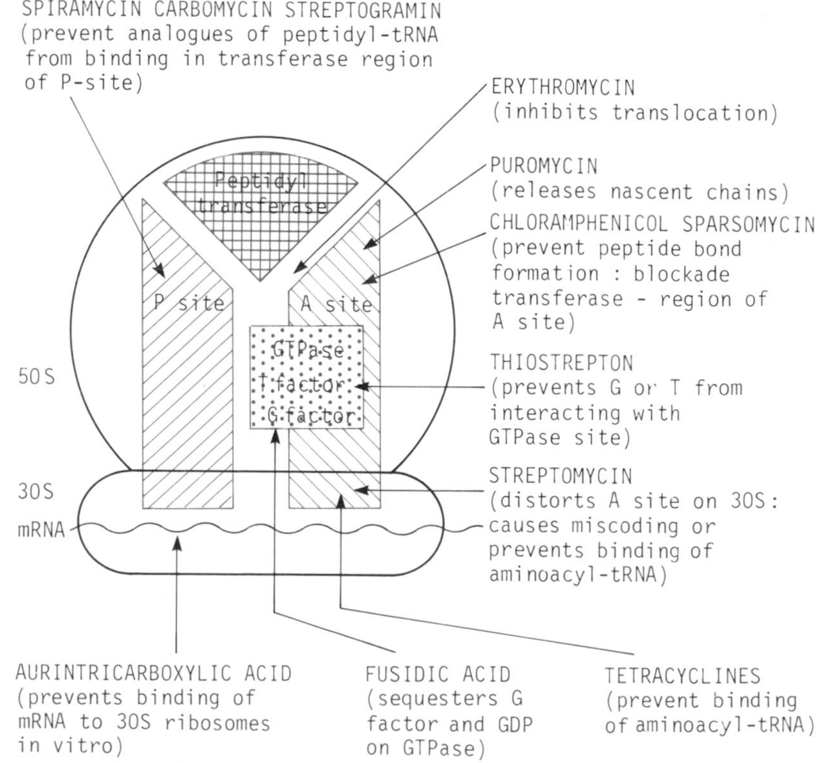

A summary of the site of action of the many drugs that inhibit DNA synthesis and replication, transcription of DNA to RNA and translation of RNA to polypeptide chain.

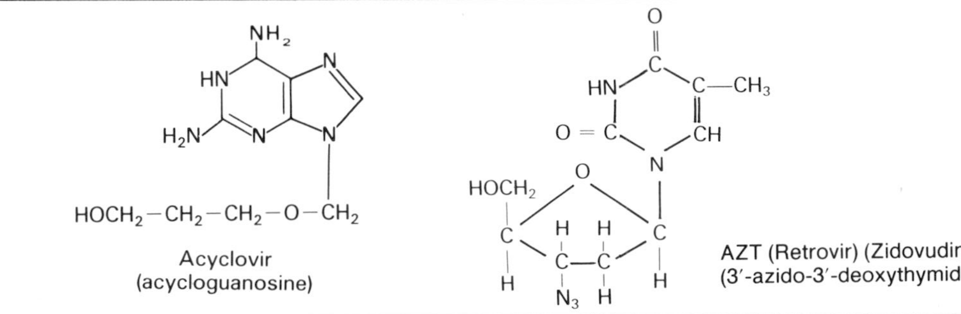

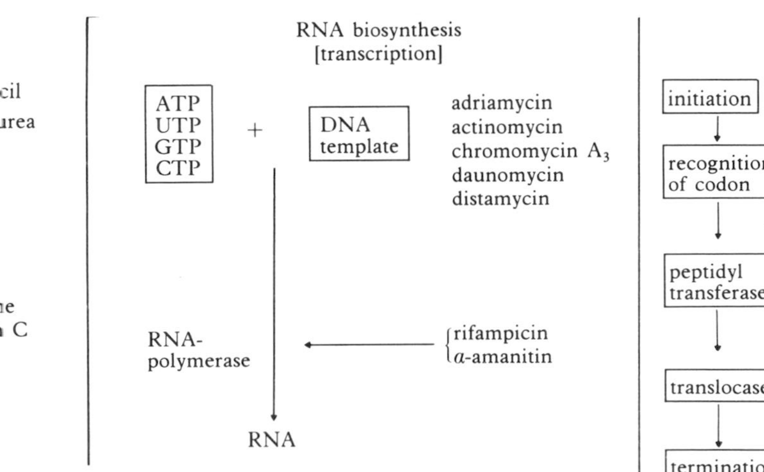

Distinction is made between four types of inhibitors of DNA biosynthesis according to site of attack:

1. Inhibitors which reversibly inhibit the DNA-polymerase reaction. Aromatic compounds which form a complex with the DNA double helix and thus inhibit replication, e.g. chloroquine. Nalidixic acid also inhibits but site of action is uncertain.
2. Inhibitors which react with the DNA with formation of covalent bonds, enter into so-called cross links with the two DNA strands necessary for semi-conservative replication, e.g. the alkylating agents including nitrogen mustard, chlorambucil, cyclophosphamide; and myleran, mitomycin C.
3. Acyclovir is a guanine derivative which is phosphorylated in herpes virus infected cells much faster than in uninfected cells; due to the action of a herpes-specific thymidine kinase. The product, acycloguanosine triphosphate, inhibits herpes virus DNA polymerase more effectively than cellular DNA polymerase.
4. Inhibitors such as N-hydroxyurea, 5-fluorouracil, aminopterin and methotrexate inhibit a certain step in the biosynthesis of deoxyribonucleosidetriphosphates, and hence indirectly DNA synthesis, by lowering the pool concentration of deoxynucleotides (see p.94)

1. Action at the bacterial ribosome (prokaryotic cell ribosome)
   a. streptomycin, neomycin, kanamycin and kasugamycin act at the 30S-subunit.
   b. chloramphenicol, erythromycin and puromycin act at the 50S-subunit.
2. Action at the eukaryotic cell ribosome by cycloheximide and emetine.

5. AZT undergoes phosphorylation in human T cells to a nucleoside 5′-triphosphate which competes with TTP and serves as a chain terminating inhibitor of HIV reverse transcriptase.

Control of protein synthesis

The scheme for the control of the *lac* operon in bacteria has formed the basis for our understanding of the mechanism of action of hormones on protein synthesis.

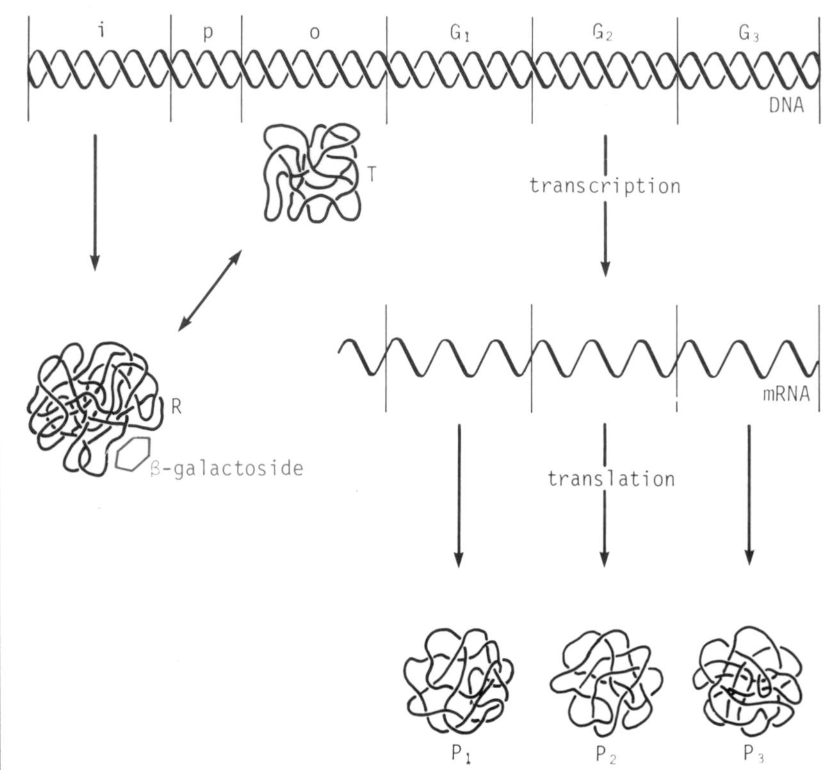

The Jacob-Monod scheme for the induction of enzymes in bacteria. R: repressor-protein in state of association with the galactoside inducer shown by the hexagon. T: repressor-protein in state of association with operator segment (o) of DNA. i: 'regulator gene' governing synthesis of the repressor. p: 'promoter' segment, point of initiation for synthesis of mRNA: G_1, G_2, G_3: 'structrural' genes governing synthesis of the three proteins in the system marked P_1, P_2, P_3. One of these proteins is β-galactosidase.

Whilst the repressor (protein) is bound to DNA, transcription cannot occur, since the site of initiation of transcription is at the promoter segment. The action of the β-galactoside is to cause the repressor to leave the DNA so that transcription and subsequently translation can occur. Since the normal state is one in which the *lac* operon is inhibited the phenomenon as described here is known as *negative control.* There are other cases in which there is *positive control* for then the phenomenon involves factors which enhance transcription above the normal level. Indeed positive control is also involved in the control of the synthesis of β-galactosidase by a mechanism which is not described here.

A

Secretory proteins

Secretory proteins are synthesized on the rough-surfaced endoplasmic reticulum from which they move through the cell. They may be modified either by the attachment of prosthetic groups or by partial proteolytic cleavage (examples of *post-translational modification*).

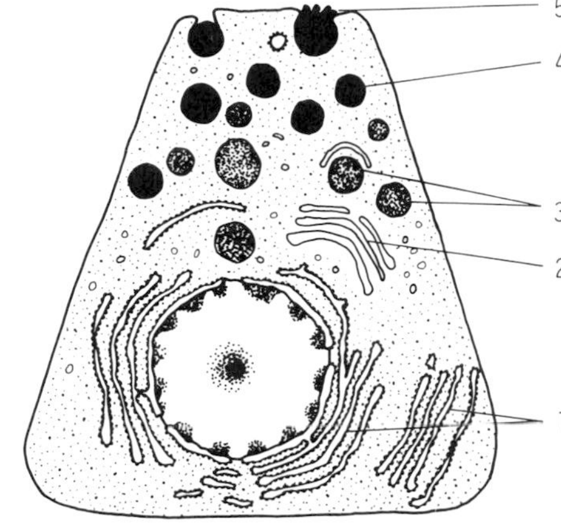

The binding of polyribosomes to the endoplasmic reticulum is an essential step in the synthesis of proteins destined for secretion.

B

The architecture of the rough-surface endoplasmic reticulum.

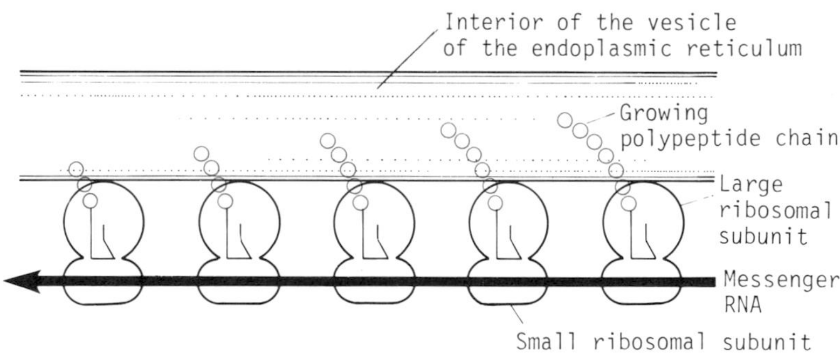

The mode of attachment of ribosomes to the membrance of the endoplasmic reticulum. The diagram shows that the larger ribosomal subunit is attached to the membrane and the growing polypeptide chain is passed into the vesicle.

In order to determine the nature of the primary translation product of the mRNA of a secretory protein, the mRNA is translated in a heterologous protein synthesizing system which lacks membranes. Such a system is one derived from wheat germ which is rich in ribosomes lacking their endogenous mRNA.

A

Polyribosomes for secretory and housekeeping proteins.

The structure of many 'signal' peptides is now known and they all contain a run of hydrophobic amino acids residues.

The mRNA in the cytoplasm associates with the membrane-free ribosomes to from polyribosomes. If the mRNA is for a secretory protein the first 20-25 amino acids from the NH_2-terminus form a signal peptide which is rich in respect to hydrophobic amino acids and has a particular affinity for the membrane of the endoplasmic reticulum. The polyribosomes then become membrane-bound. If the mRNA is for a protein which is retained by the cell, a *housekeeping protein*, there is no signal peptide and the polyribosome does not attach to membrane.

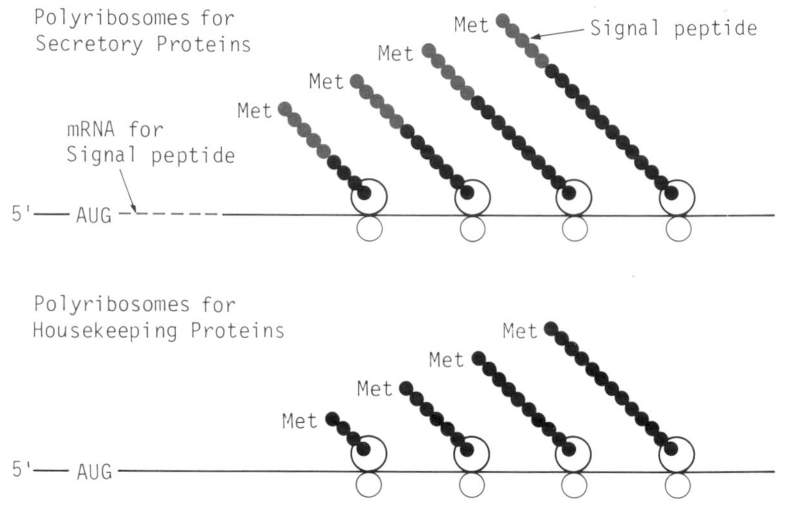

B

The signal hypothesis.

The original scheme centred on the role of the signal peptide.

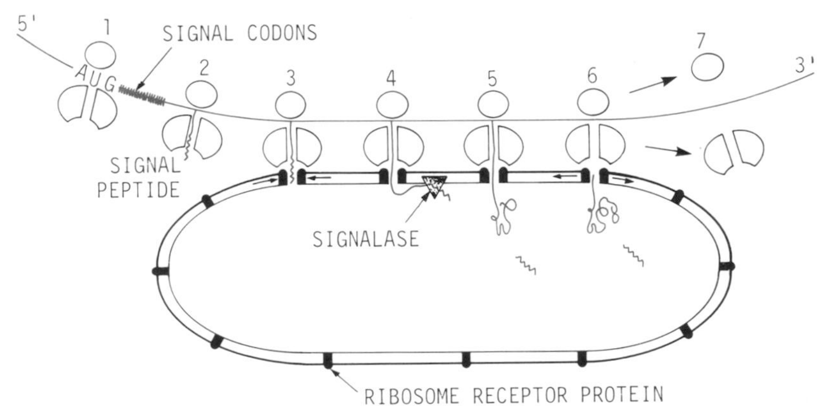

The signal peptide has an affinity with the membrane of the rough e.r.* and as a result the free polyribosomes translating the mRNA for secretory proteins become membrane bound. As the signal peptide emerges into the cisternae it is removed by a membrane-bound peptidase (*signalase*). Hence the mature protein or proprotein is found in the cisternae. The primary translation product is known as a *preprotein*. Hence preproinsulin, preproparathyroid hormone, preproalbumin (serum) are similarly named.

*See page 5

A

More about the biosynthesis of secretory proteins.

Since the emergence of the *signal hypothesis* there have been further elaborations.

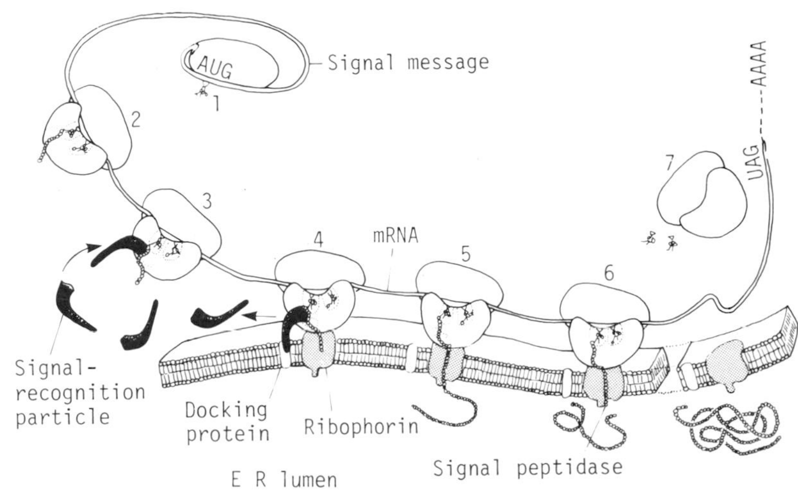

The role of various proteins has been elucidated. Thus the *signal recognition particle* which consists of several different proteins and 7S RNA binds to the signal peptide and arrests further translation until it locates the ribosome on the endoplasmic reticulum (e.r.) by the *Docking protein.* Thus secretory proteins are only made when the polysomes are bound to membrane. Various proteins contribute to the formation of the membrane pore, amongst them being the *ribophorins.* The signal recognition particle is released to the cytosol and translation by the membrane bound ribosome is resumed. (The affinity of the signal peptide is for signal recognition particle and not for the membrane of the e.r.).

B

Collagen synthesis.

In the course of the biosynthesis of the insoluble protein, collagen, a precursor protein procollagen is synthesized, which has extra peptides at each end.

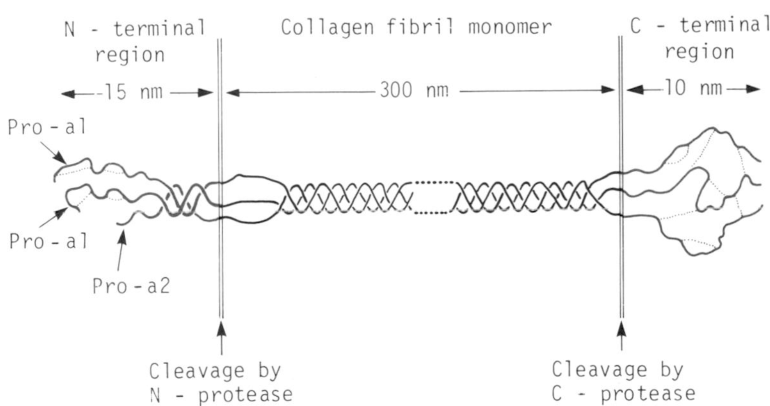

The extra peptides contain S–S bonds (indicated . . . . . .) and serve to assist in the process of helical formation of the three chains of collagen in this case shown as two identical chains ($a1$) and one different ($a2$). The peptides are lost either as the *procollagen* is secreted or just after secretion to give the insoluble protein *collagen.*

A

The formation of lysosomal enzymes.

The *mucolipidoses.*

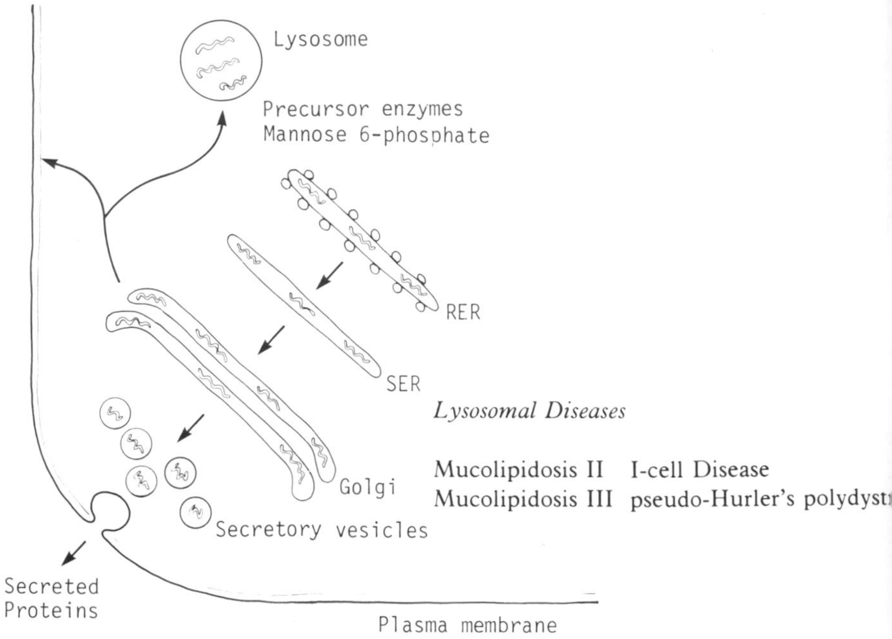

Lysosomal Diseases

Mucolipidosis II I-cell Disease
Mucolipidosis III pseudo-Hurler's polydyst

The lysosomal enzymes pass through the membranes as large precursor proteins and are glycosylated. Mannose residues are phosphorylated by a specific enzyme; an essential step in directing the enzyme to the lysosomes. It is this enzyme that is missing in the mucolipidoses in which the fibroblasts are characterized by their lack of lysosomal enzymes. Instead the enzymes are secreted extracellularly.

B

The role of membrane free polyribosomes in protein synthesis.

Th synthesis of mitochondrial, *peroxisome* and *Golgi* proteins.

Apart from membrane bound polyribosomes there exist in the cell membrane free polyribosomes. These are responsible for the translation of proteins for the cytosol, peroxisomes and mitochondria. The proteins are made as larger precursors which are degraded to mature protein as they pass through the membranes. It is *not* a process of *co*-translational insertion. The origin of the enzymes specific to the Golgi membrane is not certian. They may arise from either type of polyribosome.

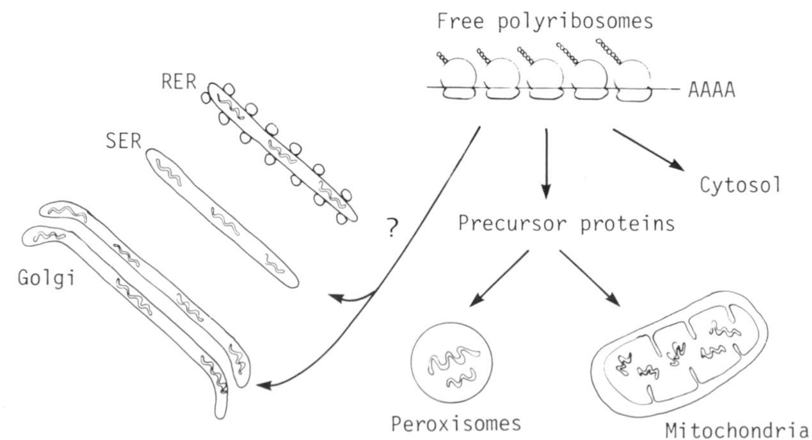

Molecular variants of proteins.

A summary showing their origin.

Molecular variants may arise normally by gene duplication from a single ancestral gene, e.g. the milk protein α-lactalbumin and enzyme lysozyme (page 248), or by alternative processing e.g. proopiomelanocortin (p. 38) or differential splicing of the primary product of the gene e.g. antibody and in particular IgM sec and IgM mem (see p. 45,116).

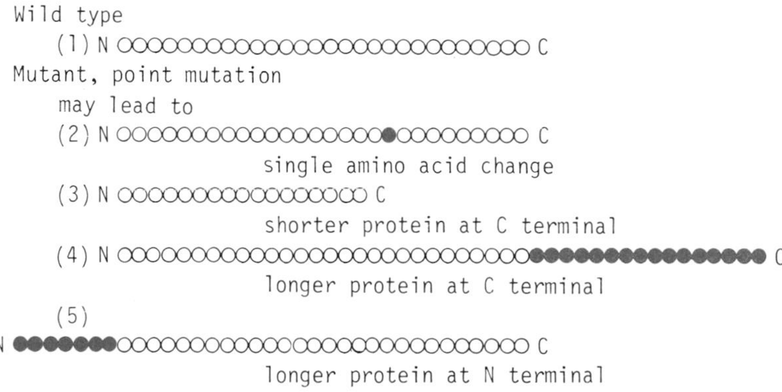

Primary structure of molecular variants of proteins

(2) Examples are the numerous haemoglobinopathies (see p. 26)
(3) A mutation of a *sense* codon to a nonsense codon.
(4) A mutation of a nonsense codon to a *sense* codon.
(5) A mutation leading to substitution of a basic amino acid (Lys or Arg) so that cleavage of proprotein does not take place e.g. proalbumin Christchurch, and insulinopathies.

A summary of the multiple steps involved in gene expression.

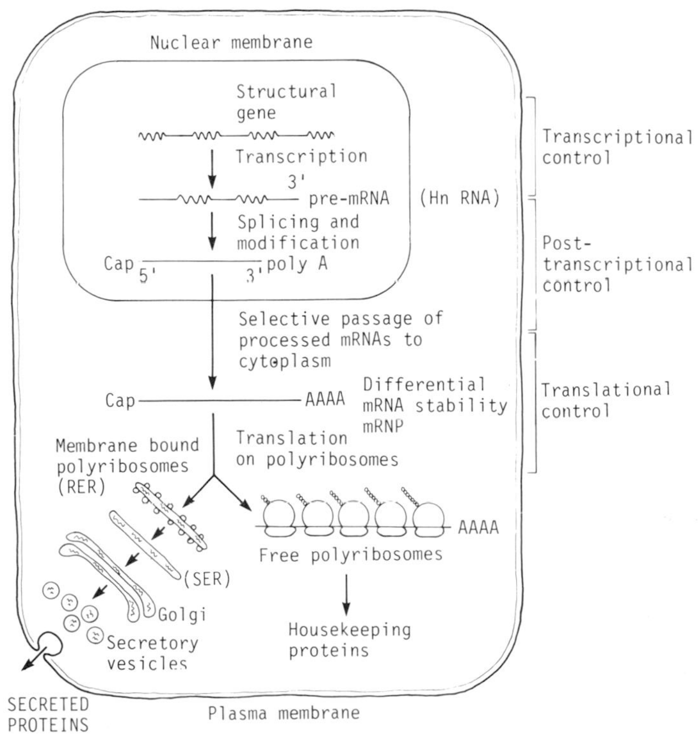

3. RECOMBINANT DNA (GENETIC ENGINEERING)

Introduction

A

Plasmids are replicated together with chromosomal DNA.

The first demonstration of the transfer of DNA from one cell type to another concerned drug resistance by bacteria. The antibiotic resistance of certain strains of *E. coli* resides in the extrachromosomal DNA of the *plasmid*. It was found that a plasmid could be transferred to a pathogenic bacterium, e.g. *Salmonella* which then became resistant to the antibiotic. The plasmid contains the gene for the synthesis of an enzyme that is secreted by the bacteria and which destroys the antibiotic.

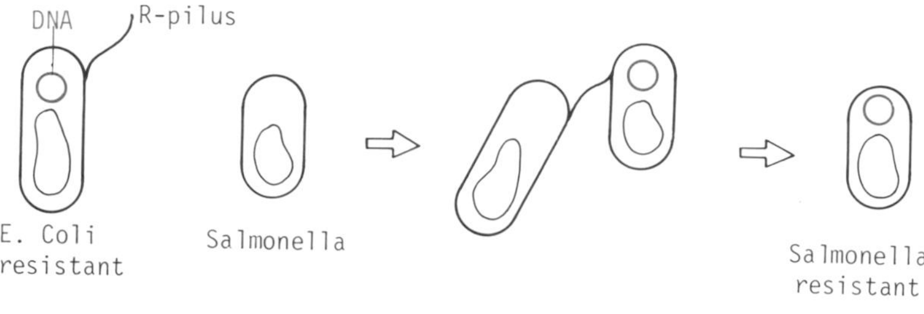

B

A first step in genetic manipulation is to insert a piece of foreign DNA into a vector (plasmid or virus) which then replicates in a suitable host. The inserted DNA can then be recovered by cleaving the vector. The recombinant DNA gene within the bacterium can be expressed in the form of protein. The host cells (e.g. bacteria) that contain the recombinant DNA are known as clones, so the process is referred to as 'cloning'.

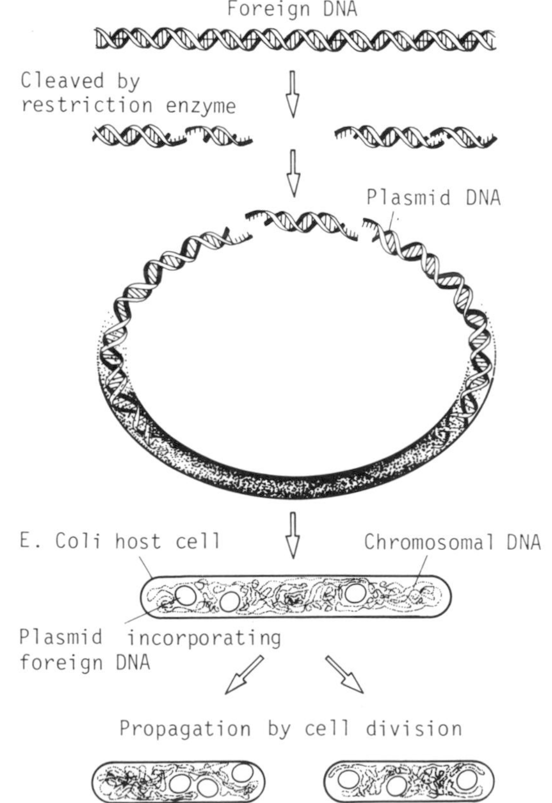

Recombinant DNA (Genetic engineering).

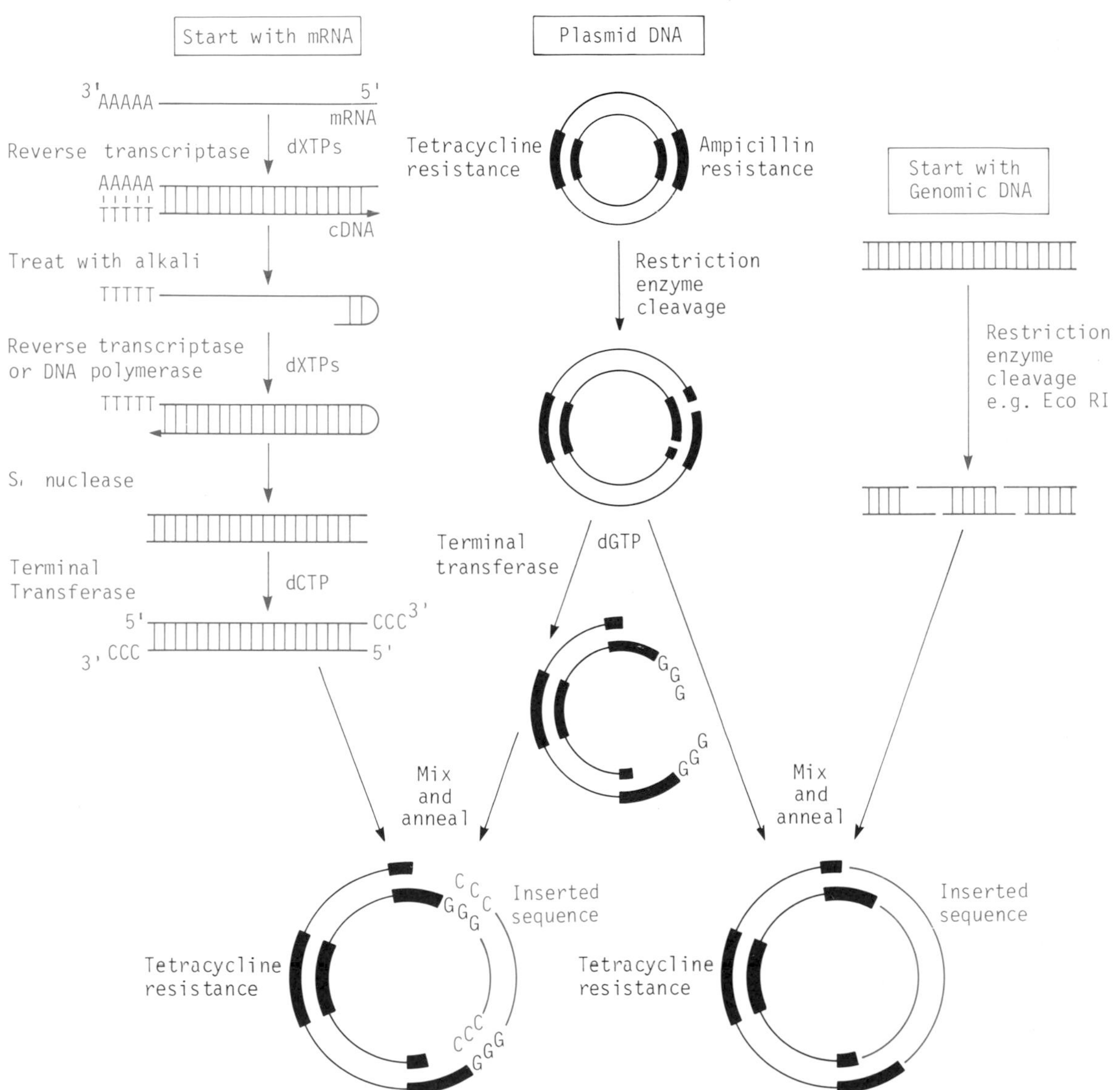

The methods used to insert the foreign DNA.

Two general methods for the construction of recombinant DNA molecules are illustrated. On the left one starts with a partially purified mRNA and double-stranded cDNA, with oligo(dC) tails, is formed. On the right genomic DNA is degraded by a restriction enzyme into roughly 'gene' size pieces. This is known as the 'shot-gun' approach. The plasmid or vector has two antibiotic resistant sites one of which is cleaved by the same restriction enzyme as is used in the 'shot-gun' approach. The 'gene-sized' piece can be annealed into the cleaved plasmid DNA. In the other approach the cleaved DNA is tailed with oligo(dG) by means of terminal transferase and annealed to the cDNA. The host is transformed and cells selected for the appropriate insert. Antibiotic resistance is useful here, since, if the bacterial colonies are grown on a medium containing tetracycline, only those colonies (clones) that contain the plasmid will grow.

A

Restriction and other enzymes

The foreign DNA to be inserted into the *plasmid* is specially prepared by recombinant techniques.

Various enzymes are used in the construction of recombinant DNA molecules. Enzymes (1), (2), (3) are involved in synthesis and enzymes (4) and (5) in degradation of DNA. By means of (2) a new copy strand complementary DNA (cDNA) of the mRNA is formed. Enzyme (5) is particularly important. The *restriction enzymes* break DNA at very specific sites. Some 200 such enzymes have been identified.

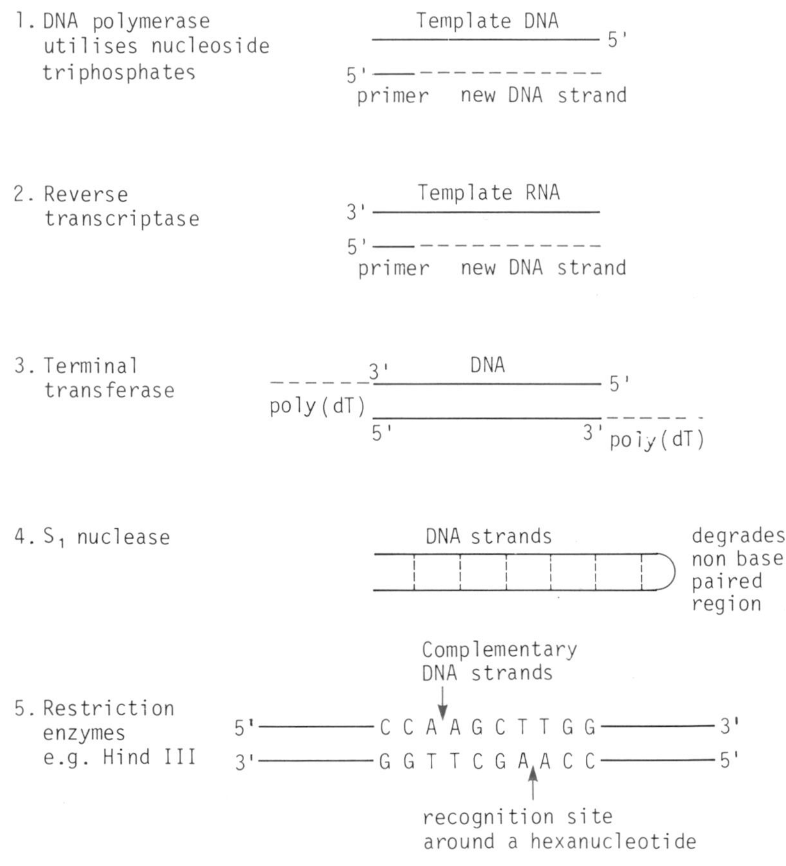

B

Nomenclature of restriction enzymes

Restriction endonucleases are the means whereby a host bacterium protects itself from foreign DNA. An infecting phage is said to be 'restricted' by the host.

The restriction enzymes that recognise a particular target sequence in a duplex DNA molecule are known as type II enzymes. Several hundred such enzymes have now been at least partially characterized having been isolated from a wide variety of bacteria. The nomenclature for these enzymes is as follows. The species name of the host organism is identified by the first letter of the genus name and the first two letters of the specific epithet to a three-letter abbreviation. Thus *Escherichia coli* = *Eco* and *Haemophilus influenzae* = *Hin*. Strain or type identification is given by a further letter hence *Hind*. Because of the symmetry of the recognition sequence the restriction enzyme may generate fragments with mutually cohesive termini as shown at (5) above. Other enzymes may generate fragments with flush ends.

A

Alternative vectors

Although until recently *E. coli* and its plasmids have been the preferred vectors there is increasing interest in alternative systems. Thus λ phage may be used as a vector in *E. coli*. This allows the use of larger inserts. *B. subtilis* may be used as an alternative bacterium and yeast protoplasts with their own plasmids also show promise. Foreign DNA may be inserted into animal cells in culture either directly as a Ca^{2+} precipitate or incorporated into a virus. Such systems hold great promise where there are problems with secretory proteins.

B

Split genes, introns and exons

The transcription and translation of the ovalbumin gene.

The primary transcription product of the DNA is known as *heterogeneous nuclear RNA* (HnRNA).

An important application of recombinant DNA has been to examine the organisation of the genomic DNA since the cDNA will hybridize with it. As indicated below, the bases which code for the egg-white protein ovalbumin are not continuous in the genome. This so-called split gene phenomenon is now commonly found, e.g. for haemoglobin α and β chains and other proteins. The intervening sequences in the genome are known as *introns* while the sequences which code for protein are known as *exons*.

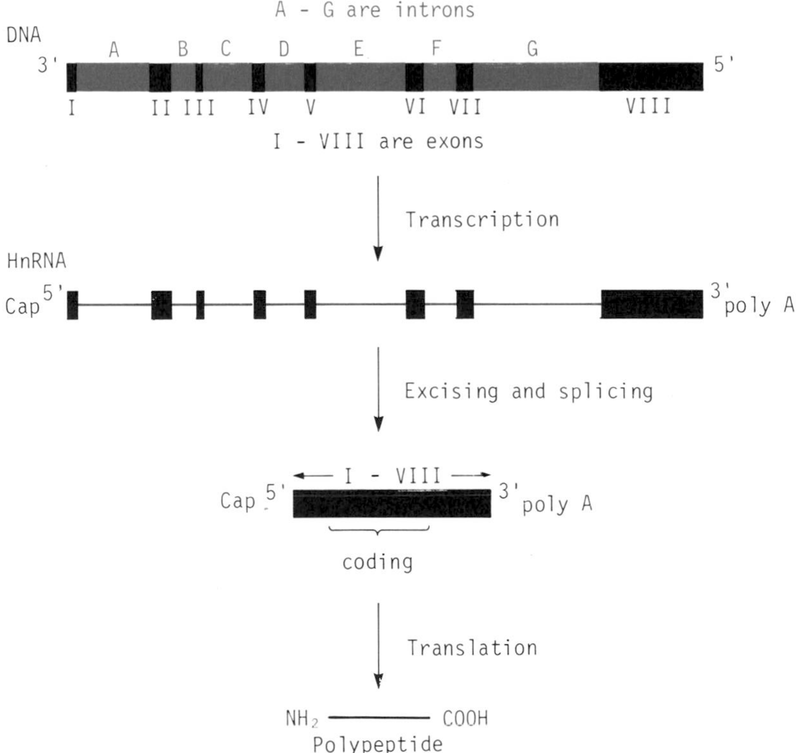

Most eukaryotic mRNA has a poly(A) tail at the 3′ end added to the primary transcription product. The poly(A) tail is not translated.

A

Electron microscopy reveals the intervening sequences.

The mRNA for the chick protein *ovomucoid* was hybridized to the genome and the DNA/RNA hybrid visualized in the electron micrograph. The DNA is shown in black and the mRNA in red. The loops are lettered and show the lengths of DNA which do not hybridize to the mRNA and hence represent the intervening sequences.

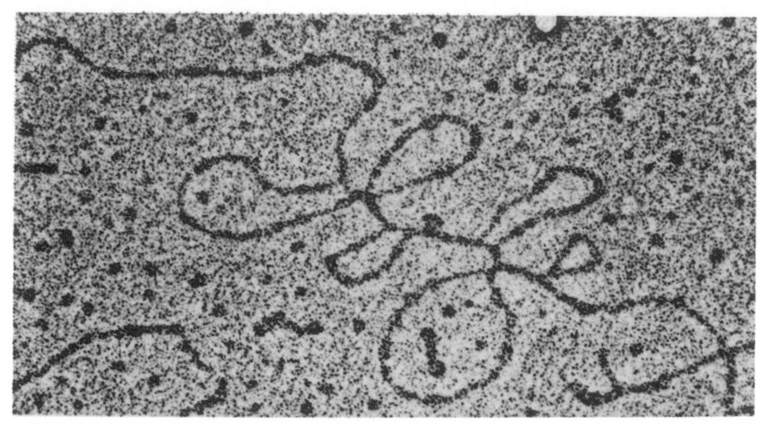

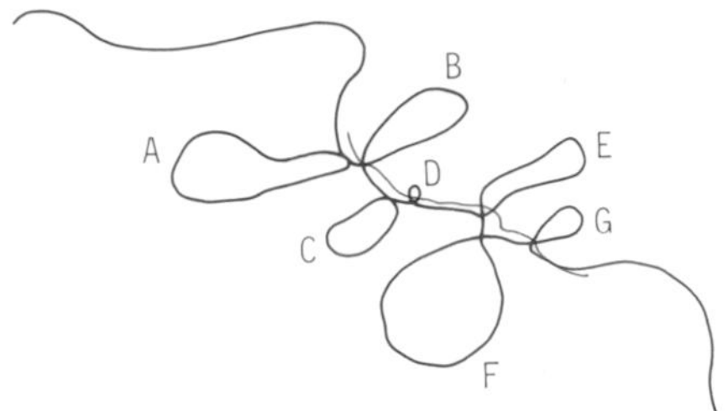

B

Differential splicing

Splicing may be used for the generation of molecular variants of proteins. In the formation of IgM for secretion and membrane insertion (see also p 45), different exons are selected for splicing to form a mature mRNA.

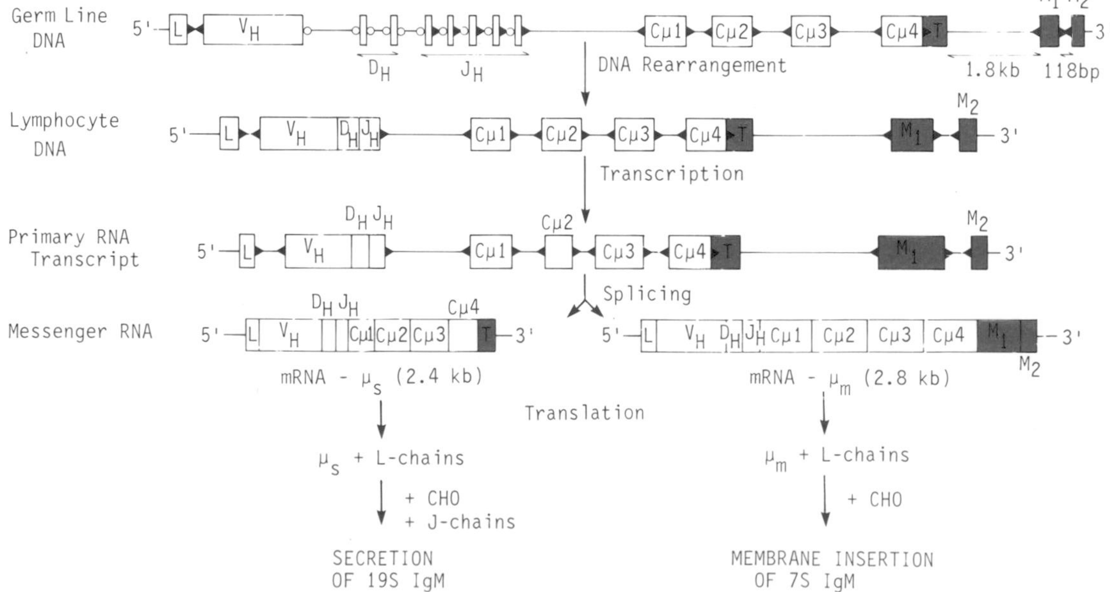

V_H, Cμ1 etc. represent the different domains of IgM shown on p 41

A

The use of radioactive 'probes' to detect gene sequences

The *Southern Blot.*

A similar method for RNA is known as the *Northern Blot.*

In order to detect fragments of DNA in an agarose gel that are complementary to a given RNA or DNA sequence Southern devised a method for transferring denatured DNA to cellulose nitrate (nitrocellulose). The DNA fragments can be permanently fixed to the cellulose nitrate by heating at 80°C. Subsequently, a radioactive probe can be hybridized to the complementary DNA on the cellulose nitrate.

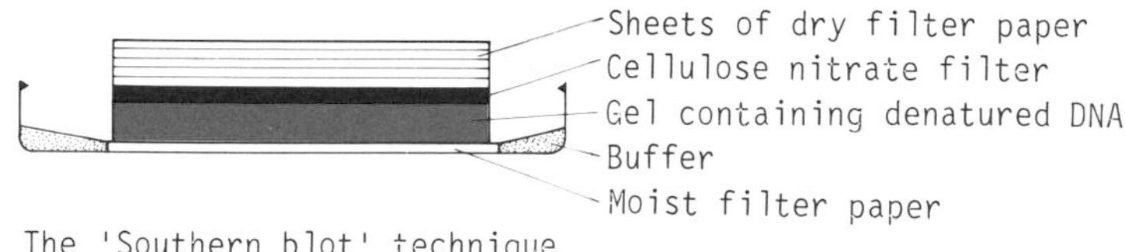

The 'Southern blot' technique

The dry filter paper draws the buffer solution up from the gel, carrying the DNA with it into the cellulose nitrate.

B

Mapping restriction sites around a hypothetical gene sequence in total genomic DNA.

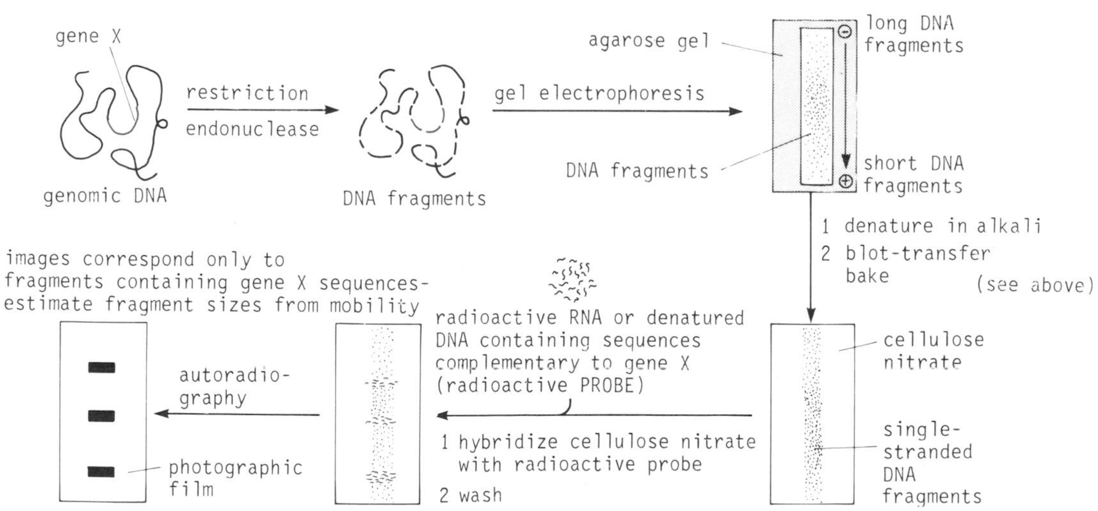

Techniques for the determination of the base sequence of DNA

Maxam and Gilbert sequencing procedure

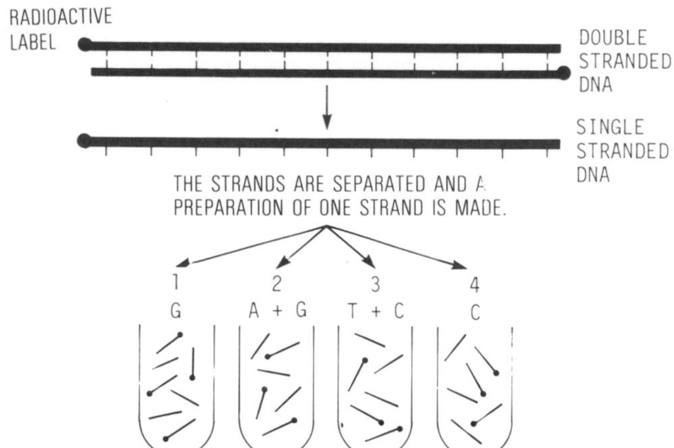

A CHEMICAL AGENT DESTROYS ONE OR TWO OF THE FOUR BASES AND SO CLEAVES THE STRANDS AT THOSE SITES. THE REACTION IS CONTROLLED SO THAT ONLY SOME STRANDS ARE CLEAVED AT EACH SITE, GENERATING A SET OF FRAGMENTS OF DIFFERENT SIZES.

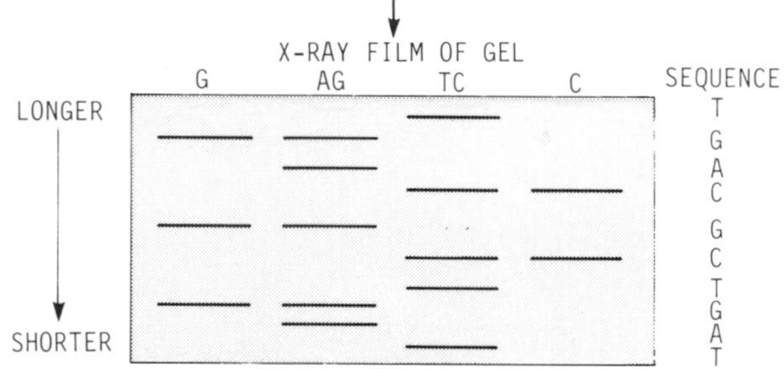

THE FRAGMENTS ARE SEPARATED ACCORDING TO SIZE BY GEL ELECTROPHORESIS AND THE RADIOACTIVE FRAGMENTS PRODUCE IMAGES ON AN X-RAY FILM. THE IMAGES ON THE X-RAY FILM DETERMINE WHICH BASE WAS DESTROYED TO PRODUCE EACH RADIOACTIVE FRAGMENT.

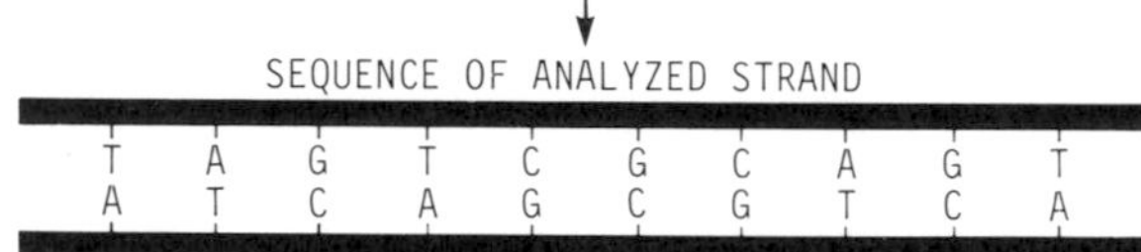

THE SEQUENCE OF THE DESTROYED BASES YIELDS THE BASE SEQUENCE OF THE ANALYZED STRAND.

A segment of DNA is labelled at one end with ^{32}P. The labelled DNA divided into four samples is treated with a chemical that specifically destroys one or two of the four bases. Only a few sites are nicked in any one DNA molecule. When the nicked molecules are treated with piperidine the fragments break up. A series of labelled fragments is generated, the length of which depend on the distance of the destroyed base from the labelled end of the molecule. The sets of labelled fragments are run side by side on an acrylamide gel that separates DNA fragments according to size and the gel is autoradiographed. The pattern of bands on the X-ray film is read to determine the base sequence of DNA.

Sanger sequencing procedure

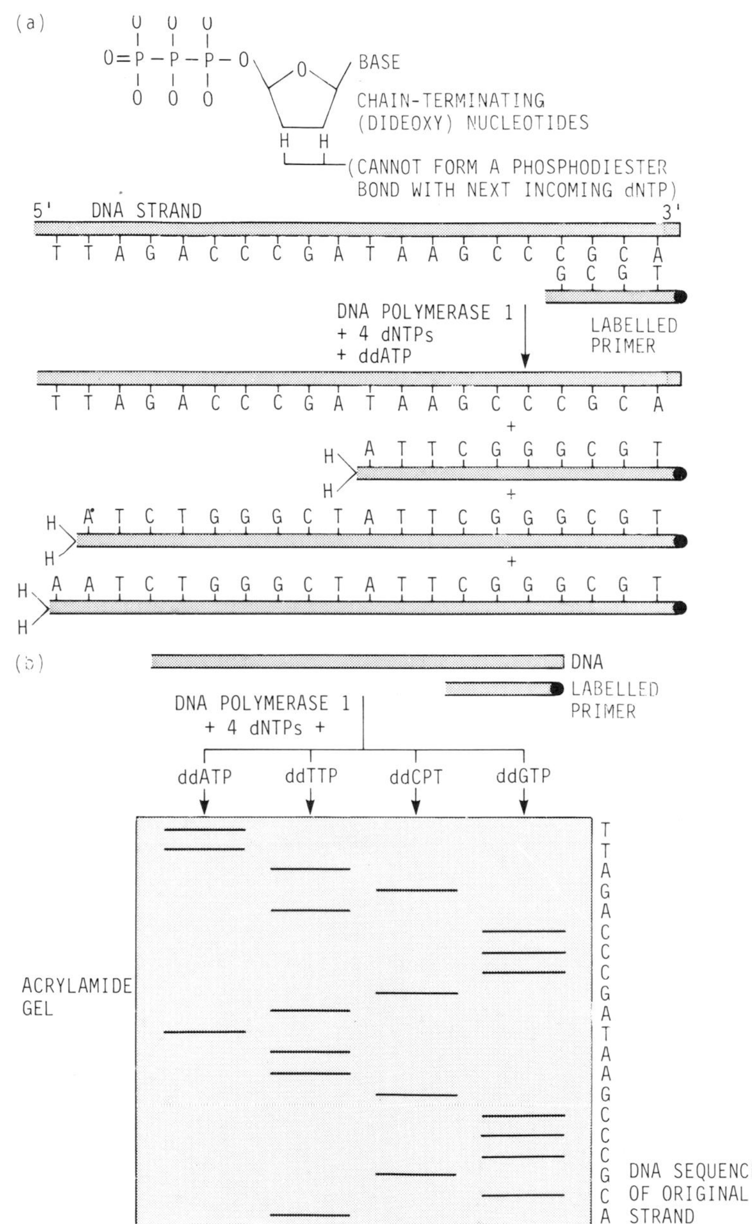

(a) In the Sanger procedure 2′, 3′-dideoxynucleotides of each of the four bases are prepared. These molecules are incorporated into DNA by *E. coli* DNA polymerase but once incorporated, the dideoxynucleotide (dd NTP) cannot form a phospohodiester bond with the next incoming dNTP and so the growth of the chain stops. A sequencing reaction consists of a DNA strand to be sequenced, a short labelled piece of DNA (the primer) that is complementary to the end of that strand, a carefully controlled ratio of one particular ddNTP with its normal dNTP and the three other dNTPs. If the correct ratio of ddNTP:dNTP is chosen a series of labelled strands will result, the lengths of which depend on the location of a particular base relative to the end of the DNA.

(b) A DNA strand to be sequenced, along with labelled primer, is split into four DNA polymerasé reactions, each containing one of the four ddNTPs. The resultant labelled fragments are separated by size as for Maxam and Gilbert.

A

Antenatal diagnosis

An example of the role of recombinant DNA technology in antenatal diagnosis

A particular application has been to devise means whereby it is possible to determine the genetic make-up of the foetus being carried by the mother. With certain knowledge of the status of the foetus the mother may in some societies be able to choose to have an abortion. An example is the problem of sickle cell anaemia and methods to determine whether the foetus is homozygous for the sickle cell trait. The genetic change is located in the β chain of HbA. As shown below, the foetus typically contains HbF, but after a few weeks there is some HbA, which may be analysed. Blood is withdrawn by a process known as foetoscopy, but this carries some dangers.

B

Haemoglobin polymorphism in the human.

The diagram shows how the different types of haemoglobin present in the normal human arise from the gene loci.

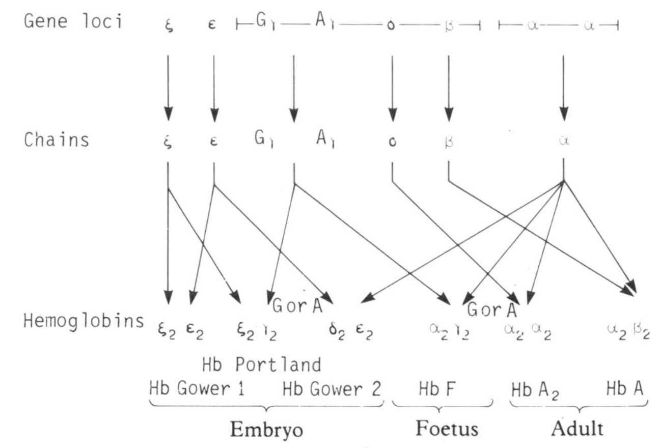

C

The analysis of the genome.

The use of a *gene probe* for the detection of sickle cell disease.

Illustrated is an indirect method which depends on a mutation in the non-coding DNA which is often linked to the point mutation leading to HbS.

An alternative technique is to withdraw some of the amniotic fluid (amniocentesis) which contains some of the fibroblast cells from the embryo. The DNA of the fibroblast is treated with a restriction enzyme (Hpa I). The fragments are separated by the Southern blot technique and probed with a [^{32}P] cDNA for β-globin. The size of the fragments (shown as thousands of bases) differs in HbA and HbS. The change in the restriction site is frequently associated with the sickle cell mutation in certain populations.

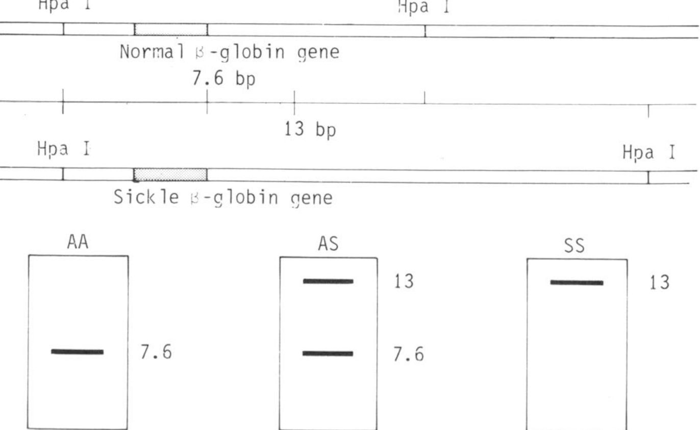

The association of a restriction fragment length polymorphism with the HbS mutation.

A brief glossary of terms that have not been previously explained.

'Alu' family DNA
The most abundant family of repetitive sequences in mammal, present throughout the genome. The human Alu sequence consists of 300 bp and appears about 300 000 times. The Alu sequence is released by the action of the Alu restriction enzyme, hence its name.

Amplification
1. Treatment designed to increase the proportion of plasmid DNA relative to that of bacterial DNA. 2. Replication of a gene library in bulk.

CAP
Not to be confused with cap; CAP is catabolite gene activator protein (sometimes CRP or CGA); it participates in the initiation of transcription in prokaryotes.
cap The structure found at the 5′ end of many eukaryotic mRNA's; it consists of 7′-methyl-guanosine-pppX, where X is the first nucleotide encoded in the DNA; it is not present in prokaryotic mRNA's; it is added post-transcriptionally near the TATA (Hogness) box.

Cistron
a DNA fragment or portion that specifies or codes for a particular polypeptide.

Cosmid cloning
A technique for cloning large eukaryotic fragments in *E. coli.* It employs the single stranded 'cos' sites at each end of λ phage which are required for packaging in phage heads.

Episome
A circular gene fragment.

Frame shift
A mutation that is caused by insertion or deletion of one or more paired nucleotides, and whose effect is to change the reading frame of codons during protein synthesis, thus yielding a different amino acid sequence beginning at the mutated codon.

Fusion proteins
Hybrid proteins containing both bacterial and eukaryotic amino acid sequences used particularly for the expression of proteins detected by antibody.

Genome
All the genes of an organism or individual.

Grünstein-Hogness assay colony
Hybridization procedure for identification of plasmid clones (colonies are transferred to a filter and hybridized with a probe).

Heat shock genes
High temperatures and other stress-inducing treatments evoke the expression of heat shock genes in *Drosophila* giving rise to heat shock proteins.

Heteroduplex
A DNA molecule, the two strands of which come from different individuals so that there may be some base pairs or blocks of base pairs that do not match.

Hogness box (TATA box)
The hypothesized eukaryotic RNA polymerase II promoter analogous to the Pribnow box.

Inversion
The alteration of a DNA molecule made by removing a fragment, reversing its orientation, and putting it back into place.

Klenow polymerase
DNA polymerase I possesses exonuclease activity in the absence of dNTPs and will degrade single strands of DNA duplex in both directions. Removal of part of the polypeptide chain to produce the Klenow fragment leaves only 3′→5′ exonuclease activity.

Linker
A small fragment of synthetic DNA that has a restriction site useful for gene splicing

M13 phage
A single-stranded DNA phage. The double-stranded replicative form can be used as a cloning vector.

Metallothionine (MMT)
A metal-binding protein that is induced in animal cells by a variety of metals such as zinc. By the use of the MMT gene promoter fused to the coding region of the protein required, the expression of a fusion protein can be regulated.

Nick translation procedure
Procedure for labelling DNA in vitro using DNA polymerase I.

Open reading frames
Long stretches of triplet codons in DNA that are not interrupted by a translational stop codon.

Palindrome
A self-complementary nucleic acid sequence, that is, a sequence identical to its complementary strand (both read in the same 5′ to 3′ direction).

Pribnow box TATAATG
Consensus sequence near the RNA start point of prokaryotic promoters.

Protein Engineering
A technique which enables deliberate changes to be made in the gene coding for a given protein so that on expression the amino acid composition of the protein is altered in a specific manner.

A brief glossary of terms that have not been previously explained. Continued from previous page.

Restriction Maps
The location of the multiple sites within a DNA which are susceptible to cleavage by a variety of restriction enzymes. Such a map will give an indication of the degree of homology of different DNA molecules.

Shine-Dalgarno sequences
A ribosome binding site, about 8 nucleotides up stream from the initiation codon which is closely complementary to the 3′ end of the smaller (16S) of the two rRNA molecules in bacteria.

Transduction
The transfer of genetic material from one cell to another by means of a viral vector (of bacteria, the vector is bacteriophage)

Transfection
Infection of a cell with isolated DNA or RNA from a virus or viral vector.

Transformation
The induction of an exogenous DNA preparation (transforming agent) into a cell.

Transgenic
Animals that have integrated foreign DNA into their germ line as a consequence of experimental introduction of DNA, commonly by microinjection or retroviral infection.

Western blotting
A technique for the detection of proteins by blotting gels, that involves the use of an antibody. (Derived from Southern and Northern blotting).

5

COENZYMES AND WATER-SOLUBLE VITAMINS

A

Coenzymes and water soluble vitamins. (Compounds of differing structures that have the same vitamin action are known as *vitamers*).

Coenzymes are molecules that cooperate in the catalytic action of an enzyme. A coenzyme may be tightly bound to the protein, or it may in other cases be free to diffuse away from the enzyme, acting essentially as an additional substrate (e.g. NAD^+ in many dehydrogenase reactions). If tightly bound to the enzyme, it may sometimes be referred to as a *prosthetic group*. The term *cofactor* refers rather non-specifically to the various materials required for full enzyme activity, including cations (e.g. Zn^{2+} or Mg^{2+}).

Water-soluble vitamins function as coenzymes, or as components of coenzyme molecules.

B

Nicotinamide nucleotides

NAD^+ and $NADP^+$ and their reduced forms are involved in a great many dehydrogenase reactions in the mitochondrion, cytosol and endoplasmic reticulum of the cell. They are water-soluble, and are usually free to diffuse away from the enzyme, after conversion to the oxidized or reduced form, to take part in another dehydrogenase reaction catalysed by another enzyme.

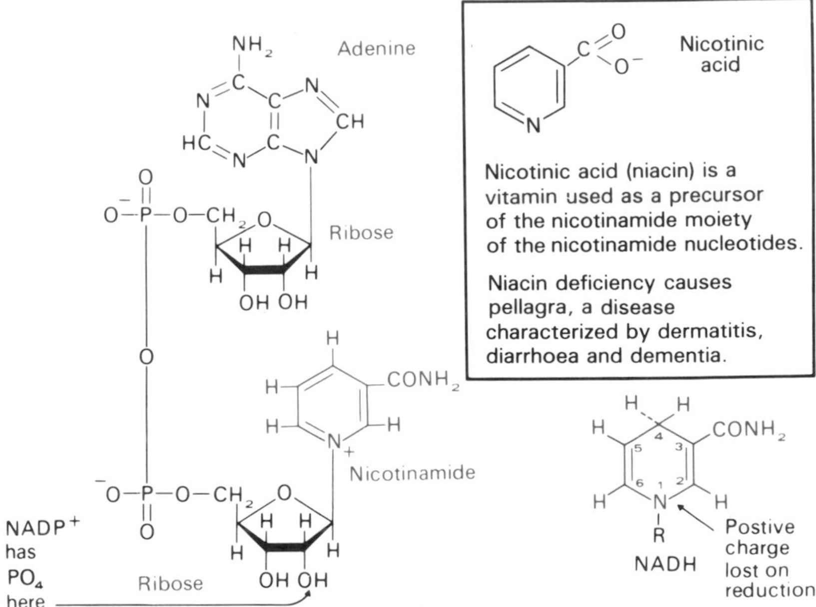

Nicotinamide adenine dinucleotide (NAD^+)
NADP is nicotinamide adenine dinucleotide phosphate.

Hydrogen received from substrate is bound to the nicotinamide moiety.

C

The spectrum of NADH differs from that of NAD^+ in that there is a peak of absorption around 340 nm.

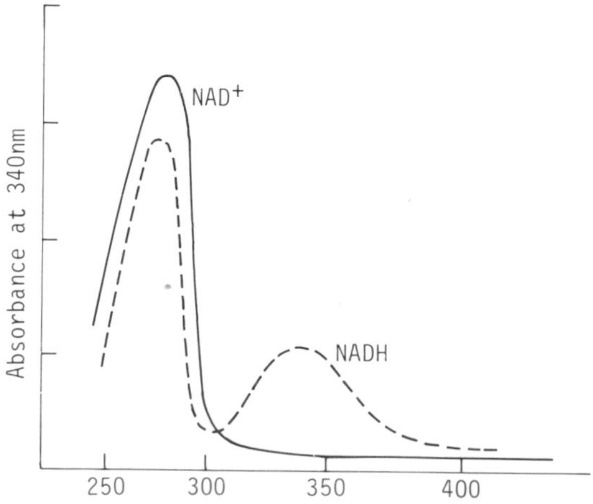

Many dehydrogenase reactions can be followed by measuring the change in absorbance at 340 nm.

A

By virtue of the fact that the concentration of these coenzymes is controlled by the respiratory state of the cell, they function to exert control from the mitochondrion on the direction of metabolism in the cell.

Some enzymes have tightly bound NAD^+ that acts as a modulator of the protein conformation. These coenzymes are allosteric effectors of many enzymes. In addition they influence enzyme activity by virtue of their relative concentrations by a mass action effect.

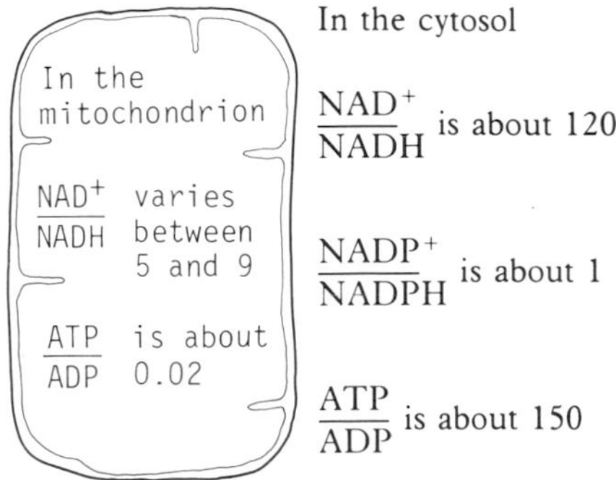

In the cytosol

$\frac{NAD^+}{NADH}$ is about 1200

$\frac{NADP^+}{NADPH}$ is about 1

$\frac{ATP}{ADP}$ is about 150

Thus in the cytosol the tendency is for reactions involving NAD^+ to proceed towards NADH (oxidation of the substrate). Because of this, NADPH is often the coenzyme used for reduction in the cytosol, which the more favourable ratio of reducing and oxidizing equivalents (*redox ratio*) facilitates.

The general form of reactions involving NAD^+ is

$$RH_2 + NAD^+ \rightleftharpoons R + NADH + H^+$$

For convenience, in writing metabolic pathways, the H^+ is omitted. However, it should be appreciated that it is always involved in the reaction.

B

Adenine nucleotide coenzymes

The ultimate purpose of tissue respiration is the phosphorylation of ADP to produce ATP. The overall direction of cellular metabolism is regulated by this process.

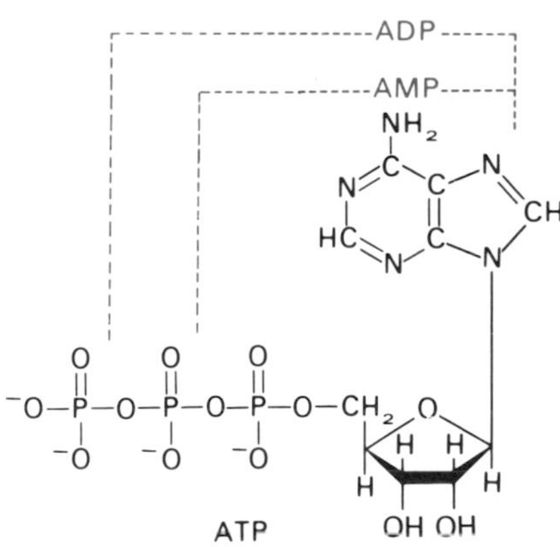

Adenosine 5′-phosphate = Adenosine monophosphate (AMP)
Adenosine 5′-diphosphate = Adenosine diphosphate (ADP)
Adenosine 5′-triphosphate = Adenosine triphosphate (ATP)

ATP, ADP and AMP form a system of coenzymes that have the function of influencing the direction of flow of metabolic pathways. These compounds are allosteric effectors for many enzymes. In addition, ATP often functions as a donor of a phosphate to other molecules (in reactions catalysed by kinases).

A

Adenine nucleotide interrelationships

The enzyme myokinase interconverts the adenine nucleotides.

$$ATP + AMP \rightleftharpoons 2ADP$$

myokinase

The metabolic status of a cell is indicated by the relative concentrations of the adenine nucleotides. A useful index is the *energy charge*, defined as

$$\frac{[ATP] + \tfrac{1}{2}[ADP]}{[ATP] + [ADP] + [AMP]}$$

B

Thiamin pyrophosphate

Thiamin (Vitamin B_1) is the precursor of thiamin pyrophosphate, the coenzyme for some important oxidative decarboxylation reactions (pyruvate dehydrogenase, oxoglutarate dehydrogenase). A deficiency of thiamin causes the disease beri-beri, a disease associated with neuropathy and cardiopathy. Experimental deficiency causes neurological symptoms (pigeons fail to hold their head erect) that can readily be reversed by administering the vitamin. In oxidative decarboxylation, loss of CO_2 is accompanied by oxidation of an aldehyde to an acid.

Thiamin pyrophosphate is important in the pyruvate dehydrogenase and similar reactions.

$$\text{Pyruvate} + \text{CoA} \xrightarrow{NAD^+ \rightarrow NADH} \text{Acetyl CoA} + CO_2$$

pyruvate dehydrogenase

The pyruvate dehydrogenase enzyme has a complex mechanism involving binding sites for pyruvate, NAD^+, thiamin pyrophosphate and lipoic acid.

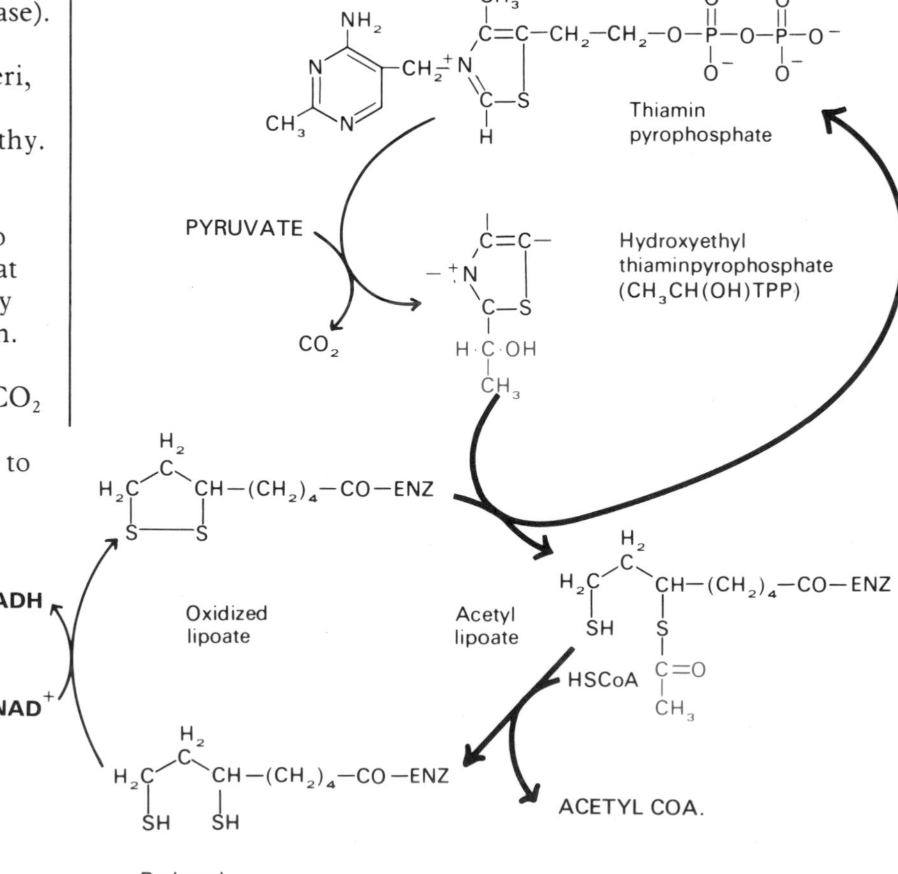

A

Flavin nucleotide coenzymes

FAD is the coenzyme of a class of dehydrogenases known as flavoproteins. The flavin moiety of the molecule is derived from riboflavin (vitamin B_2).

Flavin mononucleotide (FMN) is an important coenzyme in some flavoproteins, including NADH coenzyme Q reductase (page 175).

FMN consists of riboflavin phosphate (i.e. FAD without the adenosine monophosphate moiety.)

Riboflavin (in red)

Flavin adenine dinucleotide (FAD)

Substituted isoalloxazine

(R = remainder of molecule, above)

$FADH_2$

Reduction of FAD or FMN involves the two unsubstituted N atoms of the isoalloxazine structure, as shown in red above.

B

Flavin coenzymes remain tightly bound to the enzyme protein throughout the reaction, in contrast to nicotinamide coenzymes that bind reversibly to the enzyme.

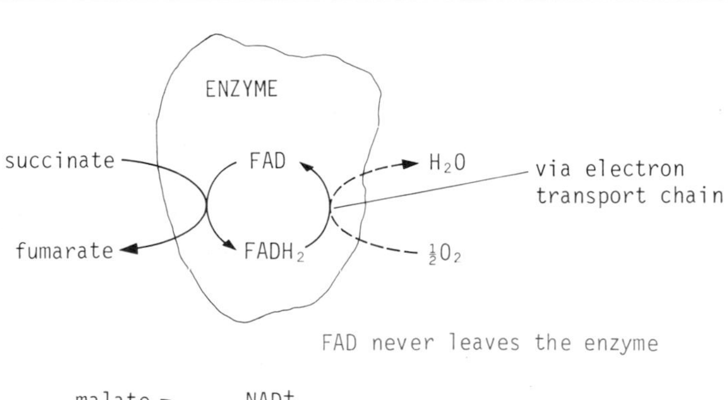

A number of mitochondrial dehydrogenases are flavoproteins; for example, NADH dehydrogenase, succinate dehydrogenase and fatty acyl CoA dehydrogenase.

A

Coenzyme A

Coenzyme A is a complex molecule which contains a free sulphydryl (—SH) group. This group can react with a carboxyl group to form a thioester.

Pantothenate is an essential food factor which forms part of coenzyme A. The free sulphydryl group can react with a carboxyl group to form a thiol ester. Such thiol esters are involved in many transfer reactions involving acyl groups, including acetyl and fatty acyl groups.

A thioester is in some ways analogous to the ester formed between a carboxylic acid and an alcohol. However, it is considerably more reactive. In acetyl CoA, the thioester linkage can activate the methyl carbon as well as the acyl carbon.

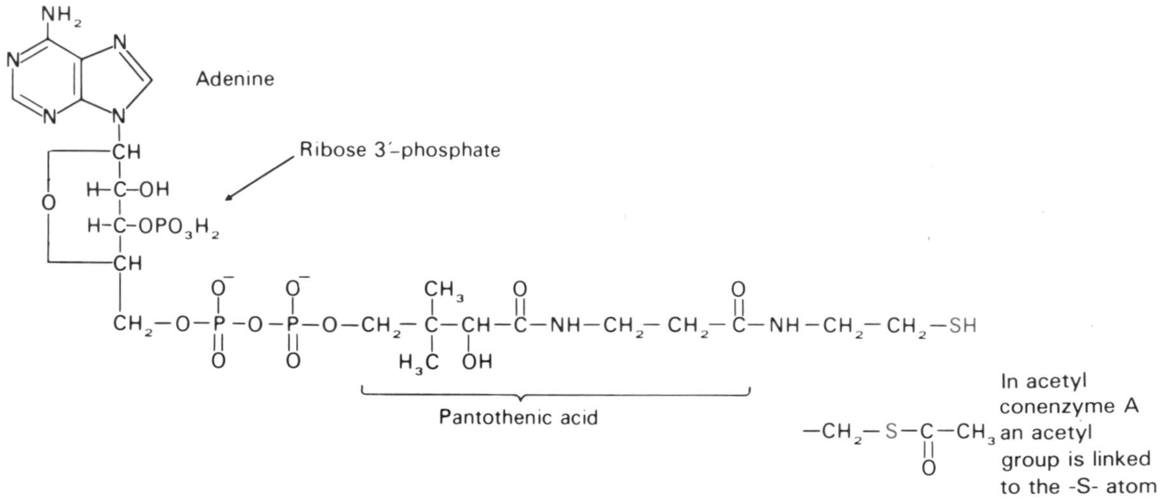

Adenine

Ribose 3′-phosphate

$CH_2-O-P(O^-)(=O)-O-P(O^-)(=O)-O-CH_2-C(CH_3)(H_3C)-CH(OH)-C(=O)-NH-CH_2-CH_2-C(=O)-NH-CH_2-CH_2-SH$

Pantothenic acid

$-CH_2-S-C(=O)-CH_3$ In acetyl conenzyme A an acetyl group is linked to the -S- atom

Coenzyme A and Acetyl Coenzyme A

B

Biotin

Biotin is an essential food factor and is a coenzyme for carboxylation reactions (e.g. pyruvate carboxylase). These reactions involve ATP, which is necessary in the first step of the reaction; this is the conversion of biotin to carboxybiotin by the addition of CO_2 to C-1 of biotin. Biotin is covalently bound to the enzyme by its carboxyl group.

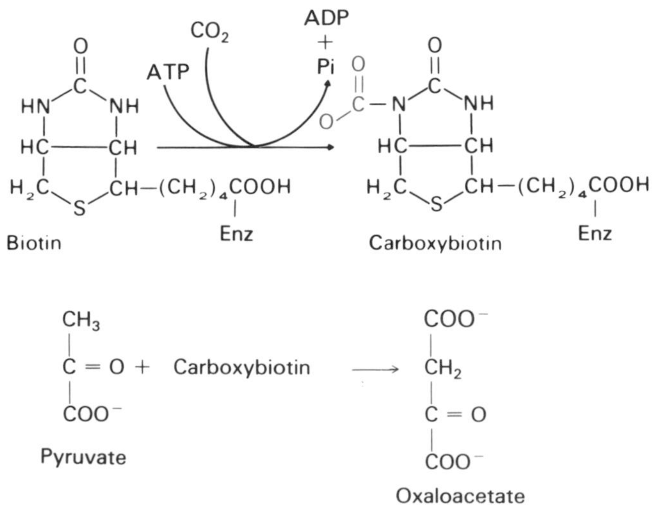

Biotin + CO_2 + ATP → Carboxybiotin + ADP + Pi

(Biotin and carboxybiotin: $-CH-(CH_2)_4COOH$, bound to Enz)

Pyruvate ($CH_3-C(=O)-COO^-$) + Carboxybiotin → Oxaloacetate ($COO^--CH_2-C(=O)-COO^-$)

A

Folate coenzymes

The transfer of a single methylene or methyl group often involves folic acid, a water-soluble vitamin, in one of its several substituted forms.

2-Amino-4-hydroxy-6-methylpteridine | *p*-Aminobenzoic acid | Glutamic acid

Folate

Pteroic acid

Pteroylglutamic acid (folic acid)

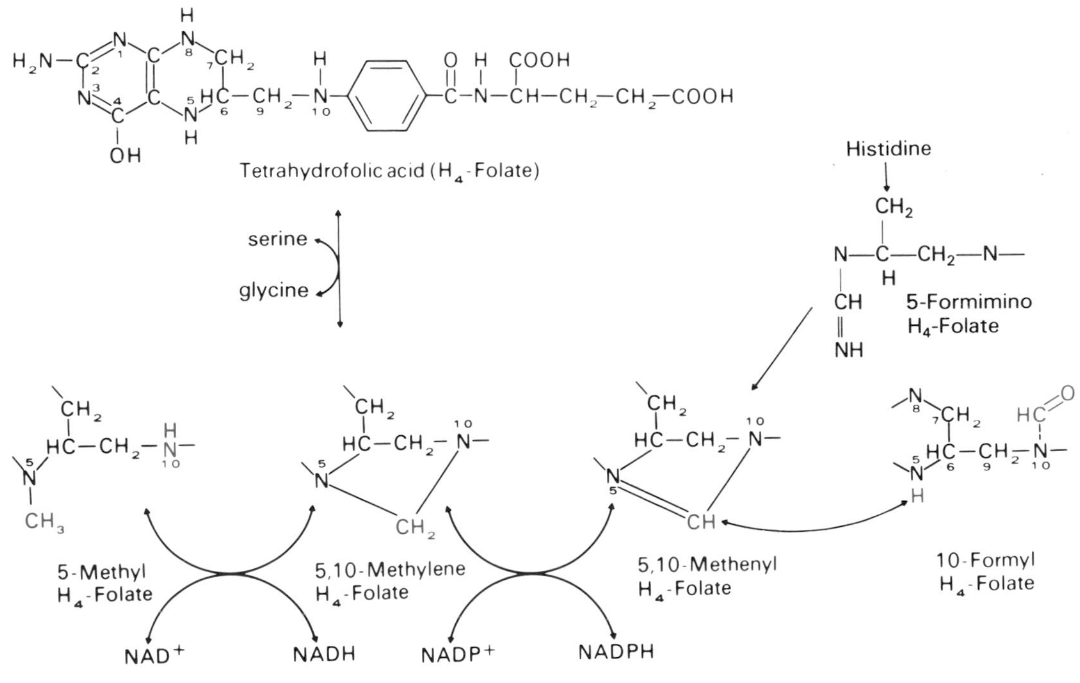

B The different folate coenzymes are specific for particular reactions as summarized in the diagram.

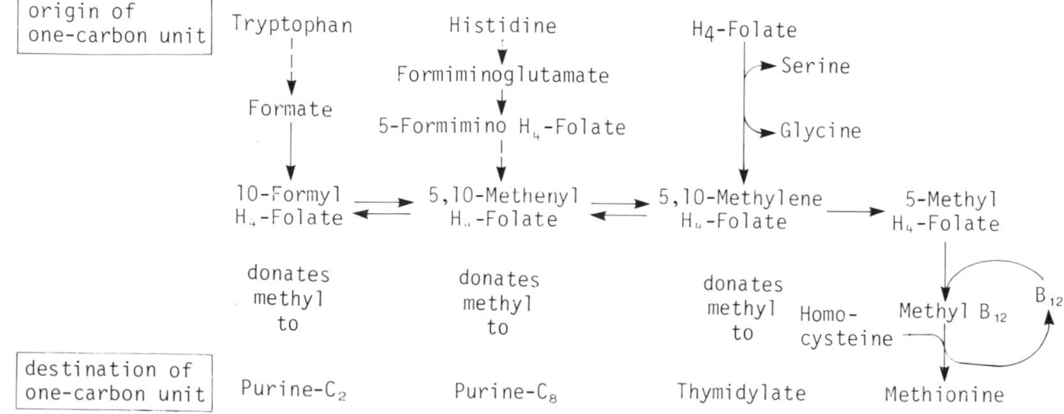

A

Pyridoxal phosphate

Pyridoxal phosphate (derivative of vitamin B_6) acts as coenzyme in transamination and decarboxylation reactions. In a transamination reaction the aldehyde group of pyridoxal phosphate first forms a Schiff base with the amino group of the amino acid, which is then converted to keto acid. Pyridoxal phosphate is thereby converted to pyridoxamine phosphate which transfers the amino group to another keto acid to form an amino acid.

X = (ring with HO, CH_2OP, H_3C)

Pyridoxal phosphate (CHO, HO, CH_2OⓅ, CH_3, N)

Pyridoxine (Vitamin B_6 derivative) (CH_2OH, HO, CH_2OH, H_3C, N)

$R{-}CH(NH_2){-}COOH$ + $X{-}CH{=}O$ $\xrightarrow{-H_2O}$ $R{-}CH(COOH){-}N{=}C(H){-}X$ $\longrightarrow$ $R{-}C(COOH){=}N{-}CH_2{-}X$ $\xrightarrow{+H_2O}$ $R{-}C(=O){-}COOH$ + $NH_2{-}CH_2{-}X$

amino acid A + pyridoxal-P — Schiff bases — keto-acid A + pyridoxamine-P

$R_1{-}C(=O){-}COOH$ + $NH_2{-}CH_2{-}X$ $\xrightarrow{-H_2O}$ $R_1{-}C(COOH){=}N{-}CH_2{-}X$ $\longrightarrow$ $R_1{-}CH(COOH){-}N{=}CH{-}X$ $\xrightarrow{+H_2O}$ $R_1{-}CH(NH_2){-}COOH$ + $X{-}CH{=}O$

keto-acid B + pyridoxamine-P — Schiff bases — amino acid B + pyridoxal-P

Pyridoxal phosphate also acts as a coenzyme in decarboxylation reactions of amino acids such as

$$^{-}OOC{-}CH_2{-}CH_2{-}\overset{+}{H_3}NCHCOO^{-} \xrightarrow[CO_2]{H^+} {}^{-}OOC{-}CH_2{-}CH_2{-}\overset{+}{H_3}NCH_2$$

glutamate → γ-aminobutyrate (GABA, a neurotransmitter)

Vitamin B_6 deficiency is rare in man, because the vitamin is widely distributed in common foodstuffs and in addition is synthesized in appreciable quantities by intestinal flora. The main abnormality seen in B_6 deficiency is a dermatitis which is readily cured by administration of the vitamin. In addition, however, laboratory animals made B_6-deficient may be susceptible to spontaneous epileptiform convulsions, and are more susceptible to convulsive drugs, which act as GABA antagonists.

A

Vitamin B_{12}

Vitamin B_{12} was first isolated as the factor in raw liver that would alleviate the symptoms of pernicious anaemia. Four pyrrole rings form a porphyrin-like structure around a cobalt atom. The vitamin is known to act as coenzyme in the methylation of homocysteine, and in a reaction in which methylmalonyl CoA is converted to succinyl CoA. Both of these reactions involve a methyl group transfer.

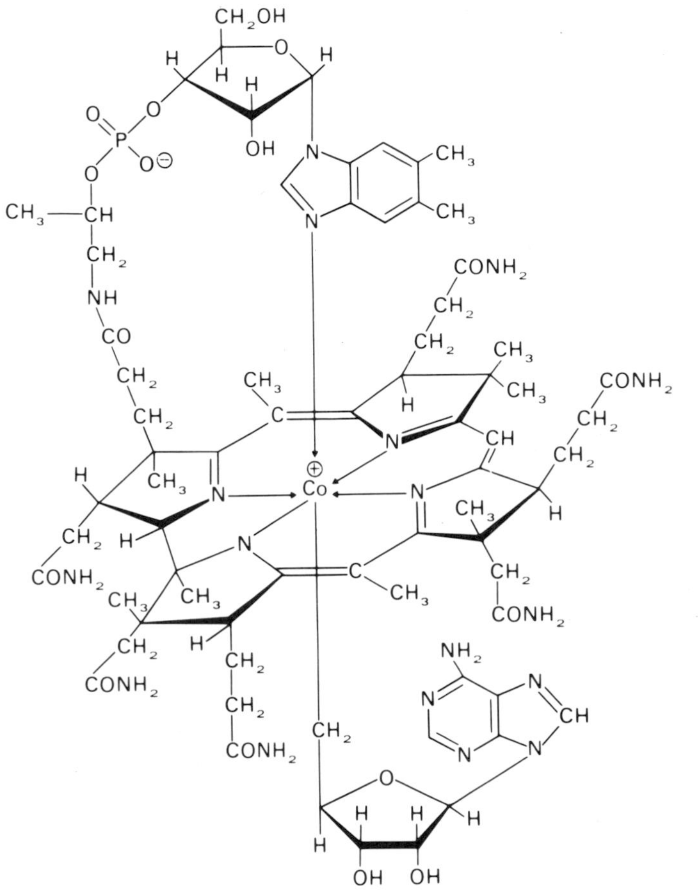

Structure of the 5,6-dimethylbenzimidazole cobamide coenzyme.

B

Ascorbic acid (vitamin C)

The action of ascorbic acid in preventing poor wound healing in scurvy may be explained by its involvement in the hydroxylation of proline.

Deficiency of ascorbic acid, found in fresh vegetables and fruit, causes scurvy, a disease characterized by defective connective tissue, with bleeding gums and loss of teeth.

Proline is hydroxylated to hydroxyproline after being incorporated into collagen precursor peptides (see p. 109B)

—Pro—Gly—+ α-oxoglutarate→
—Hypro—Gly— + CO_2 + succinate + H_2O

Fe^{3+} and ascorbic acid appear to be required in this reaction. In the absence of ascorbic acid an abnormal collagen is formed.

Ascorbic acid undergoes ionization and can act as a reducing agent, being oxidized to dehydroascorbate.

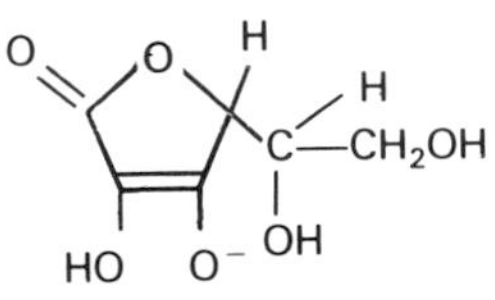

Ascorbate 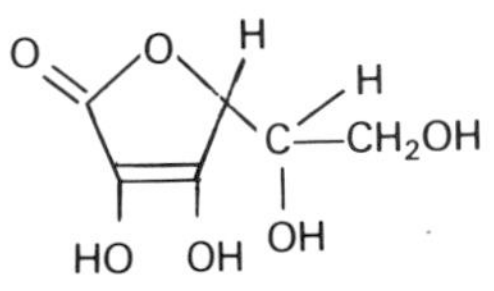 | Ascorbic acid 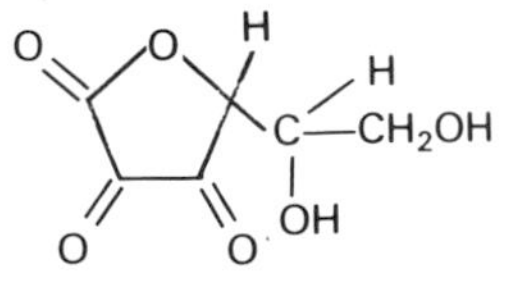 | Dehydroascorbic acid

6

CARBOHYDRATE CHEMISTRY AND INTERCONVERSIONS OF MONOSACCHARIDES

A

Chirality of Sugars

D-forms of the monosaccharides predominate in mammalian carbohydrates.

Simple sugars contain centres of asymmetry and are thus chiral molecules (see p. 10). Every sugar exists in a D- or L- form, each being a mirror image of the other.

D-Glucose		L-Glucose
CHO	1	CHO
H—C—OH	2	HO—C—H
HO—C—H	3	H—C—OH
H—C—OH	4	HO—C—H
H—C—OH	5	HO—C—H
CH_2OH	6	CH_2OH

The configuration on carbon 5 determines whether the sugar is D- or L- (the relationship can be traced back to D- or L- glyceraldehyde). Since these two sugars are mirror images, the configuration of all the other hydroxyls is reversed also in the Fischer projection shown on the left.

Note: The configuration at each carbon can also be denoted R or S (p. 10). This notation applies to individual carbons, not the whole molecule.

B

Monosaccharides normally adopt a ring conformation.

A six-membered ring structure may adopt either a chair or boat conformation, shown below for cyclohexane. Bonds perpendicular to the plane of the molecule are termed axial, whilst those in the plane of line molecule are equatorial.

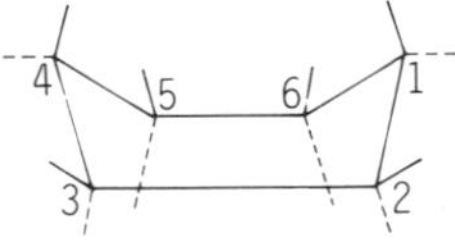

Boat

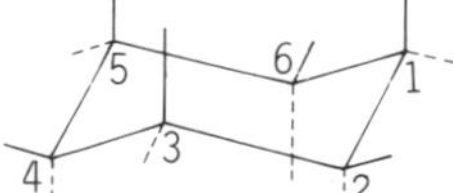

Chair

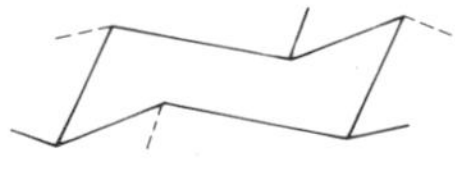
Equatorial bonds

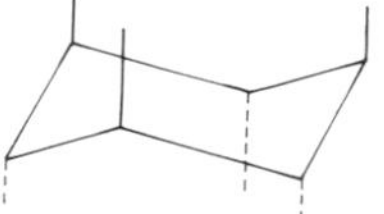
Axial bonds

When D-glucose adopts a six-membered ring structure a *hemi-acetal* bond is formed between the aldehyde carbonyl group of carbon 1 and the hydroxyl group on carbon 5.

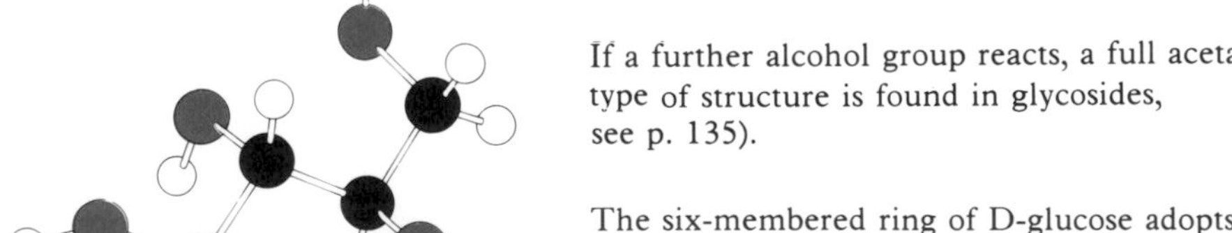

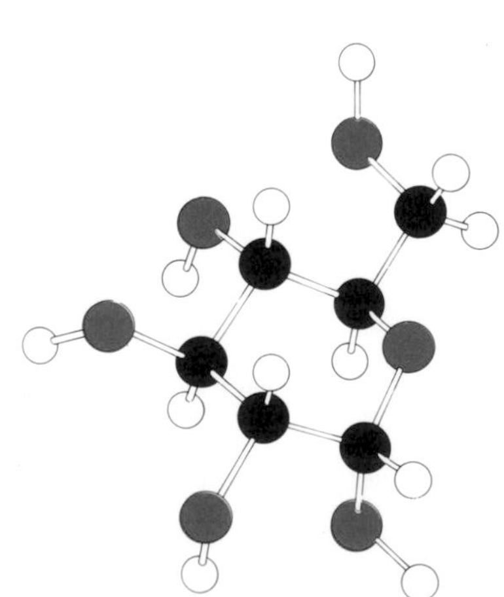

If a further alcohol group reacts, a full acetal —C(OR)(OR)H is formed (this type of structure is found in glycosides, see p. 135).

The six-membered ring of D-glucose adopts a chair configuration.

The hemi-acetal group (but not a full acetal) reacts readily with certain oxidizing agents such as Fehling's solution and this forms the basis of the well-known 'reducing test' for sugars with a hemi-acetal group, such as glucose or mannose. Some disaccharides (see p. 136) possess one hemi-acetal group, but sucrose has no hemi-acetal group, so does not react with Fehling's solution.

A

Monosaccharides

Carbohydrate structures

Simple sugars are termed monosaccharides.

The α- and β- forms of glucose spontaneously undergo interconversion (*mutarotation*) in water.

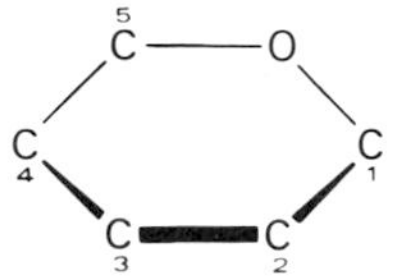

numbering for a monosaccharide in the pyranose form

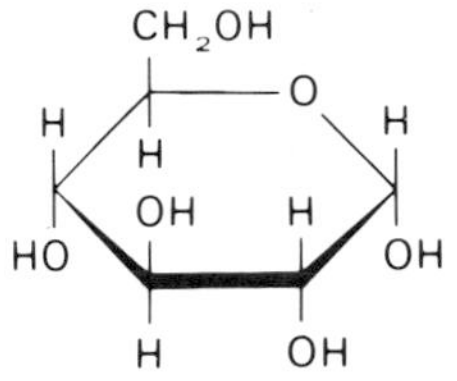

α-D-Glucose

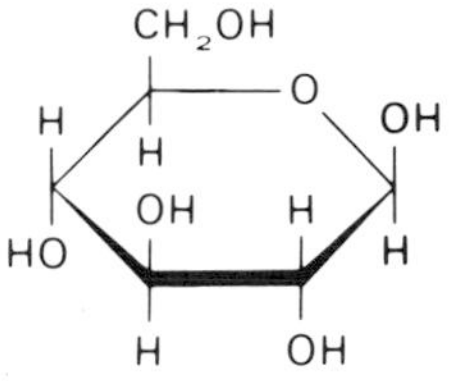

β-D-Glucose

Two conformations of D-glucose

The *pyranose* form of a sugar has a 6 membered ring
The *furanose* form of a sugar has a 5 membered ring

α-D-glucose and β-D-glucose differ only in the conformation of the hydroxyl group on carbon 1. These and other pairs of sugar molecules that differ only in this respect are termed *anomers*.

B

The ring form of the carbohydrate is in equilibrium with an open chain form.
The open chain form aids understanding of mutarotation, and of the difference between glucose, which has an aldehyde group on C-1 and is thus called an aldose, and fructose, which has a ketone group on C-2 and is called a ketose sugar. A sugar with 6 carbons is a hexose, one with 5 carbons is a pentose.

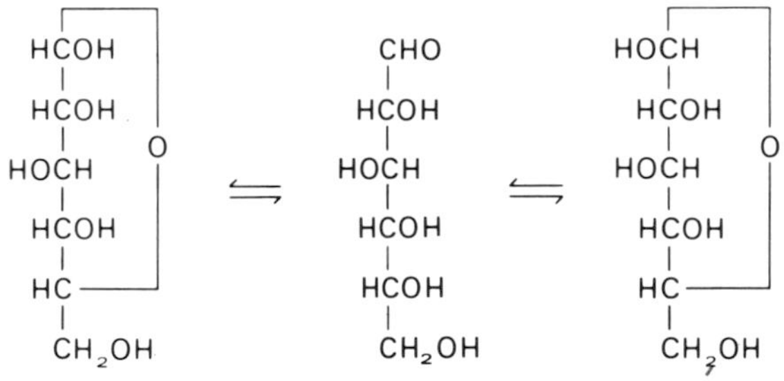

α-D-Glucopyranose

β-D-Glucopyranose

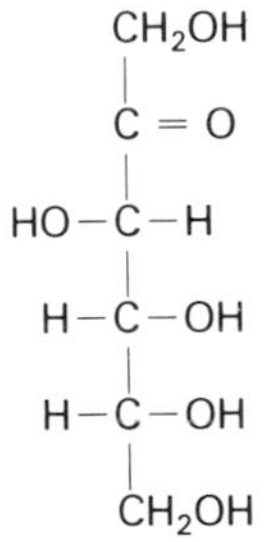

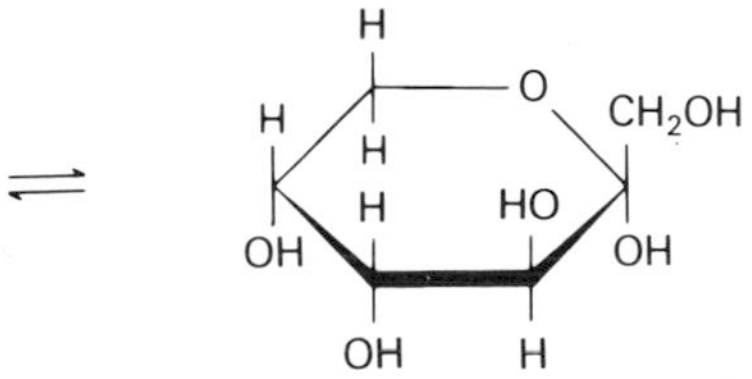

α-D-Fructopyranose

C

Enzymes that bring about reactions at C-1 between the sugar and an alcohol yield a glycoside that has either the α- or the β-configuration.

Note: Methyl glucoside is a glycoside of glucose.

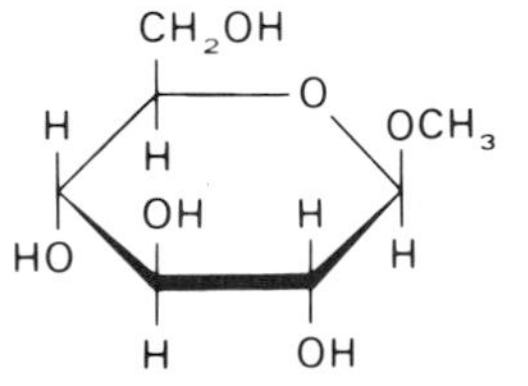

Methyl β-D-glucoside

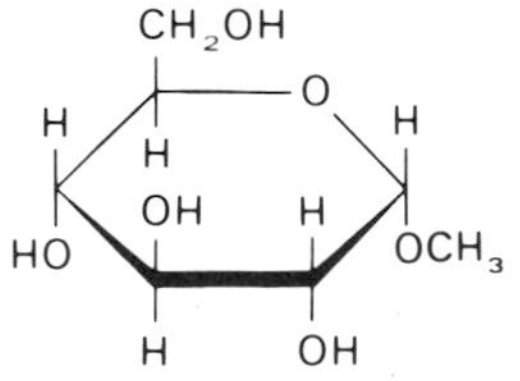

Methyl α-D-glucoside

There is no mutarotation of these compounds. Enzymes that act on glycosides are specific for the α or the β form, and are termed *glycosidases*.

A

Other monosaccharides (simple sugars) differ from one another by virtue of the configuration of one of the optically active hydroxyl groups.

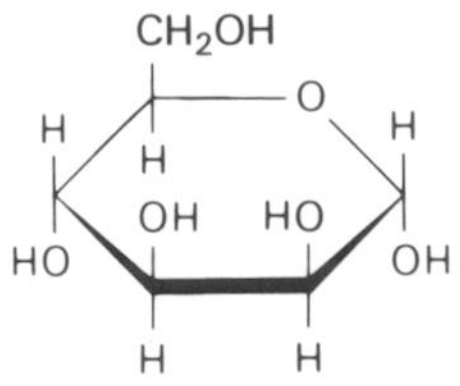

α-D-Mannose

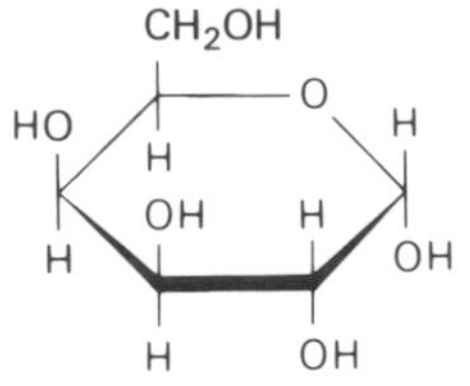

α-D-Galactose

Isomers of this type are termed *epimers*, and enzymes that change the configuration on one hydroxyl group to convert one sugar into another are termed *epimerases*.

B

Important pentoses are ribose and xylulose.

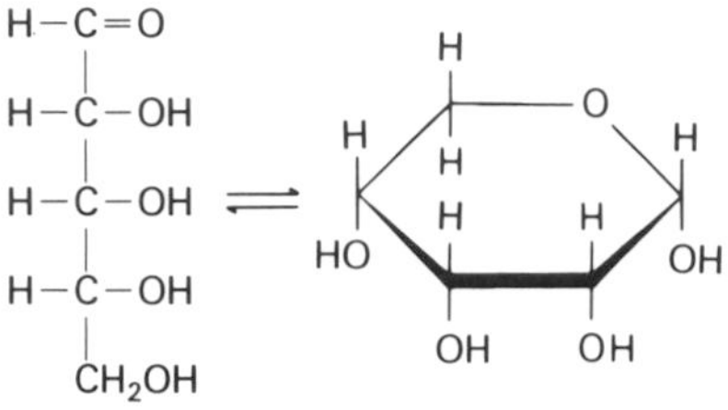

D-Ribose

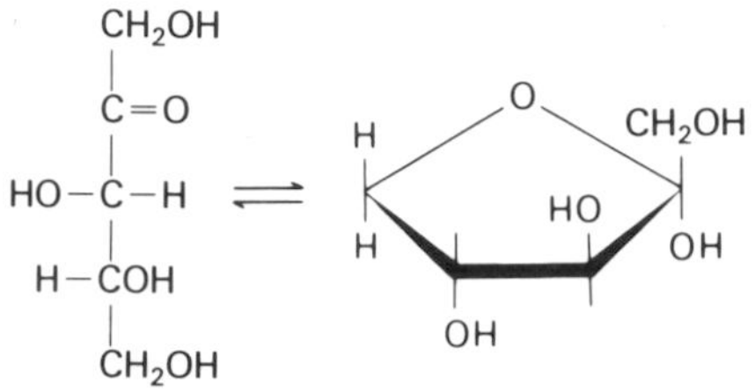

D-Xylulose

C

Amino sugars have a nitrogen on C-2 that is often acetylated.

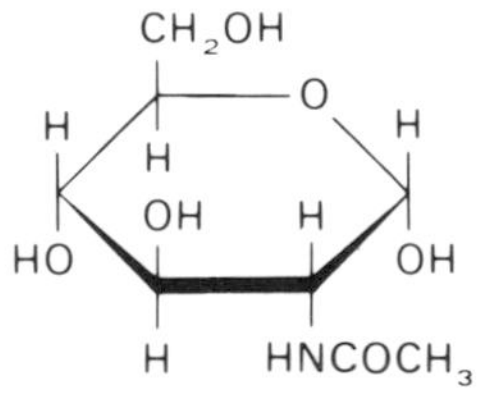

N-Acetyl α-D-glucosamine

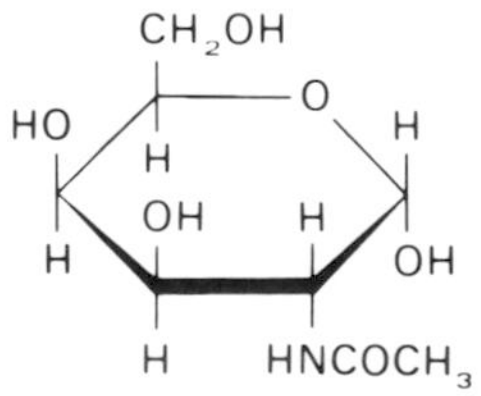

N-Acetyl α-D-galactosamine

D

Disaccharides

Maltose is a *disaccharide* formed from 2 molecules of glucose, linked between C-1 of one glucose and C-4 of the other. The link at C-1 is in the α configuration, so that an *α1,4 link* results.

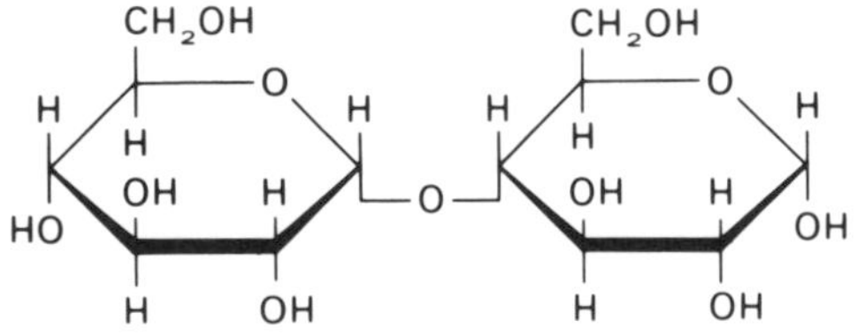

(Glucose) (Glucose)

Maltose

(α-D-glucosyl-(1→4)-α-D-glucose)

A

Sucrose and lactose have β2,1 and β1,4 links respectively.

The carbonyl group on C-1 of monosaccharides is termed a reducing group as it can be oxidized readily by weak oxidizing agents such as alkaline copper sulphate. This does not occur if C-1 is engaged in bonding to another compound. Thus sucrose has no free reducing group.

Lactose is the sugar found in milk.

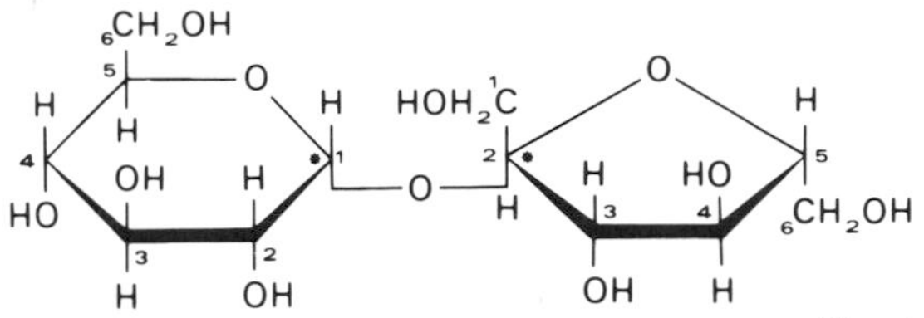

(Glucose) (Fructose)

Sucrose

β-D-fructosyl α-D-glucose

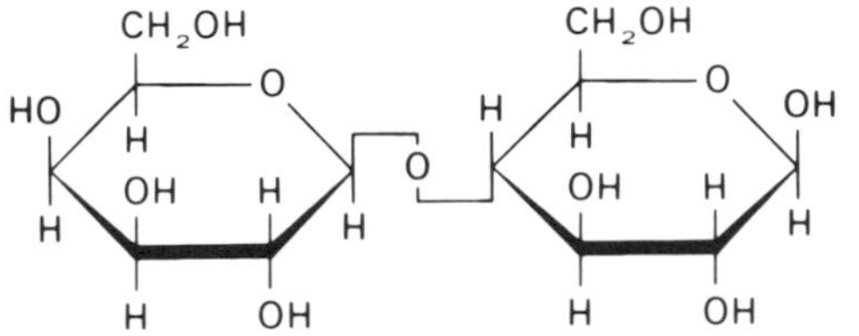

(Galactose) (Glucose)

Lactose

β-D-galactosyl-(1→4)-α-D-glucose

B

Polysaccharides (oligosaccharides)

Polysaccharides are polymers of many monosaccharide units. The term *oligosaccharide* is used, rather than *polysaccharide,* where the number of *monosaccharide* units is small.

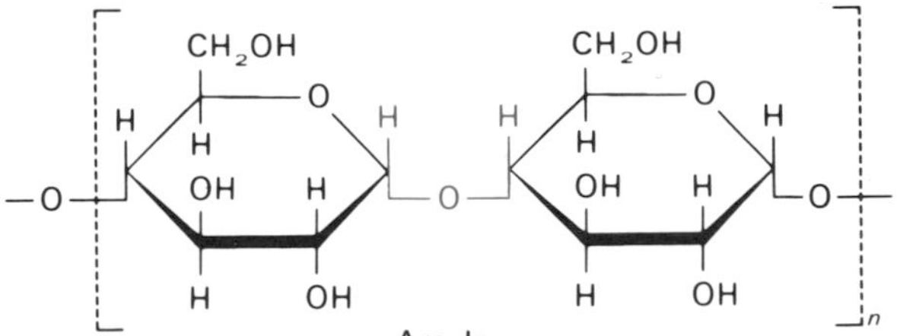

Amylose

(-α-D-glucosyl-(1→4)-α-D-glucosyl-)

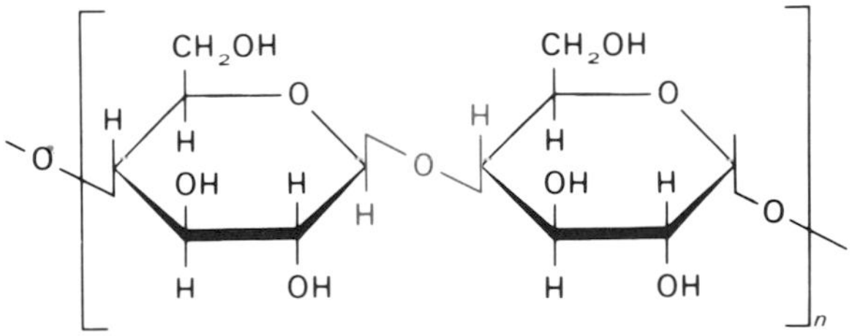

Cellulose

The repeating unit (-β-D-glucosyl-(1→4)-a-D—glucosyl-) of cellulose, leads to an insoluble product which is stabilized by intra- and inter-chain hydrogen bonds.

Starch consists of two polysaccharides, *amylose* and *amylopectin*. Amylose is a straight chain structure containing glucose linked only by a1,4 bonds, whereas amylopectin contains both a1,4 and a1,6 bonds leading to a branched structure of the type also found in glycogen in animals (p. 197, 198).

Cellulose has β1,4 bonds.

A

Glycosaminoglycans are polysaccharides important in connective tissue in which they are attached to proteins. The glycosaminoglycan-protein complexes are referred to as *proteoglycans*.

The glycosaminoglycans consist of repeating disaccharide units all of which bear negatively-charged groups. Note that in some cases these negative charges are provided by sulphate groups. The repeating units of some of the most important glycosaminoglycans are shown below.

Chondroitin 6-sulphate

Dermatan sulphate

Keratan sulphate

Hyaluronate

Heparin

B

Proteoglycans have been visualized by electron microscopy.

The scale in the photomicrograph represents 1 um.

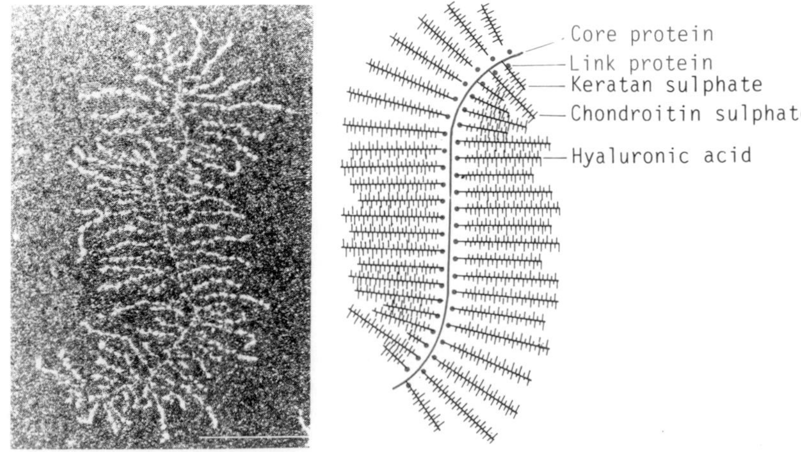

A

Glucuronic and gluconic acids

Two acids can arise from glucose. Oxidation of C-6 gives rise to *glucuronic acid*, whilst oxidation of C-1 yields *gluconic acid.*

Glucuronic acid   Glucose   Gluconic acid

Glucuronic acid is involved in the formation of glucuronides (see p. 142).

Gluconic acid is found in a phosphorylated form in the pentose phosphate pathway (6-phosphogluconic acid).

B

Uridine diphosphate glucose (UDPG)

UDPG is formed by an enzyme reaction between UTP and glucose 1-phosphate. UDPG is important in many reactions involving glucose, such as glycogen synthesis.

UDPG is an example of the activation of simple molecules by addition to a nucleotide.

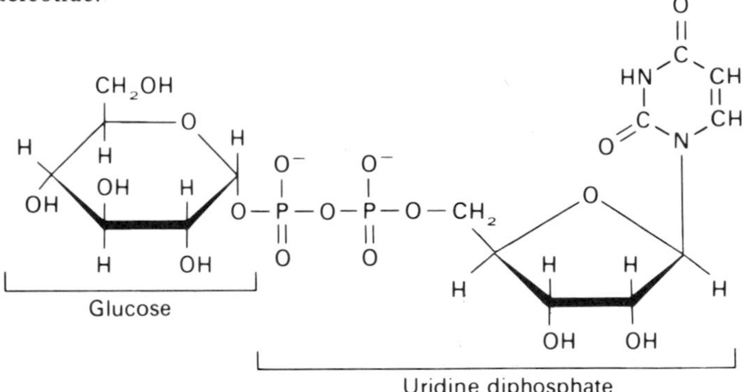

$$\text{UTP} + \text{glucose 1-phosphate} \xrightarrow{\text{UDP-glucose pyrophosphorylase}} \text{UDPG} + \text{PP}_i$$

C

Nucleotide-linked sugars include *UDP-galactose* and *GDP-mannose.*

$$\text{UTP} + \text{galactose 1-phosphate} \longrightarrow \text{UDP-galactose} + \text{PP}_i$$

$$\text{GTP} + \text{mannose 1-phosphate} \longrightarrow \text{GDP-mannose} + \text{PP}_i$$

A

Hexose interconversions

Galactose is phosphorylated by a specific kinase.

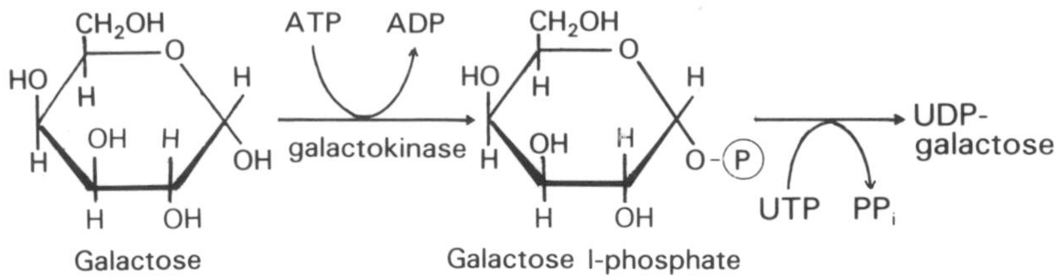

Galactose + ATP → (galactokinase) → Galactose l-phosphate + ADP

Galactose l-phosphate + UTP → UDP-galactose + PP_i

B

Reversible interconversion between UDP-glucose and UDP-galactose is important both in galactose degradation and in formation of UDP-galactose for synthesis of complex carbohydrates.

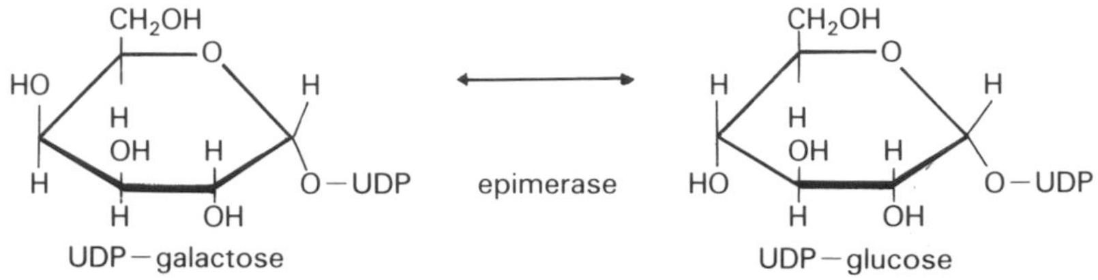

UDP—galactose ⟷ (epimerase) ⟷ UDP—glucose

UDP-galactose can also be formed by the enzyme galactose l-phosphate uridyl transferase. In this reaction galactose l-phosphate reacts with UDP-glucose, and glucose l-phosphate reacts with UDP-galactose, to give the reversible cycles shown below:

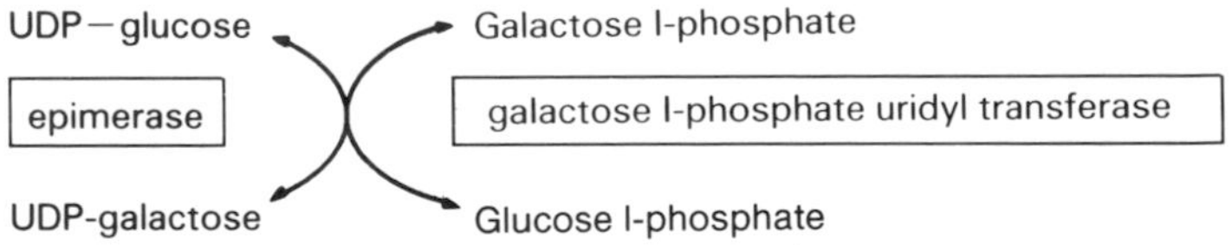

UDP—glucose ⇄ UDP-galactose (epimerase)

Galactose l-phosphate ⇄ Glucose l-phosphate (galactose l-phosphate uridyl transferase)

UDP-galactose is also formed by the reaction shown on the previous page:

Galactose l-phosphate + UTP ⟶ UDP-galactose + PP_i
UDP-galactose pyrophosphorylase

The enzyme responsible for this conversion develops only in later life, being absent in infants.

In galactosaemia, an inborn error of metabolism, galactokinase may be deficient, giving a mild form of the disease. Galactose l-phosphate uridyl transferase deficiency also occurs. In the absence of UDP-galactose pyrophosphorylase during infancy, there is thus no route for the conversion of galactose l-phosphate to UDP-galactose. This gives a more severe form of the disease.

A

The formation of UDP-glucuronic acid.

UDP-glucose $\xrightarrow{2NAD^+ \rightarrow 2NADH}$ UDP-glucuronic acid

B

Mannose metabolism
Mannose can be phosphorylated by a kinase to yield mannose 6-phosphate, which can also be formed directly from, or converted to, fructose 6-phosphate.

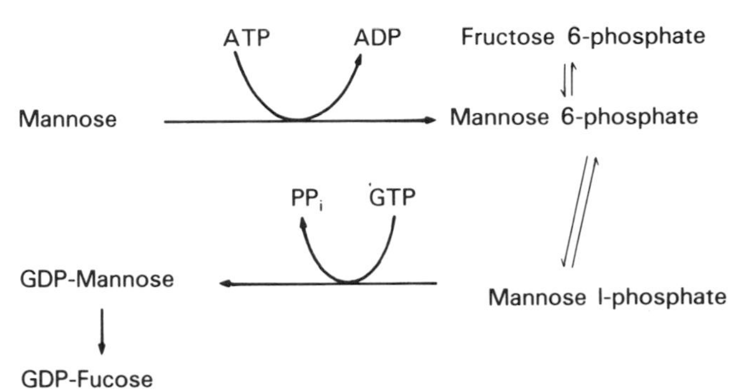

GDP-mannose and GDP-fucose are intermediates in the synthesis of complex carbohydrates.

C

Mannose derivatives participate in the formation of N-acetyl-neuraminic acid (sialic acid) but the route of formation is from glucosamine 6-phosphate, formed by reaction of glutamine with the carbonyl group of fructose 6-phosphate.

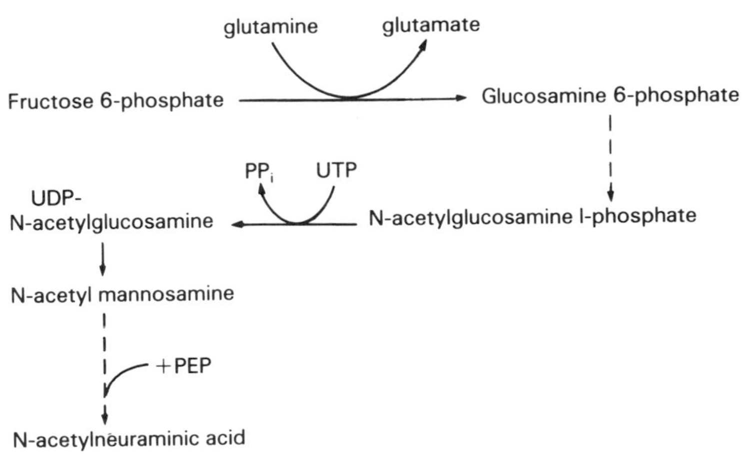

A

The role of sugars in detoxification

The introduction of a hydroxyl group into an aromatic ring in itself increases the water-solubility of the compound. There are, in the smooth endoplasmic reticulum, enzymes that catalyse addition of a single oxygen atom (i.e. mono-oxygenases). This oxidation involves a cytochrome, called cytochrome P_{450}, because it has an absorption maximum at 450 nm. The overall reaction is:

$$RH + O_2 + 2e^- + 2H^+ \longrightarrow ROH + H_2O$$

The hydrogens needed for the reaction are derived from NADPH, and are transferred by a flavoprotein called cytochrome P_{450} reductase.

These enzymes are increased 25-fold in liver by phenobarbital, which induces massive proliferation of the endoplasmic reticulum of liver.

Compounds possessing a hydroxyl group (including those formed by the reaction described above) are often converted to glucuronides (glycosides of glucuronic acid) by reaction with UDP-glucuronic acid.

B

Formation of glucuronides

The excretion of compounds that are not very soluble in water presents difficulties that can be overcome by metabolic conversions which render the material more soluble in water.

Thus, after the introduction of a hydroxyl group into an aromatic ring (see above), it can be conjugated with glucuronic acid, which also reduces the toxicity of the compound.

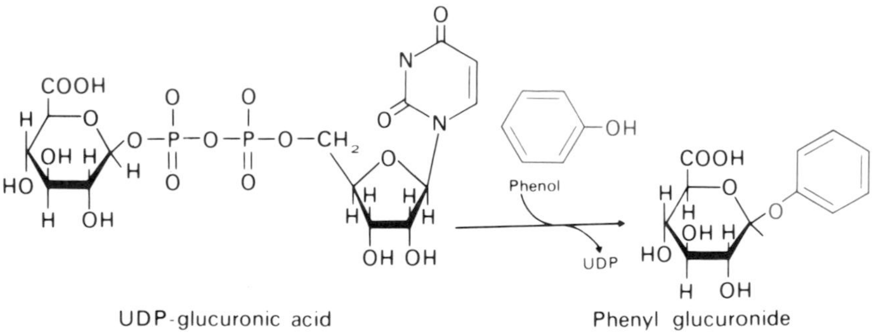

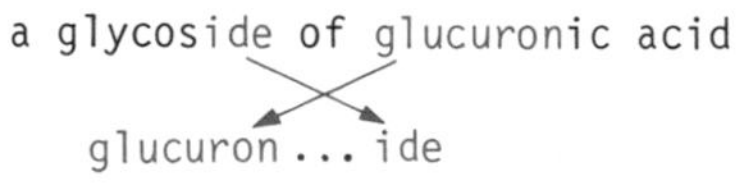

UDP-glucuronic acid + Phenol → Phenyl glucuronide + UDP

a glycoside of glucuronic acid

glucuron ... ide

α-glucuronidases and *β*-glucuronidases hydrolyse the glycosidic bond of glucuronides, and as their names imply, are specific for the configuration on C-1 of glucuronic acid.

C

Another route of detoxification involves the formation of glycine conjugates.

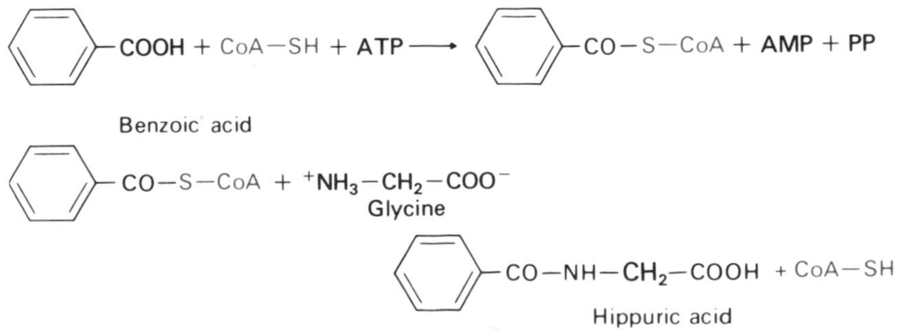

C_6H_5–COOH + CoA–SH + ATP → C_6H_5–CO–S–CoA + AMP + PP

Benzoic acid

C_6H_5–CO–S–CoA + $^+NH_3$–CH_2–COO^- (Glycine) → C_6H_5–CO–NH–CH_2–COOH + CoA–SH

Hippuric acid

Many steroid metabolites are excreted as glucuronides.
Other compounds, especially acids, may be conjugated with glycine, and excreted as glycine conjugates.

7

NITROGEN METABOLISM

A

Essential and non-essential amino acids

The amino acids that are used for protein synthesis cannot all be synthesized in the body. Those that can be synthesized are termed *non-essential amino acids*, whilst those that cannot be synthesized are termed *essential amino acids.*

Essential amino acids
Histidine
Isoleucine
Leucine
Valine
Lysine
Methionine
Threonine
Tryptophan
Phenylalanine

Non-essential amino acids synthesized directly by transamination from metabolites readily available from major pathways

Amino acid	*Metabolite from which synthesized*
Alanine	Pyruvate
Aspartic acid	Oxaloacetate
Glutamic acid	Oxoglutarate

Non-essential amino acids synthesized by special pathways

Amino acid	*Metabolite from which synthesized*
Ornithine	Glutamic acid
Proline	
Glycine	Pyruvate
Serine	
Cysteine	Methionine
Tyrosine	Phenylalanine
Arginine*	Ornithine

*Arginine may be needed in the diet for adequate growth rates in the young.

If it is to provide adequate nutrition (i.e. to have a high 'biological value'), protein must contain a balanced proportion of all of the essential amino acids. Animal proteins (meat, eggs) are a good source, but a number of plant proteins may be deficient in one or more of the essential amino acids. Corn (maize) is deficient in lysine, and rice protein is deficient in lysine and threonine. New strains are being bred with more adequate proportions of these amino acids.

B

Transamination and oxidative deamination

Amino groups can be removed by oxidative deamination or by transamination, and can be added by transamination.

The general reaction for a transamination reaction is:

amino acid A → keto acid A
keto acid B → amino acid B

R
|
C=O
|
COOH
a keto acid

Enzymes of this type are termed transaminases, or, in more recent terminology, *amino transferases.*

Individual keto acids are now designated as the oxo-derivative e.g. oxoglutarate. The term keto acid is still used to designate the class of compound.

Reactions that remove the amino group, or add it to a chain that previously lacked it, are central to amino acid metabolism.

A

Glutamate dehydrogenase is one of the most important enzymes removing the amino group.

$$\underset{\text{Glutamate}}{COO^- - CH_2 - CH_2 - CH(NH_3^+) - COO^-} + NAD^+ + H_2O \rightleftharpoons \underset{\text{2-Oxoglutarate*}}{COO^- - CH_2 - CH_2 - C{=}O - COO^-} + NADH + H^+ + NH_4^+$$

Glutamate dehydrogenase

*α-ketoglutarate, the original name for 2-oxoglutarate, is still frequently used.

B

A transaminase reaction of central importance, alanine transaminase, transfers the amino group of alanine to oxoglutarate to form glutamate.

$$\underset{\text{Alanine}}{CH_3 - CH - COO^-} + \underset{\text{Oxoglutarate}}{COO^- - CH_2 - CH_2 - C{=}O - COO^-} \rightleftharpoons \underset{\text{Pyruvate}}{CH_3 - C{=}O - COO^-} + \underset{\text{Glutamate}}{COO^- - CH_2 - CH_2 - CH - COO^-}$$

Alanine transaminase–ALT
(or Glutamate-pyruvate transaminase (GPT))

C

The most important of the enzymes that carry out oxidative deamination is glutamate dehydrogenase. Operating in conjunction with transaminases, it can convert the amino group of most amino acids to free ammonia.

amino acid → keto acid; oxoglutarate → glutamate (Many transaminases)

glutamate + $NAD^+ + H_2O$ → oxoglutarate + $NADH + H^+ + NH_4^+$ (glutamate dehydrogenase)

D

Alanine transaminase can act as a link between glutamate dehydrogenase and amino acids that do not react directly with oxoglutarate.

amino acid → keto acids; pyruvate → alanine (Many transaminases utilize alanine as substrate)

alanine → pyruvate; oxoglutarate → glutamate (alanine transaminase)

glutamate + $NAD^+ + H_2O$ → oxoglutarate + $NADH + H^+ + NH_4^+$ (glutamate dehydrogenase)

A

Transaminases that transfer amino groups to oxaloacetate to form aspartate are important as this is a route for incorporating the $-NH_2$ group into urea.

glutamate → oxaloacetate

oxoglutarate ← aspartate

Aspartate transaminase—AST

(or Glutamate-oxaloacetate transaminase (GOT))

B

Another source of free ammonia is the monoamine oxidase reaction.

Vanillylmandelate in the urine can be measured as an index of catecholamine metabolism.

Biological amines e.g. adrenaline —monoamine oxidase (MAO)→ NH_4^+ + aldehydes

For example:

Noradrenaline —MAO→ (HO)₂C₆H₃–CH(OH)–CHO ↓ methylation and oxidation → Vanillylmandelate

$$\text{Noradrenaline: } (HO)_2C_6H_3-CH(OH)-CH_2-NH_2 \xrightarrow{MAO} (HO)_2C_6H_3-CH(OH)-CH=O$$

$$\xrightarrow{\text{methylation and oxidation}} \text{Vanillylmandelate: } HO(OCH_3)C_6H_3-CH(OH)-COOH$$

C

Cooperation between the various transaminase and deaminase enzymes leads to the formation of free ammonium ion and net formation of aspartate. The ammonium ion and the amino group of the aspartate provide the two nitrogens of urea and are thus eliminated from the body.

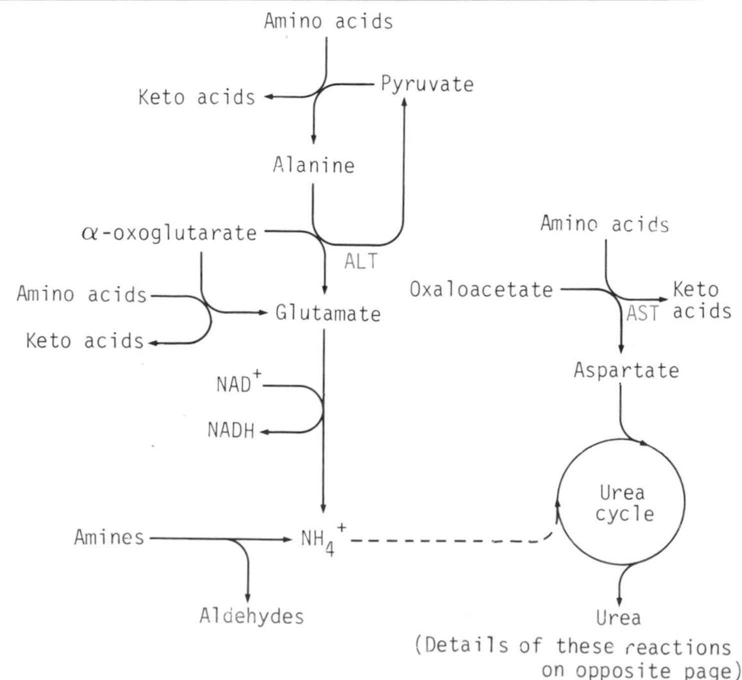

The urea cycle

The formation of urea takes place through a complex series of reactions known as the urea cycle. The start of the urea cycle may be regarded as the reaction between the amino acid ornithine and carbamoyl phosphate, to form citrulline. Ornithine is finally regenerated in the reaction that forms urea, thus giving rise to the cyclic form of the overall reaction.

Free ammonia is needed for the synthesis of carbamoyl phosphate. This reaction, and the formation of glutamine (p. 157) are important in maintaining a low concentration of free ammonia, which is toxic,

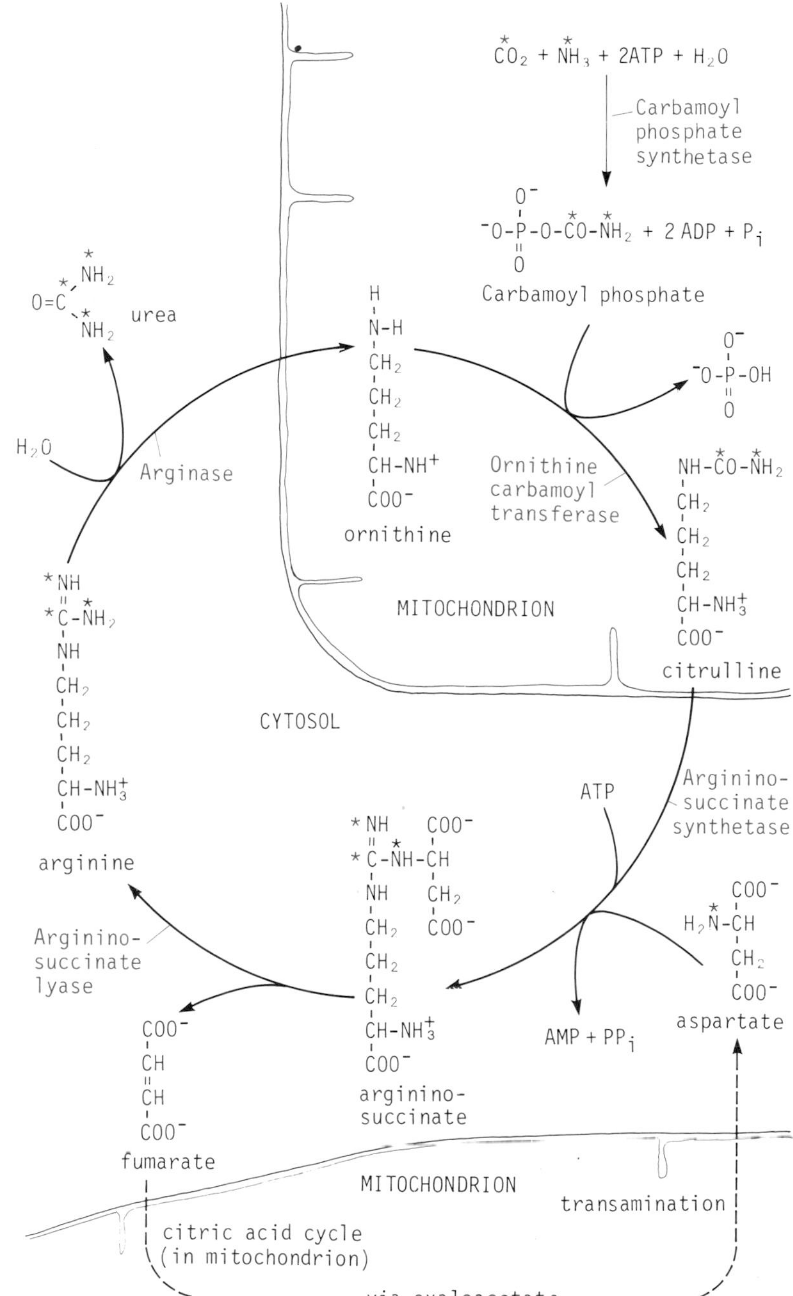

The importance of aspartate in providing one of the nitrogens of urea is noteworthy. This nitrogen is derived from transamination reactions that culminate in the aspartate transaminase reaction.

A

Inherited defects of urea cycle enzymes

Deficiency of urea cycle enzymes leads to metabolic disorders, some of which are listed in the table.

The number of cases is, of course, approximate, and increases with time, but gives an indication of relative incidence. The prognosis varies with the severity of the disease. For example, a neonatal form of carbamoyl phosphate synthetase deficiency is associated with complete or almost complete absence of the enzyme and normally results in death in the neonatal period. However, partial deficiency of the enzyme also occurs, with correspondingly longer survival to early adulthood or later.

Enzyme deficient	No. of cases reported	Type of inheritance	Characteristic features
Carbamoyl phosphate synthetase	26	autosomal recessive	Hyperammonaemia and aminoacidaemia without orotic aciduria (see page 94)
Ornithine carbamoyl transferase	110	X-linked dominant	Hyperammonaemia and aminoacidaemia with orotic aciduria
Argininosuccinate synthetase	53	Autosomal recessive	Citrullinaemia and citrullinuria
Argininosuccinate lyase	60	Autosomal recessive	Mild argininosuccinic acidaemia with argininosuccinic aciduria
Arginase	13	Autosomal recessive	Argininaemia (variable with dietary load) and argininuria

B

An important regulatory mechanism involves the compound *N–acetylglutamate.* The synthesis of this is regulated by the concentration of arginine.

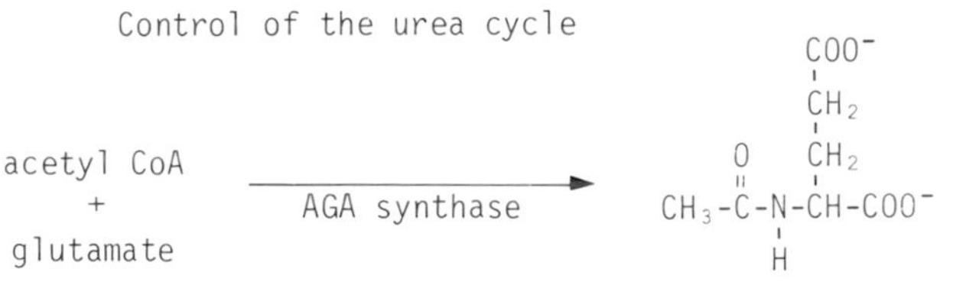

AGA synthase activity is stimulated by high protein intake.

C

Carbamoyl phosphate synthetase requires N-acetylglutamate as an allosteric effector. Thus indirectly it is controlled by the level of arginine, which stimulates AGA synthase.

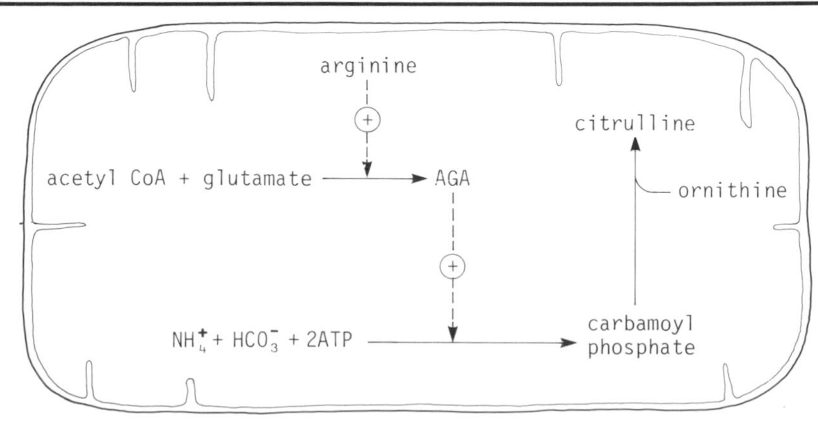

A

The synthesis of non-essential amino acids

Synthesis of *alanine, aspartate* and *glutamate.* These three amino acids are synthesized by direct transamination from pyruvate, oxaloacetate and 2-oxoglutarate reacting with another amino acid.

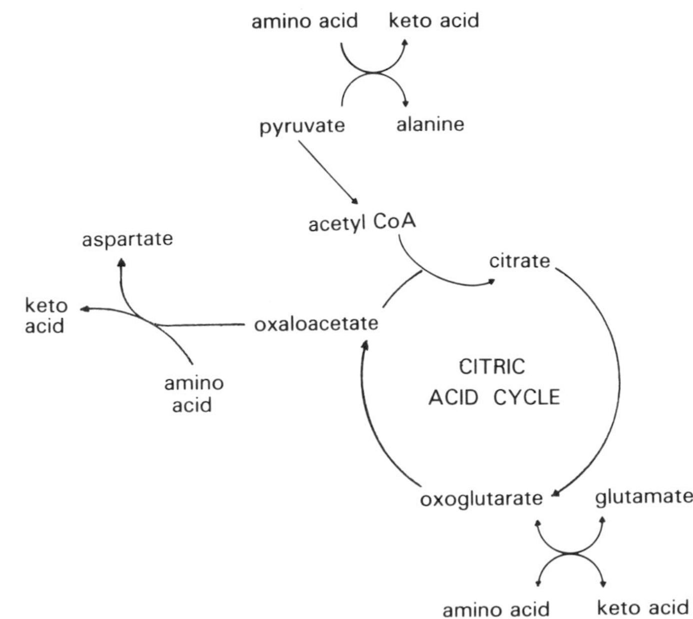

B

Synthesis of *glycine* and *serine*

Both serine and glycine are synthesized from 3-phosphoglyceric acid, derived from glucose as an intermediate in the glycolytic pathway.

$$\begin{matrix} CH_2O\text{Ⓟ} \\ | \\ CHOH \\ | \\ COO^- \end{matrix} \xrightarrow[NAD^+ \; NADH]{\text{3-phosphoglyceric acid dehydrogenase}} \begin{matrix} CH_2O\text{Ⓟ} \\ | \\ CO \\ | \\ COO^- \end{matrix}$$

3-Phosphoglyceric acid → 3-Phosphohydroxypyruvic acid

transamination

$$\begin{matrix} CH_2O\text{Ⓟ} \\ | \\ CH{-}NH_3^+ \\ | \\ COO^- \end{matrix} \xrightarrow[P]{\text{phosphoserine phosphatase}} \begin{matrix} CH_2OH \\ | \\ CH{-}NH_3^+ \\ | \\ COO^- \end{matrix} \longrightarrow \text{glycine}$$

(Loss of formaldehyde)

3-Phosphoserine → Serine

C

Glycine and *serine* are interconvertible in a reversible reaction involving a folate coenzyme.

$$\begin{matrix} CH_2NH_3^+ \\ | \\ COO^- \end{matrix} + \text{5,10-methylene } H_4\text{-Folate} \rightleftharpoons \begin{matrix} CH_2OH \\ | \\ CH{-}NH_3^+ \\ | \\ COO^- \end{matrix} + H_4\text{-Folate}$$

Glycine → Serine

A

The synthesis of *cysteine*

The synthesis of cysteine from methionine involves *S-adenosylmethionine.*

S-Adenosyl methionine is an important intermediate in the transfer of a methyl group from methionine to a variety of acceptor molecules.

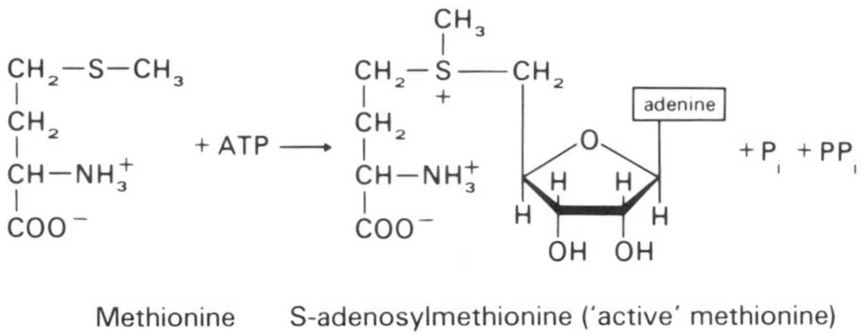

Methionine + ATP ⟶ S-adenosylmethionine ('active' methionine) + P_i + PP_i

Methionine: $CH_2{-}S{-}CH_3$ / CH_2 / $CH{-}NH_3^+$ / COO^-

S-adenosylmethionine: $CH_2{-}S^+(CH_3){-}CH_2$–ribose–adenine / CH_2 / $CH{-}NH_3^+$ / COO^- (ribose ring: O, H, H, H, H, OH, OH)

B

S-adenosylmethionine is converted to homocysteine in a reaction in which its methyl group is transferred to *guanidoacetic acid.* Homocysteine can then form cysteine after reaction with serine.

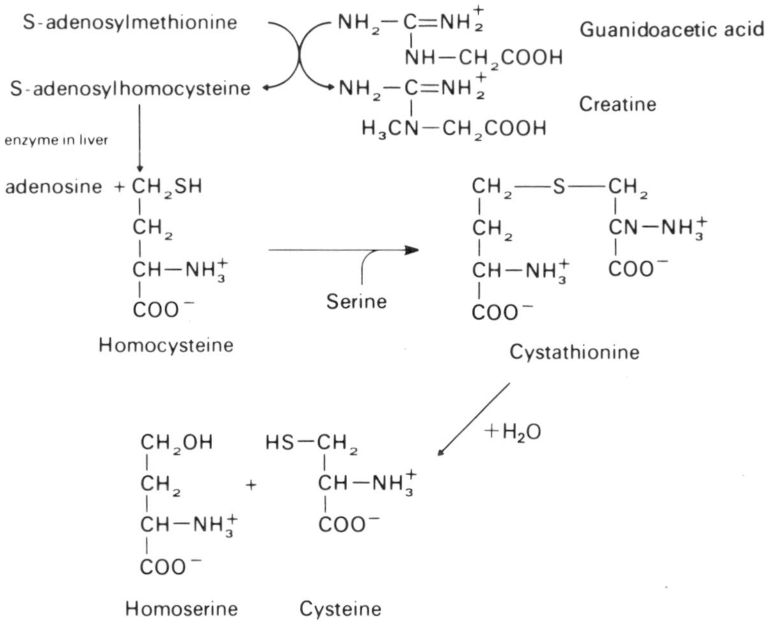

S-adenosylmethionine + Guanidoacetic acid ($NH_2{-}C{=}NH_2^+$ / $NH{-}CH_2COOH$) ⟶ S-adenosylhomocysteine + Creatine ($NH_2{-}C{=}NH_2^+$ / $H_3CN{-}CH_2COOH$)

S-adenosylhomocysteine —enzyme in liver→ adenosine + Homocysteine (CH_2SH / CH_2 / $CH{-}NH_3^+$ / COO^-)

Homocysteine + Serine ⟶ Cystathionine ($CH_2{-}S{-}CH_2$ / CH_2 / $CH{-}NH_3^+$ / COO^- ; $CN{-}NH_3^+$ / COO^-)

Cystathionine + H_2O ⟶ Homoserine (CH_2OH / CH_2 / $CH{-}NH_3^+$ / COO^-) + Cysteine ($HS{-}CH_2$ / $CH{-}NH_3^+$ / COO^-)

C

The synthesis of *proline* from *ornithine*

Ornithine is synthesized from arginine by the action of arginase. Proline can then be synthesized from ornithine.

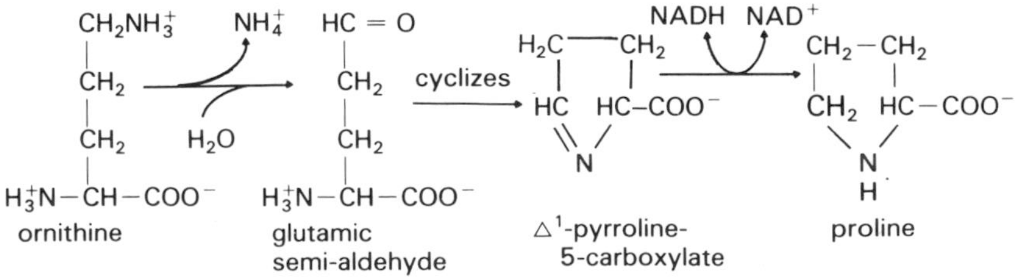

ornithine ($CH_2NH_3^+$ / CH_2 / CH_2 / $H_3^+N{-}CH{-}COO^-$) + H_2O ⟶ NH_4^+ + glutamic semi-aldehyde ($HC{=}O$ / CH_2 / CH_2 / $H_3^+N{-}CH{-}COO^-$)

glutamic semi-aldehyde —cyclizes→ Δ^1-pyrroline-5-carboxylate ($H_2C{-}CH_2$ / HC $HC{-}COO^-$ / N)

Δ^1-pyrroline-5-carboxylate + NADH ⟶ NAD$^+$ + proline ($CH_2{-}CH_2$ / CH_2 $HC{-}COO^-$ / N / H)

The catabolism of phenylalanine and tyrosine

In general the catabolic pathways that degrade the essential amino acids are complex and specialized. However, the pathways involved in the catabolism of phenylalanine and tyrosine have a special interest because deficiencies of enzymes in these pathways are associated with hereditary diseases of historical importance. *Alcaptonuria* and *albinism* were two of the first diseases to be ascribed to a genetic origin, and *phenylketonuria* was the first genetic disease for which a successful therapy was devised.

Phenylalanine → Phenylpyruvic acid → phenyl lactic acid, phenyl acetic acid

Phenylalanine —(A; H_4 – biopterin → H_2 – biopterin)→ Tyrosine

Tyrosine —(D)→ melanins

Tyrosine —(α – oxoglutarate → glutamate)→ *p*-OH-phenylpyruvic acid

p-OH-phenylpyruvic acid —(B; O_2 → CO_2)→ Homogentisic acid

Homogentisic acid —(C; O_2)→ Maleylacetoacetic acid → Fumarylacetoacetic acid → Fumaric acid + Acetoacetic acid

Enzyme A missing—*phenylketonuria* produced. Blockage of the main pathway diverts phenylalanine to phenylpyruvic acid and derivatives whose accumulation leads to mental deficiency.

Enzyme B missing—*tyrosinosis* produced. A very rare condition in which p-OH-phenylpyruvic acid and tyrosine are excreted in the urine.

Enzyme C missing—*alcaptonuria* produced. Homogentisic acid is excreted in the urine which may turn black on standing.

Enzyme D missing—*albinism* produced. The natural melanin pigments of skin, hair and eyes are not formed.

A

The hyroxylation of phenylalanine to tyrosine is carried out by an enzyme that utilizes molecular oxygen for the introduction of an oxygen atom into the aromatic ring. This enzyme is a *monooxygenase* or *mixed function oxidase*.

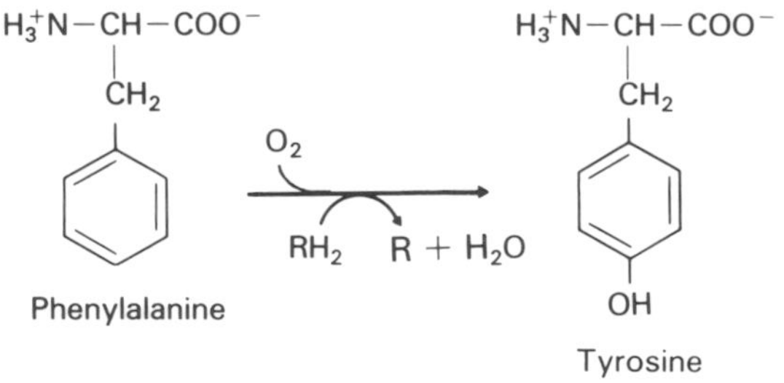

$H_3^+N{-}CH{-}COO^-$ / CH_2 / Phenylalanine —(O_2; RH_2 → $R + H_2O$)→ $H_3^+N{-}CH{-}COO^-$ / CH_2 / OH / Tyrosine

Tyrosine hydroxylase and tryptophan hydroxylase have essentially similar mechanisms to phenylalanine hydroxylase.

B

In the phenylalanine hyroxylase reaction, RH_2 is 5,6,7,8-tetrahydrobiopterin, which is oxidized to 6,7-dihydrobiopterin. This can then be reduced by dihydrofolate reductase to regenerate the reduced form.

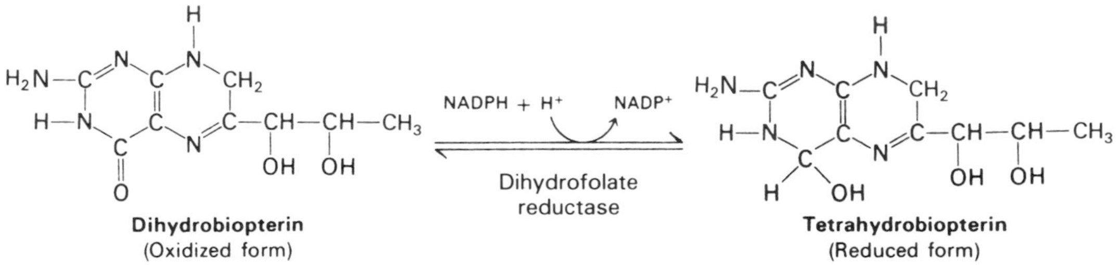

Dihydrobiopterin (Oxidized form) ⇌ (NADPH + H⁺ → NADP⁺; Dihydrofolate reductase) **Tetrahydrobiopterin** (Reduced form)

C

Phenylketonuria.

Congenital absence of phenylalanine hydroxylase occurs in phenylketonuria. In this condition the inability to convert phenylalanine to tyrosine results in metabolism of large amounts of phenylalanine by the transaminase reaction to form phenylpyruvate (the 'ketone' found in the urine).

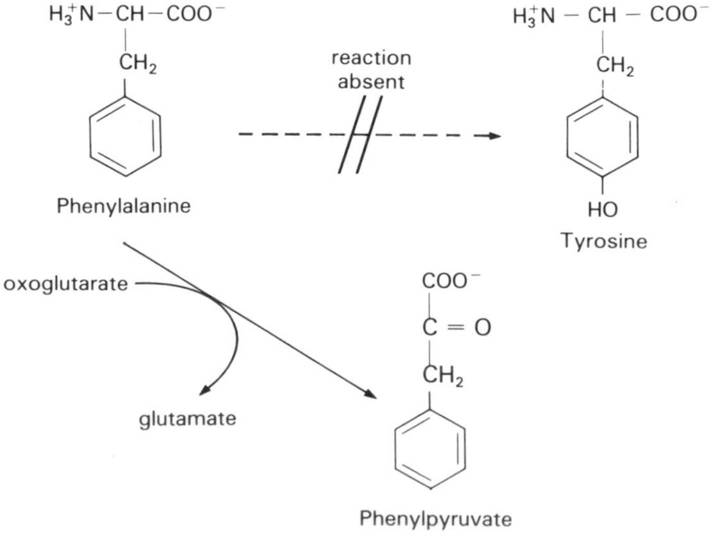

Phenylalanine --(reaction absent)--> Tyrosine

Phenylalanine + oxoglutarate → Phenylpyruvate (COO^-, $C=O$, CH_2) + glutamate

A

The formation of thyroid hormones

The synthesis of the hormone thyroxine involves the iodination of tyrosine, whilst bound as tyrosine residues of the protein, *thyroglobulin*.

$$HO-C_6H_4-CH_2-CH(NH_3^+)-COOH$$

L-Tyrosine

$$HO-C_6H_3I-CH_2-CH(NH_3^+)-COO^-$$

L-3-Monoiodotyrosine

$$HO-C_6H_2I_2-CH_2-CH(NH_3^+)-COO^-$$

L-3,5-Diiodotyrosine

$$HO-C_6H_3I-O-C_6H_2I_2-CH_2-CH(NH_3^+)-COO^-$$

L-3,5,3′-Triiodothyronine
(Triiodothyronine)

$$+\ H_2C(OH)-CH(NH_3^+)-COO^-$$

L-Serine

2 molecules

$$HO-C_6H_2I_2-O-C_6H_2I_2-CH_2-CH(NH_3^+)-COO^-$$

+ L-Serine

L-3,5,3′5′-Tetraiodothyronine
(Thyroxine)

B

Thyroglobulin, the protein containing the tyrosine residues from which thyroxine is formed is contained in the follicle within the thyroid gland, and the synthesis is at the macromolecular level.

PAS, refers to the Periodic Acid Schiff reaction, specific for carbohydrate.

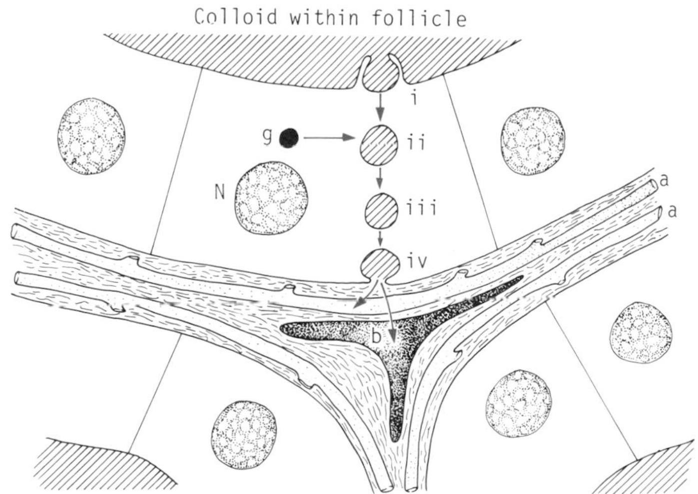

Diagrammatic representation of the breakdown of thyroglobulin in the thyroid gland. (i) Phagocytosis of colloid to form PAS-positive vesicle. (ii) Transference of hydrolytic enzymes from dense granule (*g*). (iii) Digestion of colloid with progressive loss of PAS-positive material. (iv) Extrusion of iodothyronines mainly to the thyroid blood capillary (*a*). *N*: nucleus.

Tyrosine → (Tyrosine hydroxylase) → Dopa (Dihydroxyphenylalanine) → Dopamine (Dihydroxyphenylethylamine) → Noradrenaline → Adrenaline

Catechol

A

The formation of adrenaline and noradrenaline

(the *catecholamines)*

Adrenaline is synthesized from tyrosine in the adrenal medulla.

Dopa is a neurotransmitter substance used in the treatment of Parkinson's disease

---

Arginine + Glycine → (glycine transamidinase) → Guanidinoacetic acid + Ornithine

Guanidinoacetic acid → (S-adenosylmethionine ('active' methionine) → S-adenosylhomocysteine) → Creatine

Phosphocreatine + ADP ⇌ (creatine kinase) ⇌ Creatine + ATP

B

Phosphocreatine

In muscle, during contraction, ATP is for the most part derived from glycolysis, for which muscle glycogen acts as initial substrate. However, before ATP from glycolysis becomes available, *phosphocreatine* acts as a temporary source of ATP.

Phosphocreatine — phosphate → Creatinine

Creatine — water → Creatinine

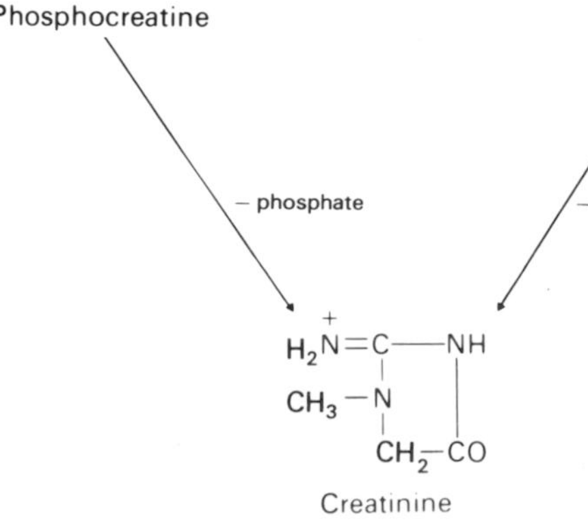

Creatinine

Phosphocreatine, and creatine, may break down to creatinine. This breakdown is fairly constant over a 24 hour period. Urinary creatinine is sometimes used as an index of urinary excretion.

The metabolism of histidine

A

Formation of histamine

The important *autacoid**, histamine, is released by *mast cells* in response to a variety of stimuli, including IgE (mediating allergic responses).

Histamine is produced as a result of the action of *histidine decarboxylase.*

Histidine $\xrightarrow[\text{histidine decarboxylase}]{-CO_2}$ Histamine $\xrightarrow[\text{monoamine oxidase}]{H_2O \;\to\; NH_4^+ + H_2O_2}$ Imidazolyl aldehyde $\xrightarrow{NADH \;\leftarrow\; NAD^+}$ Imidazolylacetate

Histamine is destroyed by monoamine oxidase, forming imidazolyl aldehyde, which is oxidized to imidazolylacetate, in which form it is excreted in the urine (as a conjugate with ribose 5′-phosphate).

*The term autacoid denotes any compound secreted into the blood by one organ to act on another.

B

Degradation of histidine

The main degradative pathway for histidine yields glutamic acid.

Formiminoglutamic acid (FIGLU) may be excreted in quantity in the urine in certain pathological conditions such as formimino transferase deficiency. As a folate coenzyme is required for its metabolism, it accumulates and is excreted in the urine in folate deficiency.

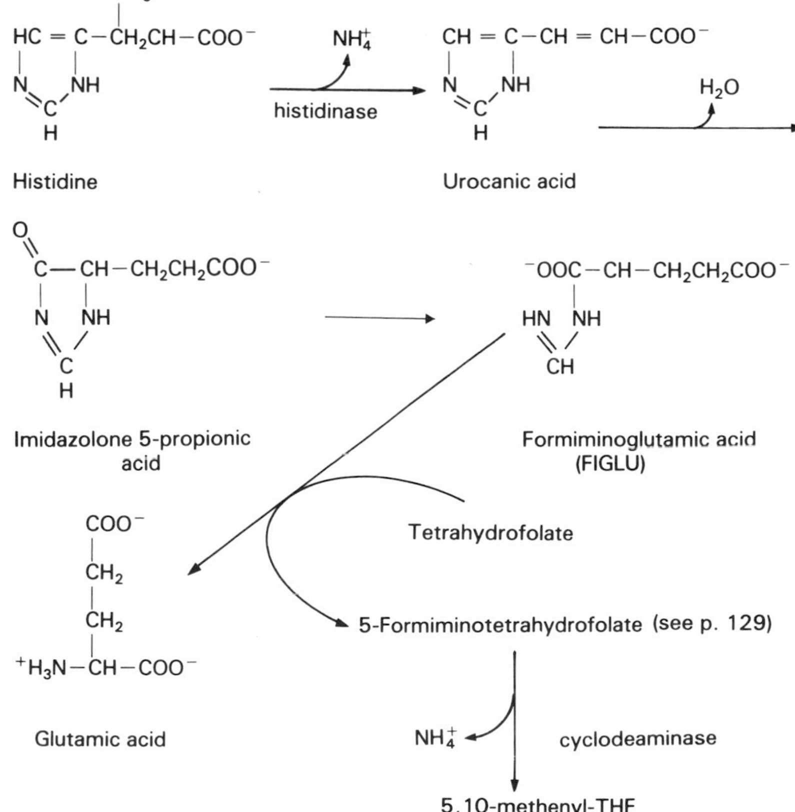

The metabolism of tryptophan

Tryptophan serves as the precursor to two important compounds, *serotonin* and *nicotinamide*. Serotonin (5-OH tryptamine) a neurotransmitter, is formed after hydroxylation and decarboxylation of tryptophan. The synthesis of the nicotinamide moiety of the nicotinamide nucleotides is initiated by the enzyme *trytophan dioxygenase* (*tryptophan pyrrolase*), which opens the indole ring.

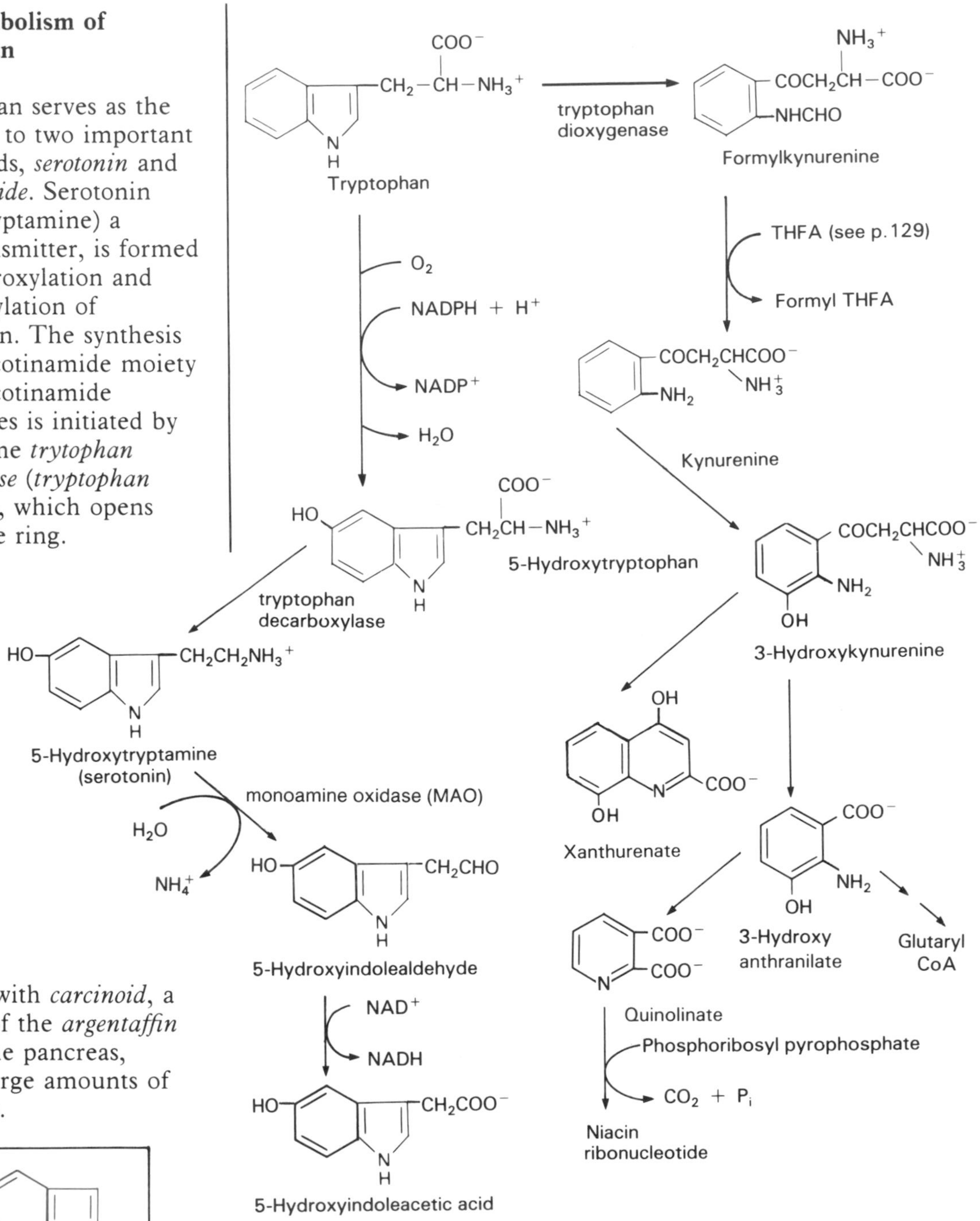

Tryptophan

tryptophan dioxygenase

Formylkynurenine

THFA (see p.129)

Formyl THFA

O_2

NADPH + H^+

$NADP^+$

H_2O

Kynurenine

5-Hydroxytryptophan

tryptophan decarboxylase

3-Hydroxykynurenine

5-Hydroxytryptamine (serotonin)

monoamine oxidase (MAO)

H_2O

NH_4^+

Xanthurenate

5-Hydroxyindolealdehyde

3-Hydroxy anthranilate

Glutaryl CoA

Quinolinate

NAD^+

NADH

Phosphoribosyl pyrophosphate

CO_2 + P_i

Niacin ribonucleotide

5-Hydroxyindoleacetic acid

Patients with *carcinoid*, a tumour of the *argentaffin cells* of the pancreas, secrete large amounts of serotonin.

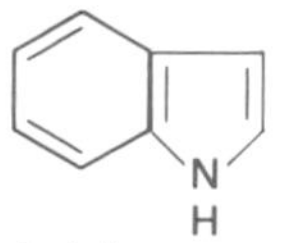

Indole

Tryptophan is indolylalanine

In B_6 deficiency, metabolism of 3-hydroxykynurenine is depressed, due to the fact that the enzyme is B_6-dependent. Under these conditions, much more tryptophan is metabolized to xanthurenic acid, an event that can be utilized to assess the adequacy of B_6 nutritional status.

A

Glutathione and glutamine

The tripeptide *glutathione* is found at a high concentration in many cells.

Compounds such as ethyl bromide can be detoxified in a reaction in which they react with the sulphydryl group of glutathione.

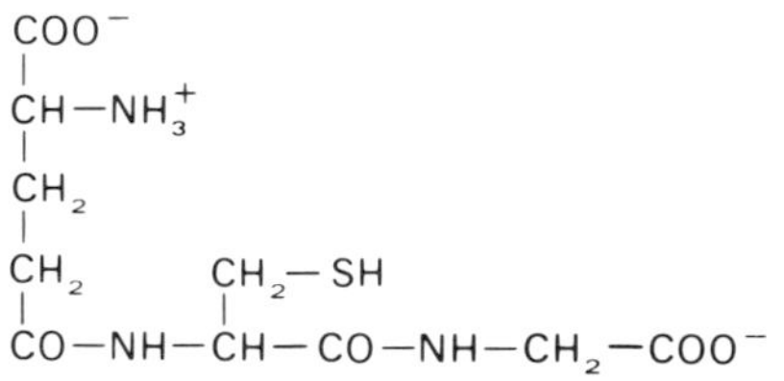

$$
\begin{array}{l}
COO^- \\
| \\
CH-NH_3^+ \\
| \\
CH_2 \\
| \\
CH_2 \qquad\quad CH_2-SH \\
| \qquad\qquad\quad | \\
CO-NH-CH-CO-NH-CH_2-COO^-
\end{array}
$$

Glutathione (γ-glutamyl-cysteinyl-glycine)

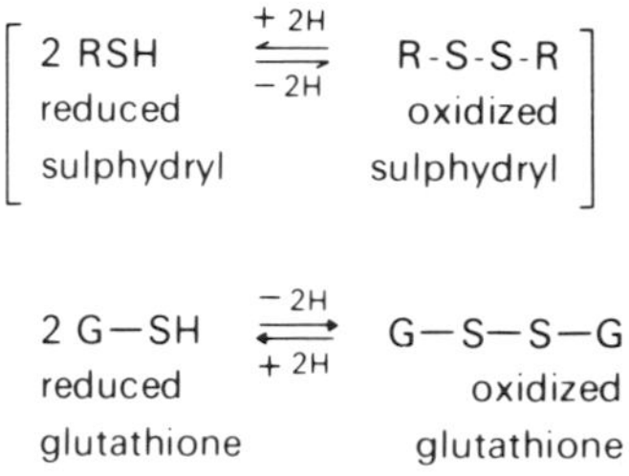

$$\left[\begin{array}{ccc} 2\,RSH & \underset{-2H}{\overset{+2H}{\rightleftharpoons}} & R\text{-}S\text{-}S\text{-}R \\ \text{reduced} & & \text{oxidized} \\ \text{sulphydryl} & & \text{sulphydryl} \end{array}\right]$$

$$\begin{array}{ccc} 2\,G-SH & \underset{+2H}{\overset{-2H}{\rightleftharpoons}} & G-S-S-G \\ \text{reduced} & & \text{oxidized} \\ \text{glutathione} & & \text{glutathione} \end{array}$$

Among a variety of functions, glutathione appears to be involved in preventing peroxide formation in membranes, in transporting amino acids through membranes, and in a number of oxidation-reduction reactions, especially those utilizing enzymes with an -SH group in the active site, which can be maintained in the reduced state by reaction with glutathione.

B

γ-glutamyltransferase, located in the plasma membrane, may be involved in the membrane transport of certain amino acids and in the salvage of glutathione.

Amino acid + glutathione ⟶ γ-glutamyl amino acid + cysteinylglycine
γ-glutamyl transferase (γGT)

γGT has diagnostic value, as it is released into the blood from liver membranes in cases of biliary obstruction, and alcoholism (See p. 73)

C

Glutamine acts as a means of trapping and transporting free ammonia (which is toxic). Hydrolysis of glutamine provides ammonia in the kidney for secretion of NH_4^+ ions into the urine, providing a means of generating additional cations to accompany excreted anions.

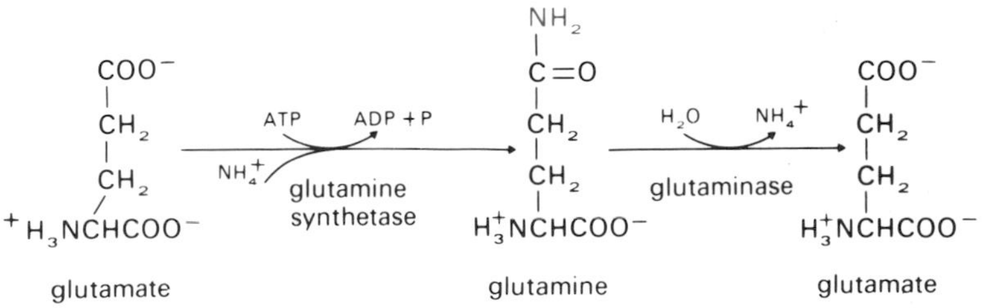

$$
\begin{array}{ccccc}
 & & NH_2 & & \\
 & & | & & \\
COO^- & & C{=}O & & COO^- \\
| & & | & & | \\
CH_2 & \xrightarrow[\text{glutamine synthetase}]{ATP,\ NH_4^+ \ \rightarrow\ ADP+P} & CH_2 & \xrightarrow[\text{glutaminase}]{H_2O\ \rightarrow\ NH_4^+} & CH_2 \\
| & & | & & | \\
CH_2 & & CH_2 & & CH_2 \\
| & & | & & | \\
{}^+H_3NCHCOO^- & & H_3^+NCHCOO^- & & H_3^+NCHCOO^- \\
\text{glutamate} & & \text{glutamine} & & \text{glutamate}
\end{array}
$$

In animals asparagine is usually formed by amide transfer from glutamine by asparagine synthase. Some tumour cells lack this enzyme so satisfy their requirement for asparagine by uptake from the blood. Certain forms of leukaemia may be treated by administering asparaginase to remove the asparagine from the blood and hence 'starve' the tumour cells. Asparaginase, an analogous enzyme to glutaminase, hydrolyses the amide group of asparagine.

A

The synthesis of haem

The synthesis begins with the compound *δ-aminolaevulinic acid* (ALA), which itself is synthesized by a condensation between glycine and succinyl CoA, with loss of CoA and CO_2, catalysed by *ALA synthase.*

Succinyl CoA + glycine

ALA synthase (CO_2, CoA)

I: COOH–CH_2–CH_2–C(=O)–CH_2–NH_2 + II: COOH–CH_2–CH_2–C(=O)–CH_2–NH_2

2 molecules of ALA

Porphobilinogen synthase

Porphobilinogen (I: COOH, CH_2; II: COOH, CH_2, CH_2; CH_2, NH_2; N–H)

Uroporphyrinogen synthase

Uroporphyrinogens I and III

Two molecules of ALA condense to form compounds that are converted to *porphobilinogen*, four molecules of which then polymerize to the tetrapyrroles, *uroporphyrinogen I and III.*

Both ALA synthase and ALA dehydratase are inhibited by haem. In acute intermittent porphyria (page 160), haem synthesis is decreased, and the feedback inhibition on ALA synthase is decreased. Certain drugs such as barbiturates and oestrogens induce this enzyme and may precipitate an attack.

Synthesis of ALA occurs in the mitochondrion whilst porphobilinogen synthesis is cytoplasmic.

Pyrrole

B

These molecules are identical except that the order of the carboxymethyl and carboxyethyl side chains on one pyrrole is reversed.

Uroporphyrinogen I is synthesized in smaller amounts and converted to several products essentially similar to those synthesized from uroporphyrinogen III. However, the I series do not form haem and are excreted in the urine.

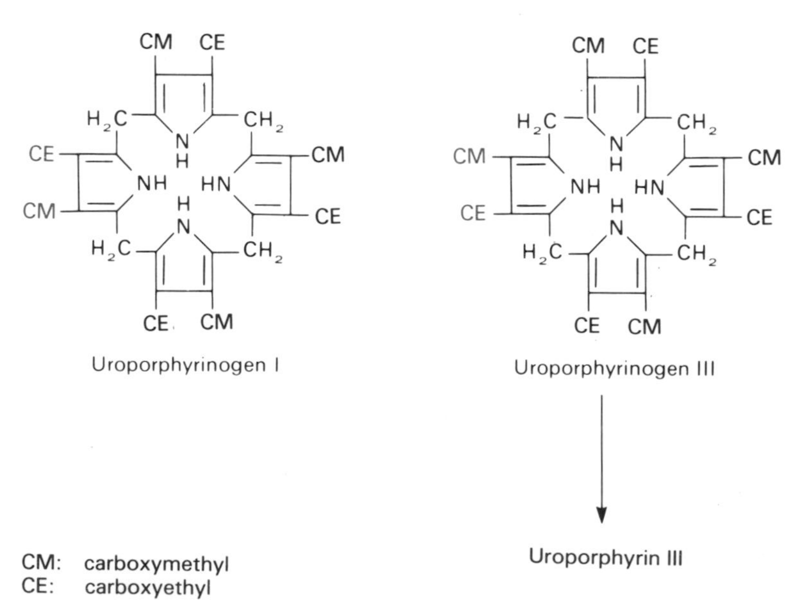

Uroporphyrinogen I

Uroporphyrinogen III

Uroporphyrin III

CM: carboxymethyl
CE: carboxyethyl

(see next page)

Uroporphyrinogen III is converted through several intermediate compounds to haem.

Metabolism of Uroporphyrinogen III to haem involves only the side chains of the pyrroles and the hydrogen atoms on the pyrrole nitrogens.

CM = carboxymethyl
CE = carboxyethyl
M = methyl
V = vinyl

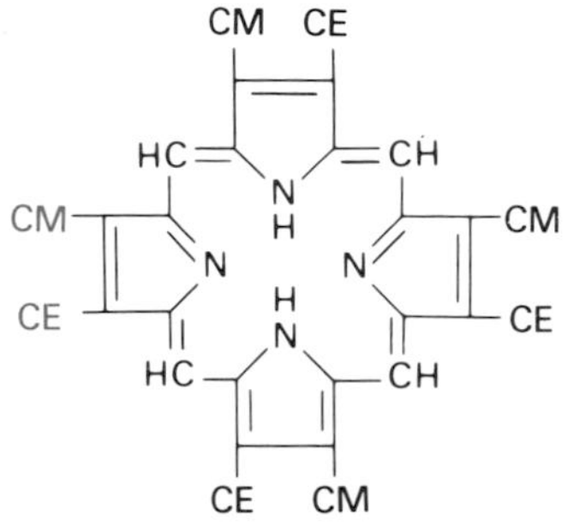

Uroporphyrin III

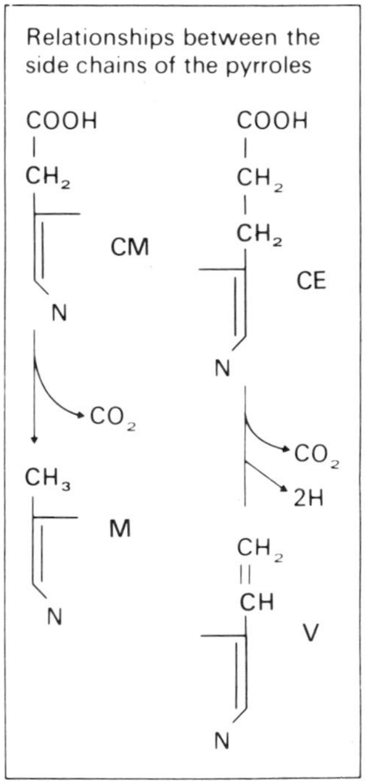

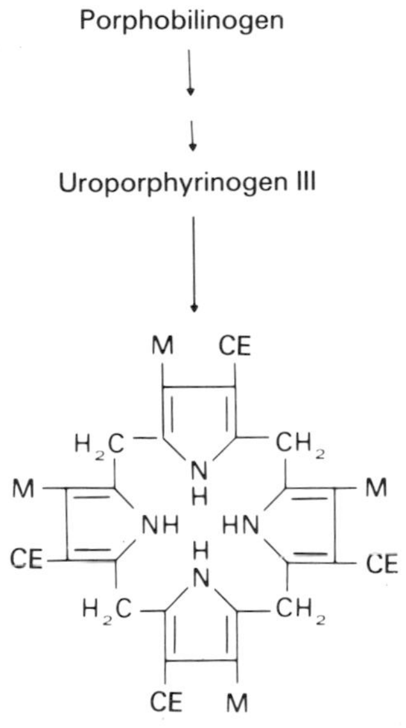

Porphobilinogen → Uroporphyrinogen III → Coproporphyrinogen III

Relationships between the side chains of the pyrroles

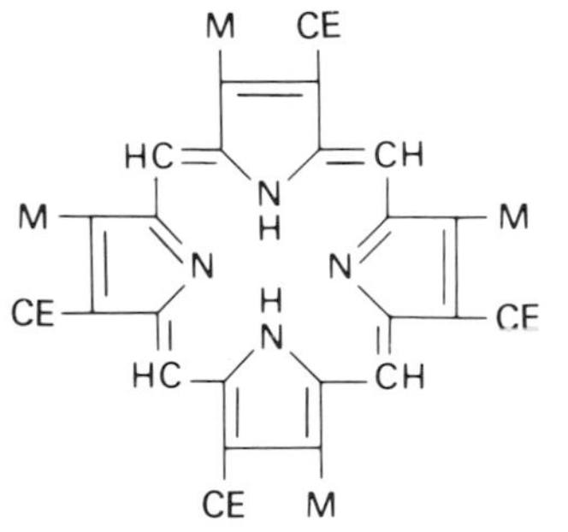

Coproporphyrin III

Coproporphyrinogen III

The synthesis of coproporphyrinogen III from porphobilinogen is cytoplasmic. The remaining steps in the synthesis of haem occur in the mitochondrion.

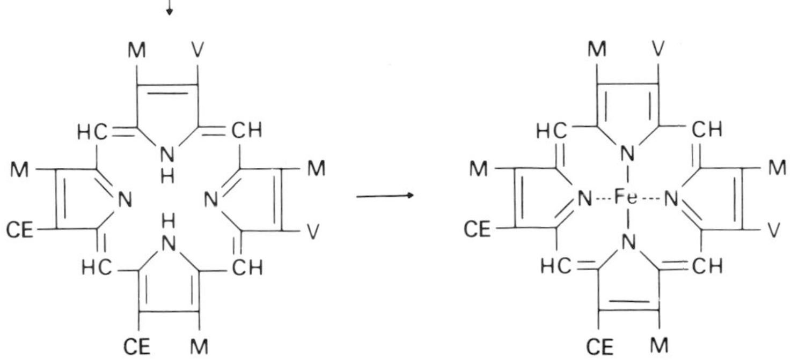

Protoporphyrin III → Haem

A

The *porphyrias* are inherited diseases of porphyrin metabolism

The porphyrias are a group of diseases that arise from disordered porphyrin metabolism.

Name:	Erythropoietic Uroporphyria	Erythropoietic Protoporphyria	Acute intermittent porphyria
Cell affected:	Erythrocyte	Erythrocyte	Liver
Type of inheritance:	Autosomal recessive	Autosomal dominant	Autosomal dominant
Accumulating materials:	Uroporphyrin I Coproporphyrin	Protoporphyrin	δ-aminolaevulinic acid Porphobilinogen
Diagnosis:	1. Both compounds in urine & faeces & red cells 2. Red cells fluoresce under u.v. light	Excretion of protoporphyrins in faeces	1. Both compounds in urine 2. Urine turns to deep red in sunlight as porphobilinogen is converted to uroporphyrin

B

Bile pigments

At the end of the life of the red cell it is broken down by cells of the reticuloendothelial system and the haem is degraded to *biliverdin*. This involves oxidation of one of the inter-pyrrole bridges. Reduction of biliverdin yields *bilirubin*. These compounds are found in the bile and are known as the bile pigments.

Biliverdin $\xrightarrow{+2H}$ Bilirubin

In the liver, bilirubin reacts with 2 molecules of UDP-glucuronate to form a bisglucuronide, which has a glucuronic acid residue in glucuronide linkage to each of the carboxyethyl residues.

C

Further metabolism in intestine and kidney yields other pigments that are found in faeces and urine.

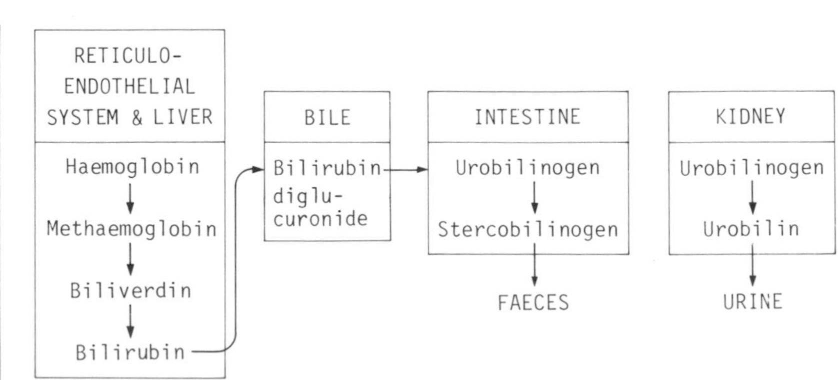

8

CARBOHYDRATE AND FAT METABOLISM

A. Oxidative catabolism

A

Forces driving metabolic processes

All life depends on the utilization of energy emitted by the sun.

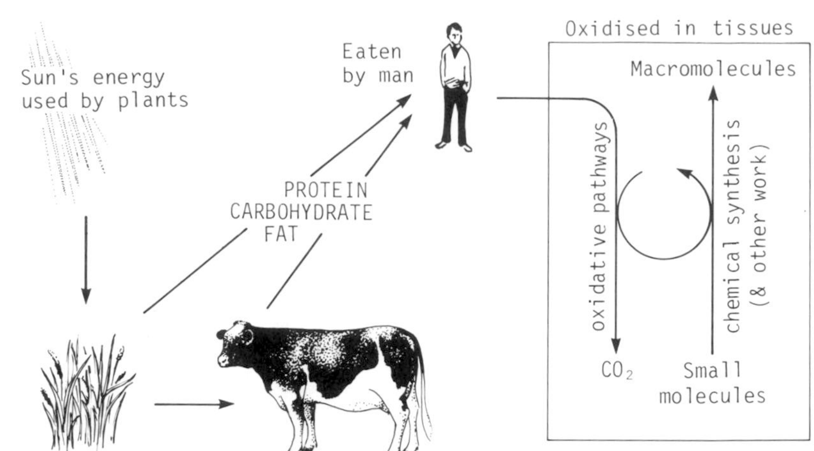

Chemical reactions tend always to proceed towards equilibrium. If a product of a chemical reaction (such as a gas) is continually removed, the reaction never reaches equilibrium and will proceed as long as reactants (substrates in an enzymic reaction) are available. Thus continual loss of CO_2 (an end product of mammalian oxidation pathways) through the lungs ensures that oxidation processes can act as a constant driving force for other reactions in the body. Although one of the reactants, O_2, is also a gas, it is in much greater proportion in the atmosphere than is CO_2 and is concentrated at the tissues by the action of haemoglobin.

B

Synthetic (*anabolic*) pathways depend on a supply of certain coenzymes generated by degradative (*catabolic*) pathways.

Note: NADPH is not itself generated by many major oxidative pathways, and is not exclusively used for reductive processes in synthetic pathways, but for simplicity only the one nicotinamide coenzyme is shown.

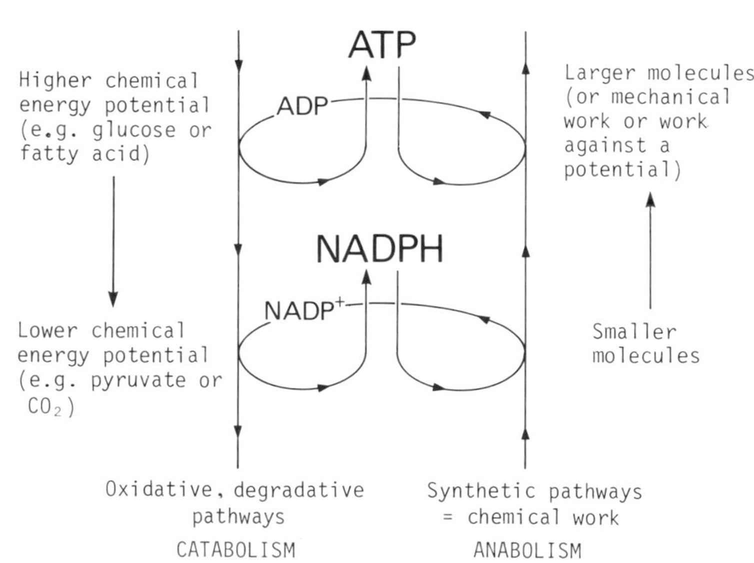

Certain coenzyme systems (ATP/ADP, NADH/NAD^+, NADPH/$NADP^+$) are involved in many reactions within a cell. Of these coenzymes, ATP, NADH and NADPH often participate as reactants in synthetic reactions (or, in the case of ATP, as an essential component of a system doing work, such as muscular contraction, or transport of molecules across membranes).

A

The pathways that generate the major part of the ATP and NADH formed in the cell are those of glycolysis, fatty acid oxidation, the citric acid cycle and the electron transport chain.

NADPH can be formed from $NADP^+$ and NADH by a mitochondrial transhydrogenase, linked to the electron transport system and essentially irreversible.

$$NADH + NADP^+ \longrightarrow NADPH + NAD^+$$

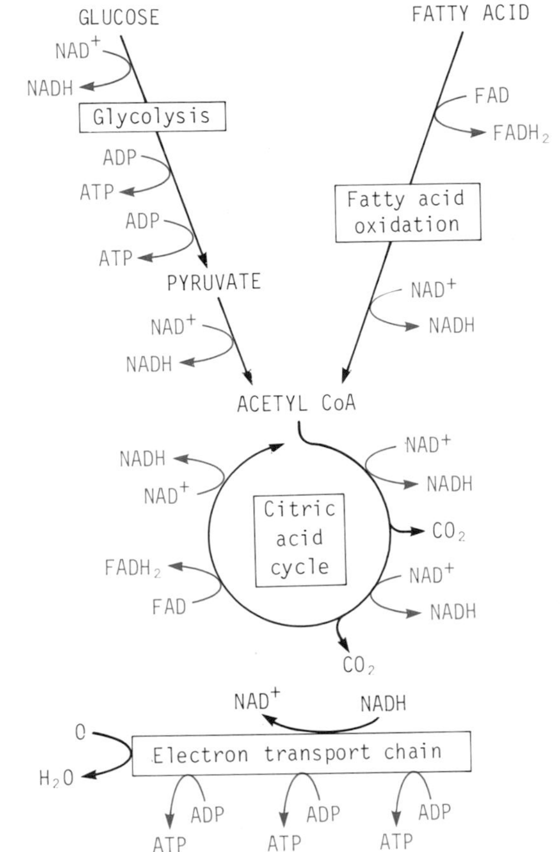

In the cytosol, transfer of hydrogen between NADH and $NADP^+$ is brought about by malate dehydrogenase and the malic enzyme (see p. 214).

B

Fat has a higher energy potential than has carbohydrate.

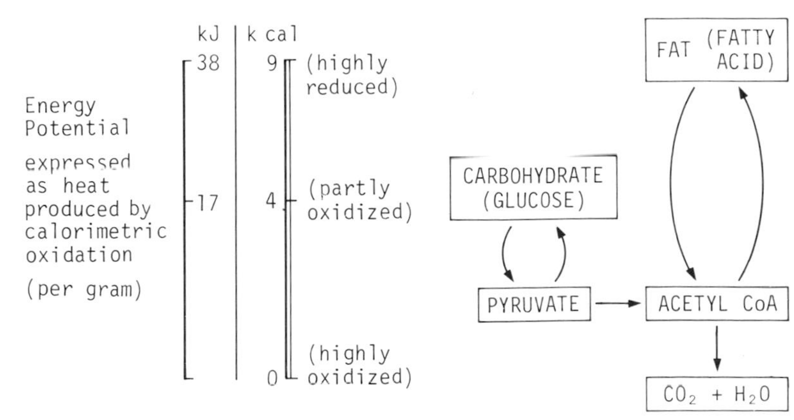

A

The carbons of glucose are already partially oxidized, whilst most of those of a fatty acid are fully reduced.

Glucose

Palmitic acid (a Fatty Acid)

$CH_3{:}CH_2{:}CH_2{:}CH_2{:}CH_2{:}CH_2{:}CH_2{:}CH_2{:}CH_2{:}CH_2{:}CH_2{:}CH_2{:}CH_2{:}CH_2{:}CH_2{:}COOH$

Most of the carbons of a fatty acid are in the fully reduced state. Thus fatty acids can potentially undergo a greater number of oxidative steps for each carbon than can carbohydrates. Thus they are an ideal substance for use as an energy store. Glycogen (a polymer of glucose) is, on the other hand, stored as an immediate reserve of blood glucose.

B

An overall plan of metabolic pathways

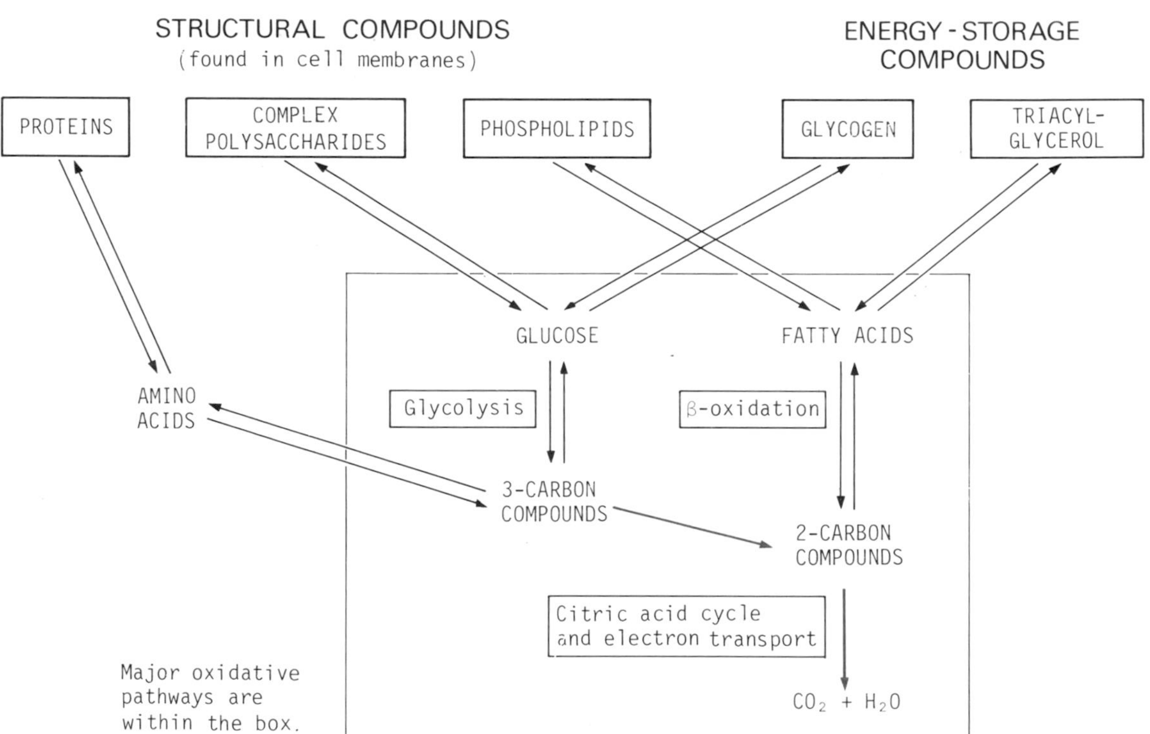

The steps shown in red are the only ones for which there is no reverse route.

A

β-oxidation of fatty acids

Fatty acids are oxidized in the mitochondria.
A fatty acid is degraded step by step by the sequential removal of 2-carbon units.
Coenzyme A is added to the carboxyl group of the fatty acid to form the fatty acyl CoA derivative.
Each 2-carbon unit is liberated from the fatty acid as acetyl CoA.
The oxidation of the fatty acid alkyl chain, by enzymes that utilize FAD and NAD^+, yields a compound that is at the oxidation state of an acid (acetyl CoA is the condensation product of acetic acid and coenzyme A).

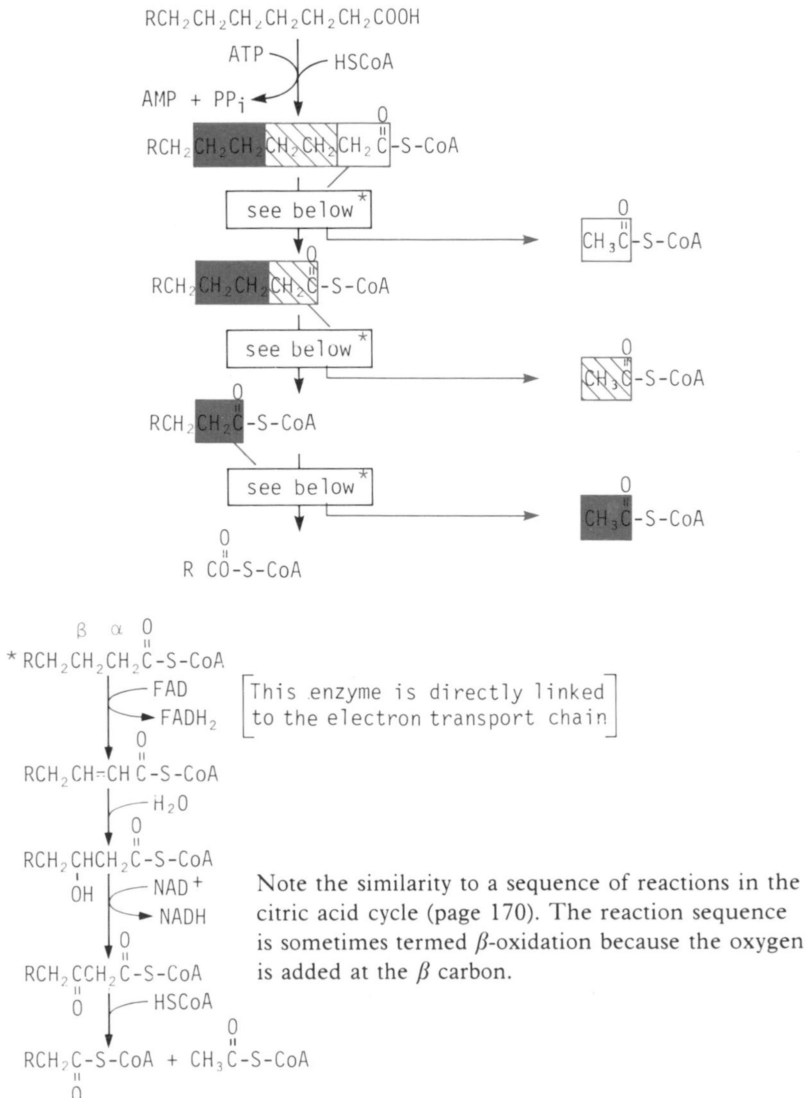

Note the similarity to a sequence of reactions in the citric acid cycle (page 170). The reaction sequence is sometimes termed β-oxidation because the oxygen is added at the β carbon.

B

A summary of fatty acid oxidation

Control of the rate of fatty acid oxidation is regulated by the rate of transport of the fatty acids into the mitochondrion. This depends on the formation of carnitine esters (see p. 203).

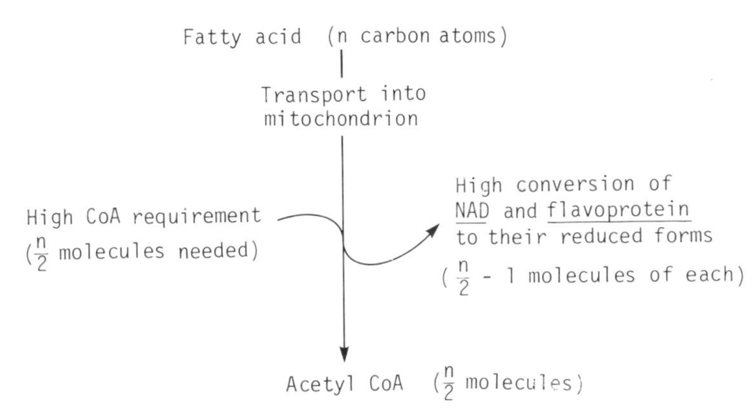

The glycolytic pathway

The oxidation of glucose to pyruvate is termed glycolysis.

Successive phosphorylation of the glucose C-1 and C-6 utilizes initially two molecules of ATP. These phosphorylations and an isomerization bring about the conversion of glucose to fructose 1,6-bisphosphate.

After the split of the 6-carbon sugar to two 3-carbon molecules (by aldolase) there is immediate oxidation of the aldehyde to an acid with simultaneous formation of an anhydride bond between the acid and a phosphate group. This complex and important reaction is achieved by glyceraldehyde phosphate dehydrogenase, utilizing NAD^+.
This formation of the phosphate anhydride by an enzyme of the cytosol is termed *substrate-level phosphorylation*, distinguishing it from oxidative phosphorylation. Conversion of the phosphate anhydride to a carboxylic acid is then

(continued on p. 167)

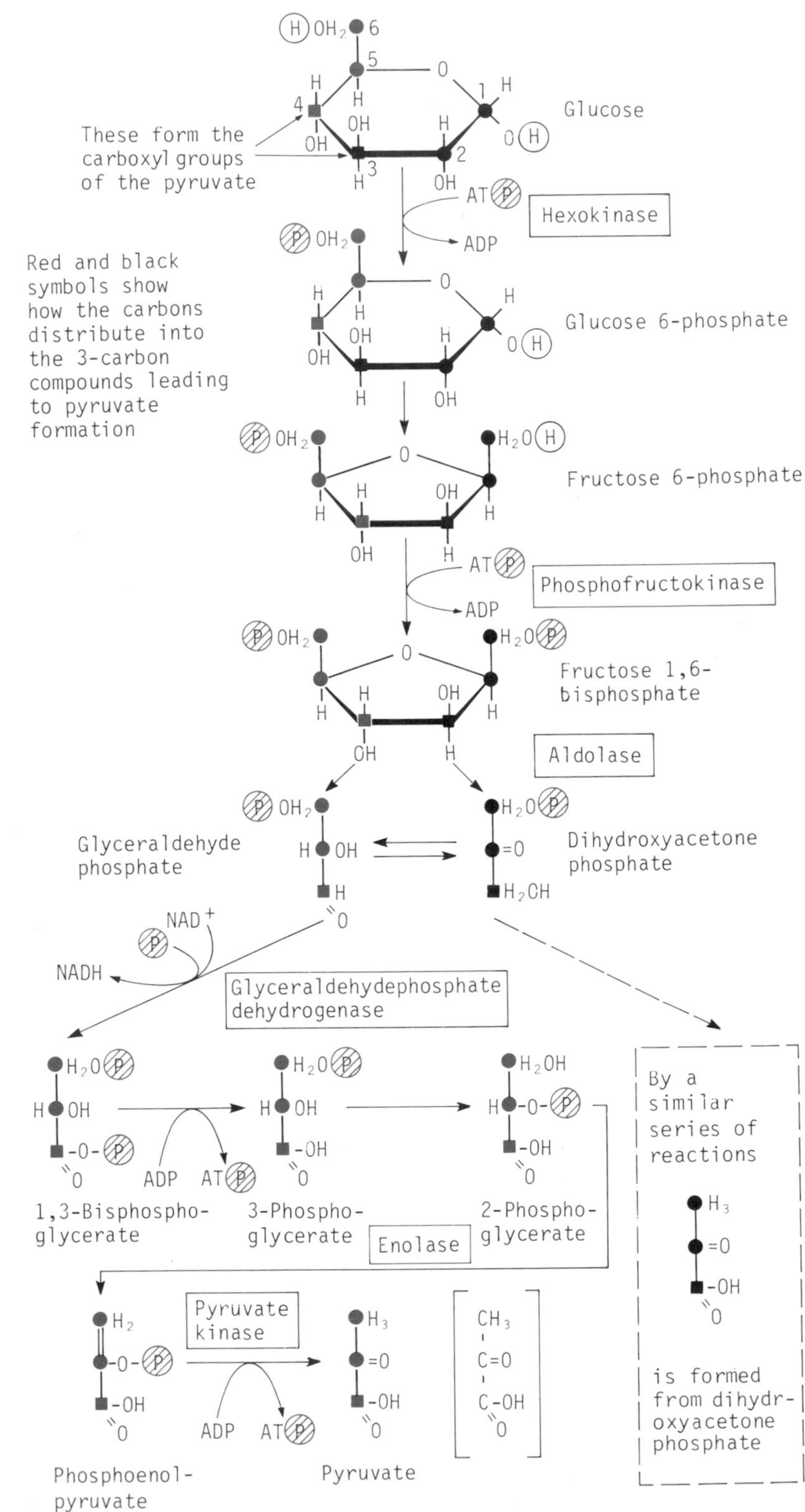

A

(cont'd from previous page)

combined with the conversion of ADP to ATP. After the move of the phosphate to C-2, enolase, by removing water, yields phosphoenolpyruvate, which is then involved in a reaction in which a further molecule of ATP is formed from ADP.
For each mole of glucose oxidized, the yield of ATP from glycolysis is two moles of ATP (4 ATP produced, 2 ATP utilized). In addition, two moles of NADH are produced (one for each of the 3-carbon products), each of which may yield a further two moles of ATP if oxidized by the electron transport chain, (see p. 172) via the glycerol phosphate shuttle.

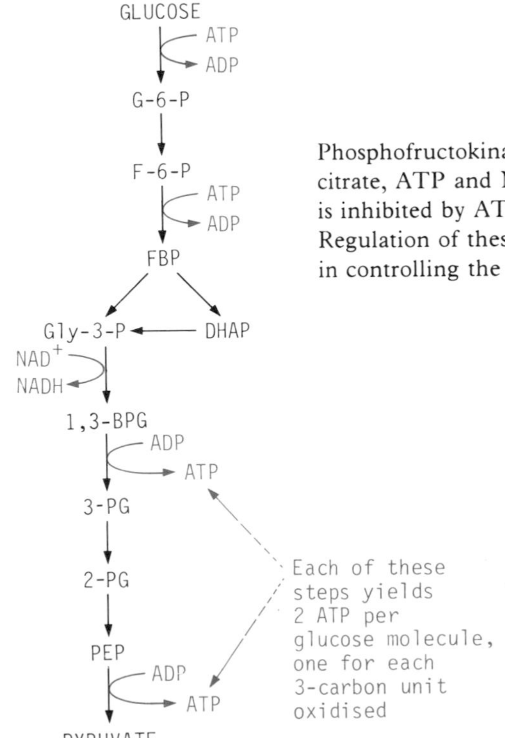

Phosphofructokinase is inhibited by citrate, ATP and NADH. Pyruvate kinase is inhibited by ATP and NADH. Regulation of these enzymes is important in controlling the rate of glycolysis.

The meaning of the abbreviations used in the scheme above can be determined by comparing the scheme with the more detailed reaction pathway shown on the opposite page. Note that these abbreviations are not those recommended for published work. They are, however, very convenient for everyday use.

B

Further metabolism of pyruvate

The pyruvate produced by the glycolytic pathway (and by other metabolic pathways) can be converted to oxaloacetate by the enzyme *pyruvate carboxylase*. Alternatively, in an oxidative reaction utilizing coenzyme A and NAD^+, it can be converted to acetyl CoA.

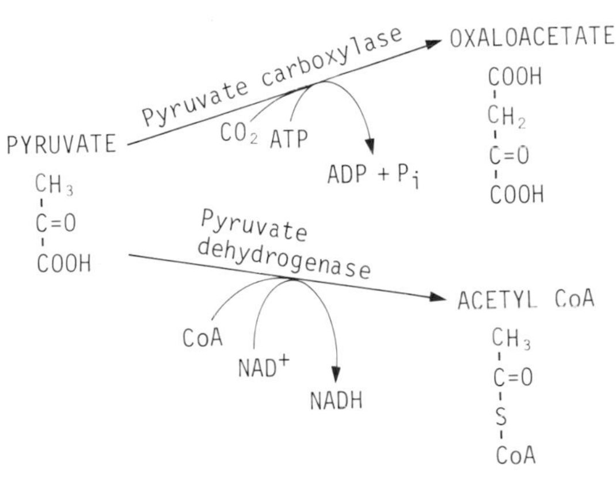

Pyruvate carboxylase has an absolute requirement for the presence of acetyl CoA as an activator. Biotin is also a coenzyme.

Pyruvate dehydrogenase exhibits product inhibition. Its activity is inhibited by a high acetyl CoA/CoA ratio.

Thiamin pyrophosphate is a coenzyme for pyruvate dehydrogenase

A

Regulation of the glycolytic pathway

The non-equilibrium reactions (see p. 65) of glycolysis are catalysed by hexokinase, phosphofructokinase and pyruvate kinase.
These are regulated by a number of allosteric modifiers.

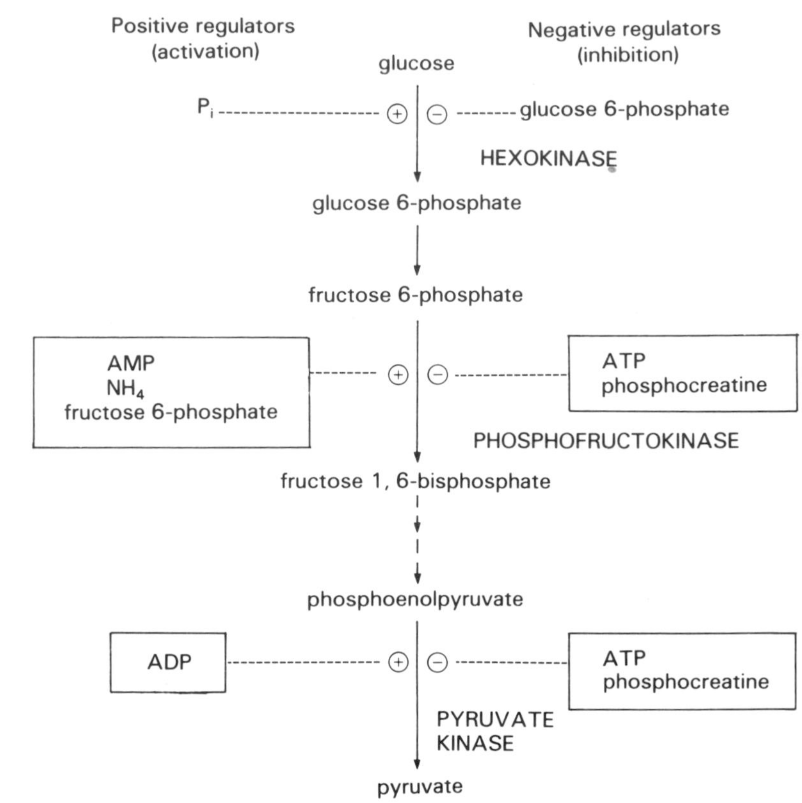

B

Phosphofructokinase (6PF1-K) is also subject to positive regulation by *fructose 2,6-bisphosphate* (F2, $6P_2$), the concentration of which is controlled by a kinase and phosphatase.

The kinase is itself regulated by covalent modification by a protein kinase which phosphorylates it, bringing about reduced activity.
This protein kinase is activated by glucagon, which thus reduces the level of F2, $6P_2$, and so also the activity of 6PF1-K.

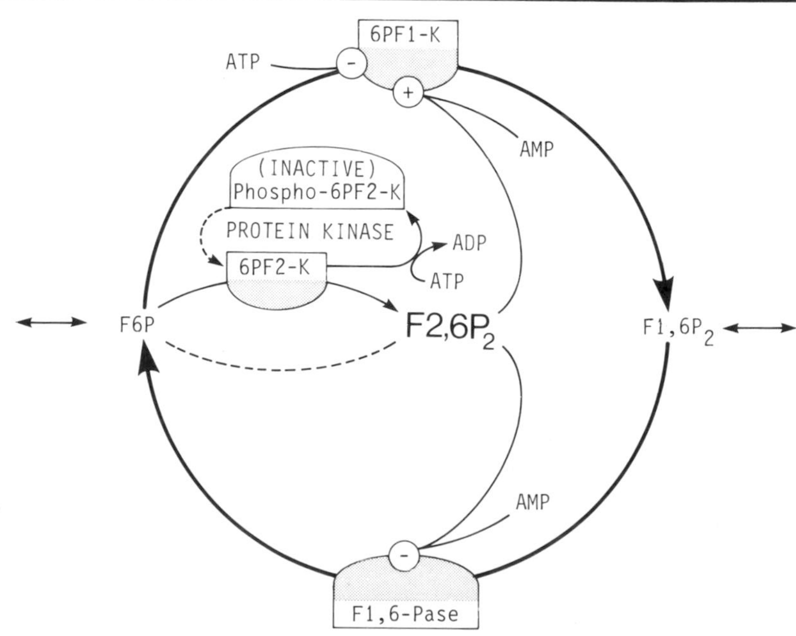

Fructose 1,6-bisphosphate (F1, $6P_2$) is hydrolysed to fructose 6-phosphate (F6P) by fructose 1,6-bisphosphatase (F1, 6-Pase — see p. 200). This enzyme is inhibited by high concentrations of F2, $6P_2$, and thus activated by glucagon which lowers the F2, $6P_2$ concentration. F2, $6P_2$ potentiates the action of AMP on phosphofructokinase and fructose 1,6-bisphosphatase, on which AMP has similar actions to F2, $6P_2$.

A

The citric acid cycle

Oxaloacetate and acetyl CoA can condense to form citrate.

The citrate that is formed is a pro-chiral molecule (see p. 10). For structure of citrate see next page.

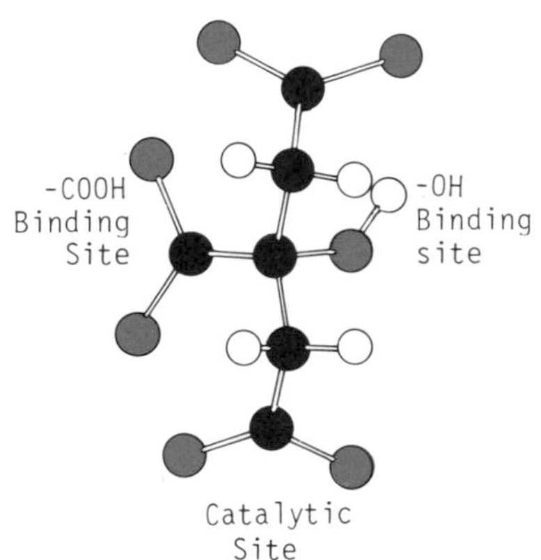

If -COOH and -OH binding sites are held on a surface, the two $-CH_2COOH$ groups are distinguishable.

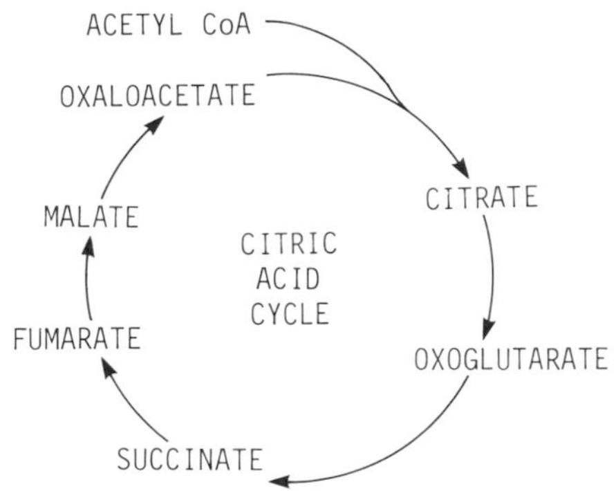

Citrate can be converted to oxaloacetate in a series of reactions in which CO_2 is lost and hydrogen removed. The oxaloacetate formed from the citrate can then react with a further molecule of acetyl CoA, and the sequence repeated an infinite number of times. In other words, the oxaloacetate is behaving as a catalyst, being regenerated in unchanged form after the reaction. This sequence of reactions, shown in detail on the next page, occurs in virtually every type of cell, with rare exceptions such as the mature erythrocyte.

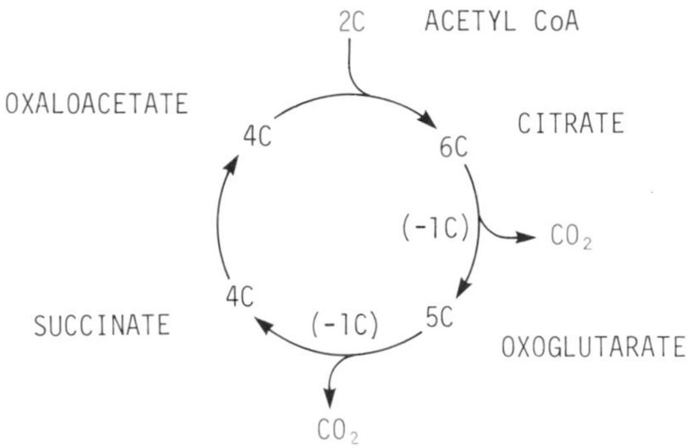

The net result of the reaction is the conversion of the acetyl residue to two molecules of CO_2 with the loss of eight hydrogen atoms to the acceptor molecules FAD and NAD^+.

B

The intermediates are free to enter and leave the cycle at any point, in which case there will be stoicheiometric conversion of molecules, one into the other, in the normal way.

As an example, glutamate can be converted into aspartate by transamination with oxaloacetate to form aspartate and oxoglutarate, which then enters the citric acid cycle to be converted to oxaloacetate, which undergoes transamination with glutamate to yield further molecules of aspartate and oxoglutarate until an equilibrium position is reached.

1. Glutamate + oxaloacetate ⟶ Aspartate + oxoglutarate
2. Oxoglutarate ⟶ oxaloacetate (Citric acid cycle enzymes)

Step 1 then repeats with a further molecule of glutamate.
The net result is the stoicheiometric conversion of glutamate to aspartate.

The citric acid cycle **cont'd.**

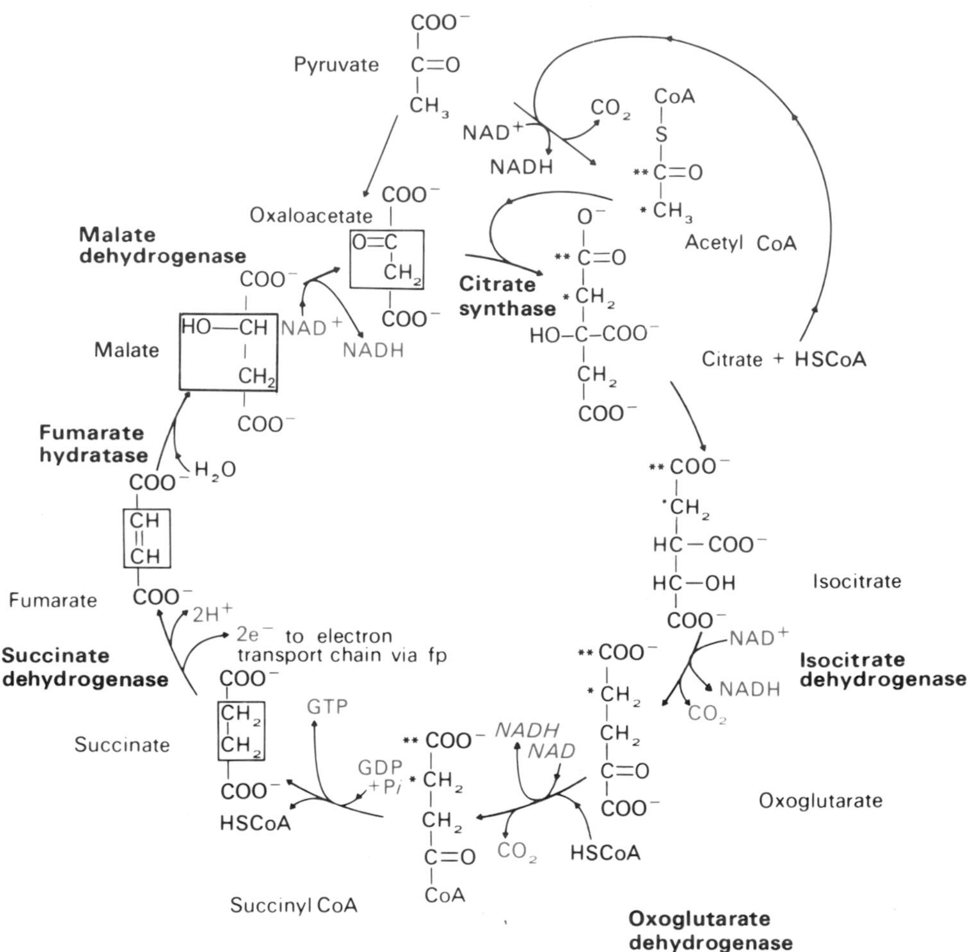

Note the reaction sequence

$CH_2-CH_2 \longrightarrow CH=CH \longrightarrow HO-CH-CH_2 \longrightarrow O=C-CH_2$

A similar series of reactions is also found in fatty acid oxidation.

* The asterisks denote the carbons from acetyl CoA. The prochiral nature of citrate enables the stereospecific oxidation of the molecule. CO_2 from acetyl CoA is not released by oxoglutarate dehydrogenase during the first turn of the cycle. At succinate, the molecule becomes completely symmetrical, and -COOH groups can then no longer be distinguished. The acetyl CoA label thus becomes randomized by the time oxaloacetate is formed, and during the next turn of the cycle 50% of the carbon from acetyl CoA will be released.

A

Pathway interactions

Major oxidative pathways may also act as routes linking other pathways.

The citric acid cycle links carbohydrate and fat metabolism, and is involved in the metabolism of amino acids. As is described in later sections, it provides an important control point for regulating metabolite flow in these pathways.

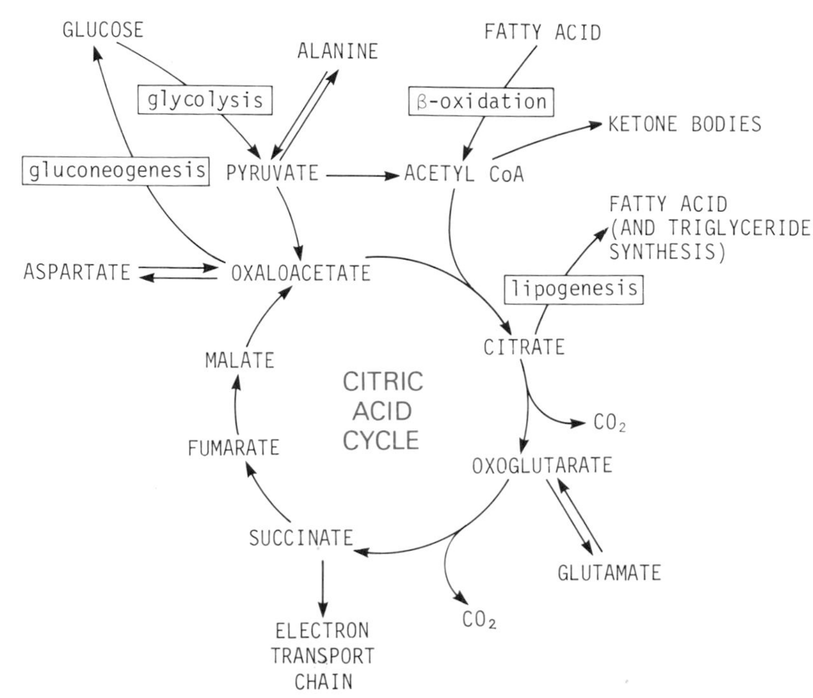

B

The enzymes of the glycolytic pathway link up the metabolism of glucose and other carbohydrates with fat and amino acid metabolism.

Substrates in boxes are important in interactions of the glycolytic pathway with other pathways.

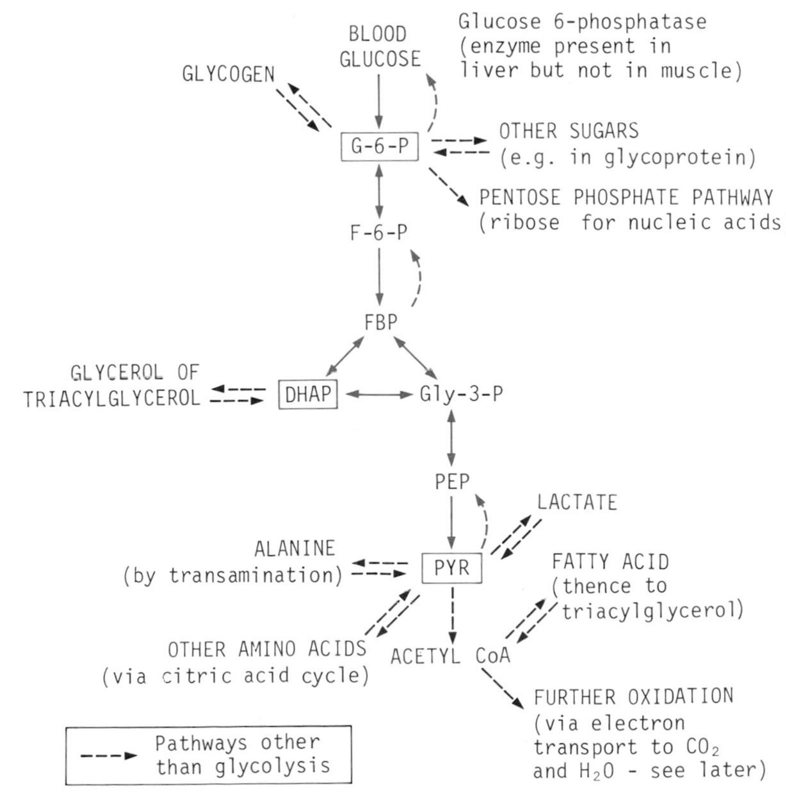

A

The electron transport chain

The oxidation of NADH and succinate.

NADH, produced by the action of citric acid cycle enzymes, diffuses to the electron transport chain, where the first step in its oxidation is the removal (by NADH dehydrogenase) of hydrogen.
Succinate dehydrogenase does the same for succinate.
Note: This succinate dehydrogenase is the same enzyme as the one shown earlier in the citric acid cycle.

NADH → NAD^+ (NADH dehydrogenase)
succinate → fumarate (succinate dehydrogenase)
Electron transport system: O_2 → H_2O

The function of the electron transport chain is to accept hydrogen atoms from substrates such as NADH or succinate and bring about their reaction with oxygen to form water.
Both NADH dehydrogenase and succinate dehydrogenase are flavoproteins (see p. 127).

These two flavoprotein dehydrogenases are linked to an electron transport system consisting of a series of enzymes termed *cytochromes*. The hydrogens removed from NADH or succinate are split to H^+ and e^-. The electrons are then transferred from one cytochrome to another until they react with O_2.

$$NADH + H^+ \rightarrow NAD^+ \; ; \; 2H \rightarrow f_p \rightarrow 2e^- \rightarrow \text{cyt } b \rightarrow \text{cyt } c_1 \rightarrow \text{cyt } c \rightarrow \text{cyt } a/a_3 \; ; \; O_2 \rightarrow O^{2-}$$

$f_p \rightarrow 2H^+$

f_p = flavoprotein dehydrogenase

B

The prosthetic group of all cytochromes is a haem molecule within which is bound an iron ion.
The iron atom is bound, within the haem group, to a porphyrin coenzyme identical with that found in haemoglobin, with the difference that in the cytochromes the iron undergoes oxidation and reduction, and is covalently linked.
The complete structure of a cytochrome (cytochrome c) is shown on page 61.
The iron at the centre of the haem group accepts and then passes on the electron, being constantly reduced and oxidized between Fe^{2+} and Fe^{3+} in the process. There are differences in the structure of the side chains on the porphyrin between different cytochromes. The structure shown is haem B, the haem of cytochrome b.

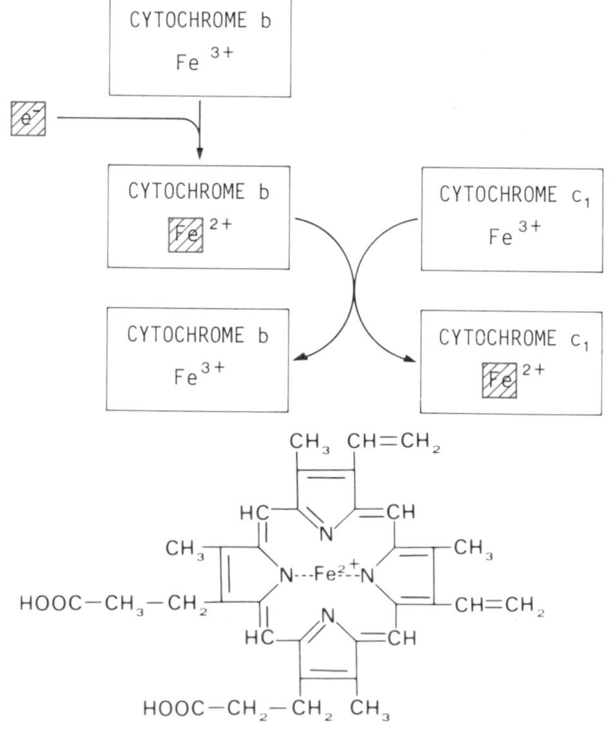

Ferroprotoporphyrin (haem)

A

The hydrogen acceptor for the flavoproteins of the electron transport chains (NADH and succinate dehydrogenases and the flavoprotein dehydrogenase of β-oxidation) is a quinone, called *ubiquinone*, or *coenzyme Q*. The electrons are then passed from coenzyme Q to the cytochromes.

H_3CO, H_3CO, CH_3, $(CH_2{-}CH{=}C(CH_3){-}CH_2)_nH$

Coenzyme Q, CoQ
(Ubiquinone) oxidized form

R_1, R_1, R_2, R_3, O˙, OH, $+H^{\cdot}$ $(H^+ + e^-)$

Semi-quinone form of ubiquinone

Coenzyme Q is a lipid-soluble compound that is not bound to protein but is free to diffuse through the hydrophobic core of the lipid bilayer.

It also has the property that it can readily form a semi-quinone, so that it can form a bridge between the two-electron reactions of the dehydrogenases and the one-electron reactions of the cytochromes.

It is at this point of the electron transport chain that the hydrogen atoms lose an electron to cytochrome b, leaving the proton to be transported by a separate transport mechanism.

B

Purification of components of the electron transport chain

The way in which the components of the chain interact can be confirmed by partially solubilizing the mitochondrial membrane and then separating out the components. When this is done four complexes of the different insoluble components can be isolated.

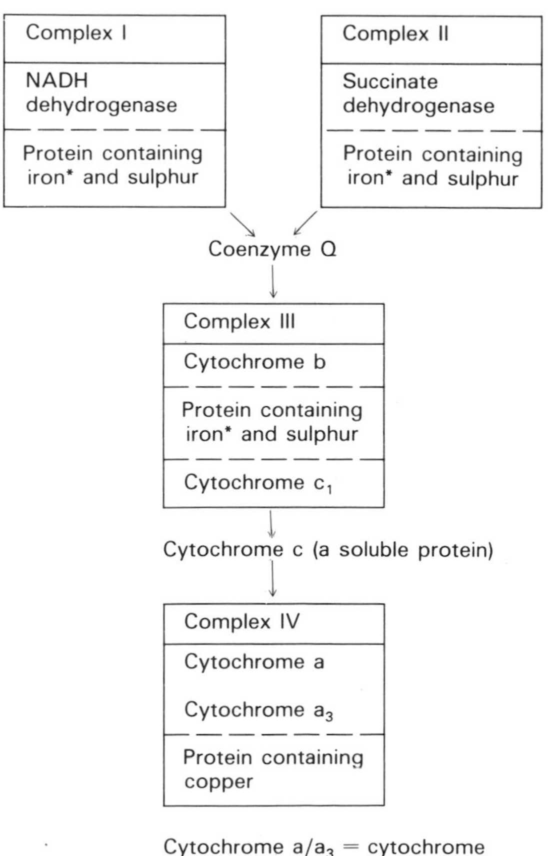

*sometimes referred to as "non-haem" iron.

Free energy change in electron transport

As with any enzyme reaction, the reaction of a cytochrome with another component of the electron transport chain involves a change in free energy. Indeed, the release of free energy as a result of the oxidation of the hydrogens of substances such as NADH or succinate, mediated through the cytochromes, yields the potential for the phosphorylation of ADP to ATP.

The reactions taking place in the electron transport chain are oxidation-reduction (redox) reactions. Redox reactions can be demonstrated in vitro by means of electrodes connecting solutions of the reactants, and the potential difference can be measured between the solutions in the two half-cells.

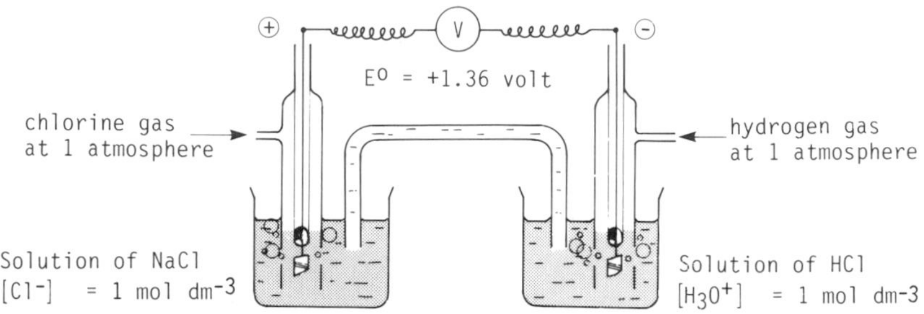

If such reactions are carried out at standard temperature (298°K) and pressure (1 atmosphere), using 1M solutions of reactants, the electromotive force, as represented by the voltage recorded on the voltmeter, is referred to as standard oxidation-reduction potential of the reaction, or standard redox potential (E_0'). By convention, the standard redox potential of the $H^+ : H_2$ couple (1M H^+ in equilibrium with H_2 gas at 1 atmosphere) is defined as 0 volts, and all other redox potentials can be related to this. For the reactions of the electron transport chain the following table can be prepared:

Oxidant	*Reductant*	*n*	$E_0'(V)$
NAD^+	$NADH + H^+$	2	−0.32
Fumarate	Succinate	2	+0.32
Ubiquinone (oxidized)	Ubiquinone (reduced)	2	+0.10
Cytochrome b (3+)	Cytochrome b (2+)	1	+0.07
Cytochrome c (3+)	Cytochrome c (2+)	1	+0.22
Cytochrome a (3+)	Cytochrome a (2+)	1	+0.29
$0_2 + 2H^+$	H_2	2	+0.82

The standard free energy change $\Delta G^{o\prime}$ is related to $\Delta E_o'$, the change in the standard oxidation-reduction potential, by the equation

$$\Delta G^{o\prime} = -n\,F\,\Delta E_o'$$

Where n = the number of electrons involved and F is the Faraday constant (96 500 coulombs). The units of F $\Delta E_0'$ are coulomb-volts, or joules, which can readily be converted to units of free energy as 4.18 joules is equal to 1 cal. The overall span of the respiratory chain is 1.14V. Thus the standard free energy change associated with oxidation of one mole of NADH can be calculated

$$\Delta G^{o\prime} = \frac{-2 * 96{,}500 * 1.14}{4.18} = -52.6 \text{ kcal/mole} = -220 \text{ kJ/mole}$$

A

Oxidative phosphorylation

Electron flow is coupled to the phosphorylation of ADP, forming ATP. During oxidation of one mole of NADH, 3 moles of ATP are formed (P : O ratio of 3.) During oxidation of one mole of succinate, 2 moles of ATP are produced (P : O ratio of 2.)
The cytochromes are termed a, b and c for historical reasons, but are thought to interact in the order b, c, a. Note that cytochromes c and c_1 are two different cytochromes and that the cytochrome which reacts with molecular oxygen, which is often called cytochrome oxidase, consists of cytochrome a and a_3 in a single enzyme complex.

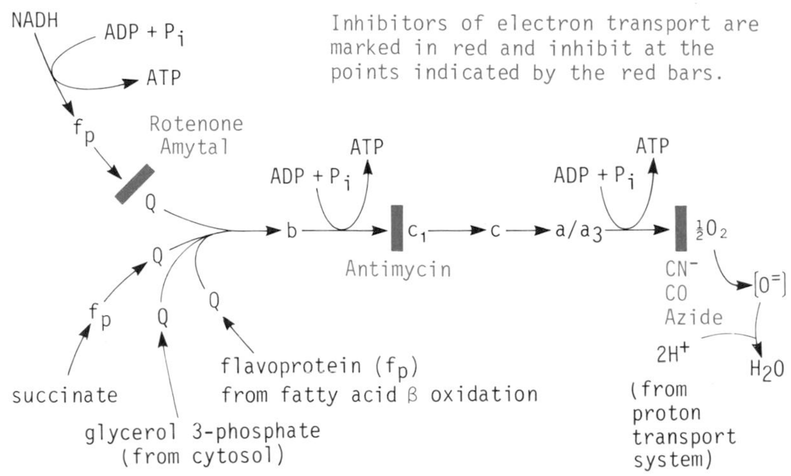

The dehydrogenases acting on NADH, succinate and fatty acyl CoA molecules are flavoproteins. They are shown as fp, and pass their hydrogens to co-enzyme Q (shown as Q above) which is 'dissolved' in the membrane lipid, and which then passes the electrons to cytochrome b, and the protons to a proton-transporting system.

Hydrogens passed to FAD during fatty acid oxidation are also passed to coenzyme Q via a flavoprotein acceptor.

B

Respiratory control and uncoupling agents

Some compounds, such as *dinitrophenol*, cause ATP synthesis to cease, and at the same time increase oxygen utilization. This is termed the *uncoupling* of ATP synthesis from electron flow, and such mitochondria are said to be *uncoupled*.
In coupled mitochondria, electron flow (oxygen utilization) is highly dependent on the relative amounts of ADP and ATP present, being high when ADP/ATP is high, and low when ADP/ATP is low. This is termed *respiratory control*.

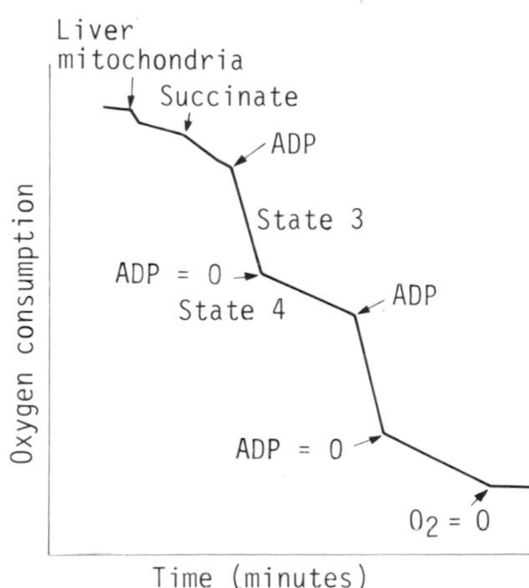

In liver mitochondria oxidizing succinate, addition of ADP elevates the ADP/ATP ratio (state 3) and increases oxygen utilization. When this ADP has been converted to ATP (low ADP/ATP ratio, 'state 4') oxygen utilization is low.

A

Structural aspects of mitochondria

Relation of structure to function in mitochondria.

A cross-sectional view of the mitochondrion portrays the membranes of the cristae (i.e. the inner membrane) as dividing the mitochondrion into compartments, termed the intermembrane space, and the matrix.

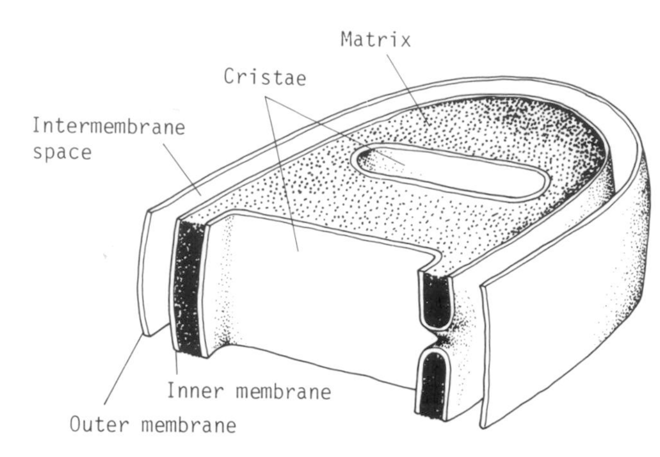

B

The inner membrane bears the enzymes of the electron transport chain and oxidative phosphorylation. The ATPase responsible for the generation of ATP is found on protrusions on the inside of the inner membrane.

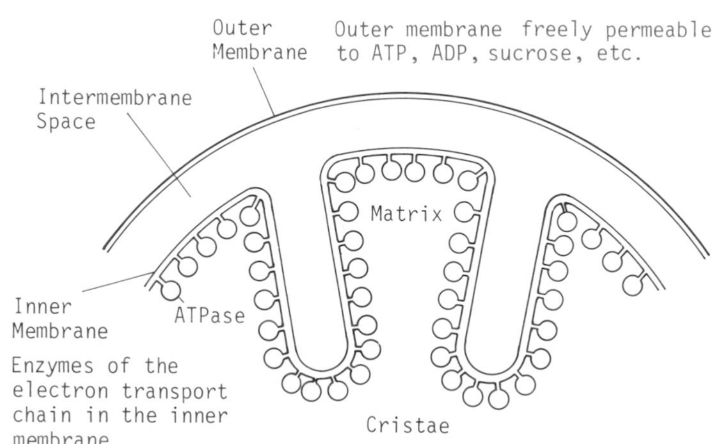

Inner membrane not freely permeable, but containing transport enzymes for transporting anions, and translocating ATP for ADP, across the membrane. The inner membrane is not permeable to NADH or other nicotinamide nucleotides.

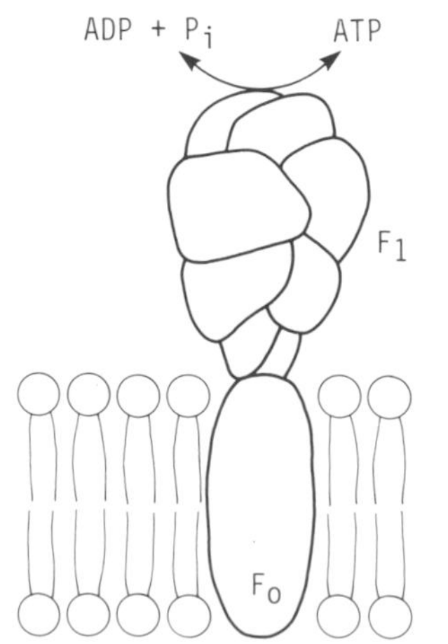

The ATPase has been purified and fractionated into several components.

The main assemblies are designated F_O embedded in the bilayer, and F_1 the ATPase itself, which is the assembly which protrudes from the bilayer and is known to be composed of several proteins.

A

Diagrammatic representations of mitochondrial structure are an attempt to reconcile the structures revealed by electron microscopy with the experimental results obtained with fractionated particles from mitochondria.

Electron micrograph of a section through part of a mitochondrion (pancreatic exocrine cell of guinea-pig).

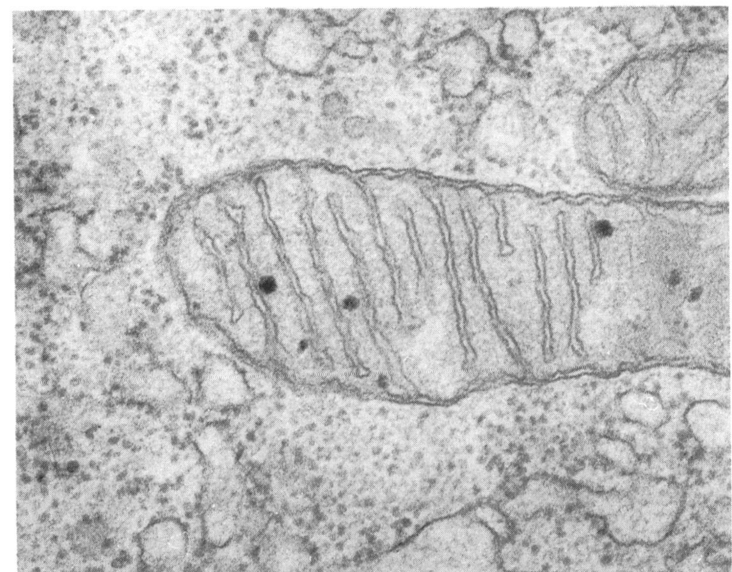

B

The *elementary particles* that contain the ATPase are readily visualized by the electron microscope, using the technique of negative staining at high magnification.

Electron micrograph of beef heart mitochondrion embedded in a thin phosphotungstate layer.

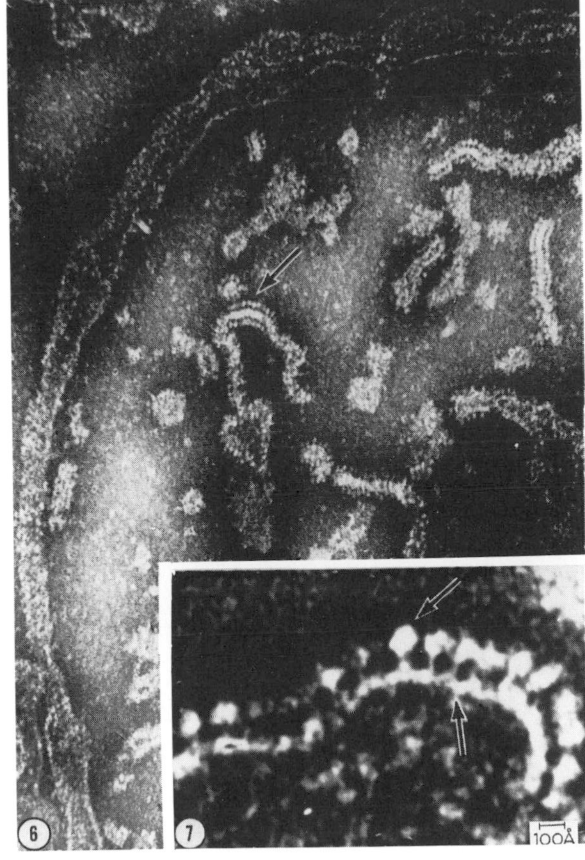

Morphology of a mitochondrial respiratory system.

From purification experiments it is now possible to build up a model for the arrangement of the respiratory system in the mitochondrial inner membrane, although the arrangement shown is highly conjectural. The diagram shows the reaction of NADH with NADH dehydrogenase, and the reaction of ADP and P_i with the ATPase complex resulting in ATP formation. Both of these reactions occur on the matrix face of the membrane. In addition to cytochromes b, c_1, c and a/a_3, proteins containing iron and sulphur and copper-containing proteins are shown. The protons from the hydrogens released by reaction of NADH with its dehydrogenase are thought to be pumped into the intermembrane space. In addition, there are other proton pumps, which are driven by the transport of electrons, one in the b/c_1 region and another at cytochrome a/a_3. The succinate dehydrogenase complex does not function as a proton pump. Thus, when succinate is oxidized, only the latter two pumps operate, whereas during oxidation of NADH these pumps in addition to NADH-coenzyme Q reductase pump electrons into the intermembrane space. As explained on the following page, this pumping of electrons creates a protonic gradient that is capable of driving the ATPase to form ATP. The potential of the gradient is proportional to the number of protons pumped, so that more ATP is synthesized as a result of NADH oxidation than when succinate is oxidized. The oxidation of one molecule of NADH results in the formation of about three molecules of ATP for each oxygen atom reduced by two electrons (*P/O ratio* of 3), whilst for succinate the yield is about two molecules of ATP (P/O ratio of 2).

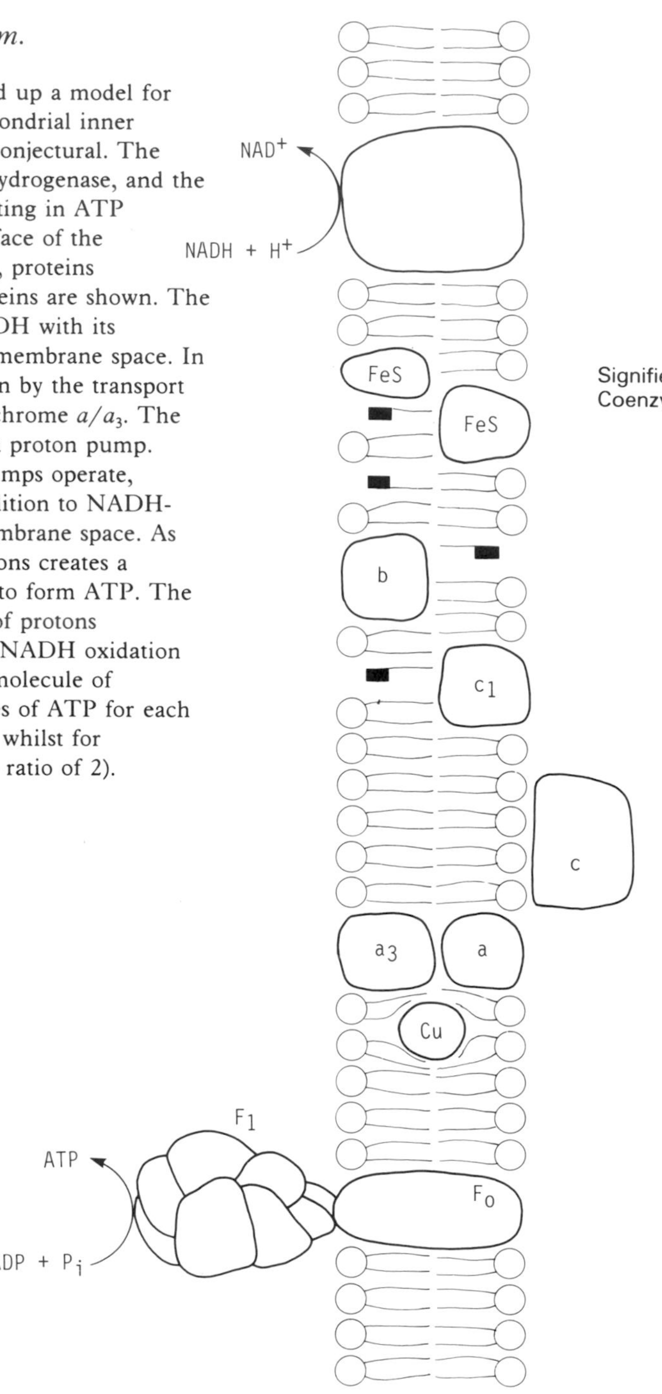

A

The way in which the free energy released by electron flow drives ADP phosphorylation is unknown, but there are several theories which attempt to explain it.

One of these is termed the *chemiosmotic theory.*

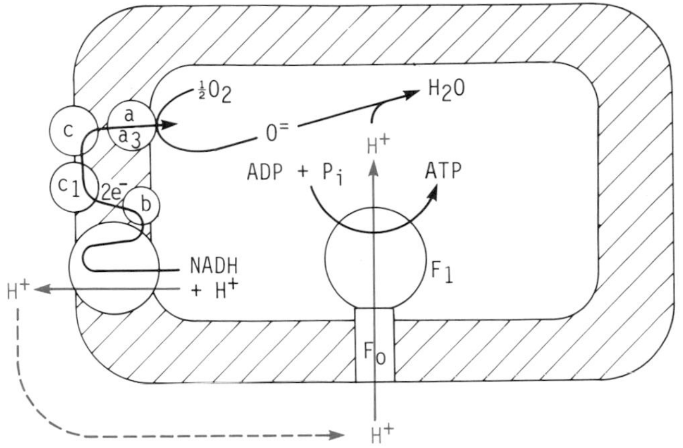

The chemiosmotic theory

1. A reversible ATPase* synthesizes ATP from ADP & P_i.
2. If, during electron transport, protons are pumped to one face of a membrane, whilst electrons pass to the other face, a protonic potential difference will be built up across the membrane.
3. The force resulting from the protons flowing back through the ATPase drives it in the direction of ATP synthesis.
4. The electrons react at cytochrome a—a_3 with oxygen, and reduce the oxygen to a form that then combines with the protons to form water.

* The enzyme responsible for catalysing the formation of ATP during oxidative phosphorylation has been termed an ATPase (i.e. the reverse reaction) because it is convenient to measure its activity by determining the phosphate formed after the hydrolysis of ATP.

B

As stated on page 175 some compounds *uncouple* oxidative phosphorylation from electron flow.

In this condition electrons flow much faster than in the coupled state, and little or no ADP is phosphorylated to ATP.

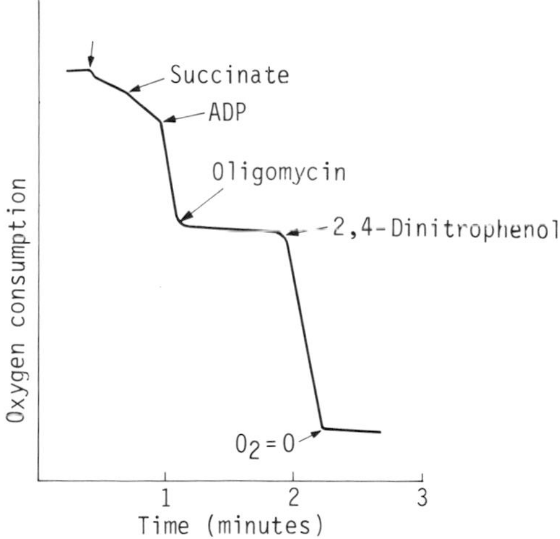

The trace from an oxygen electrode during oxidation of succinate by liver mitochondria shows the increased rate on adding ADP and arrest of oxidation by oligomycin (total inhibition of F1 protein in *tightly coupled* mitochondria stops electron flow). Then addition of 2, 4-dinitrophenol uncouples and allows rapid free flow of electrons. Tight coupling implies that electrons can only flow as ADP is phosphorylated to ATP, so that if this is totally inhibited by *oligomycin*, electron flow ceases. It is possible that 2, 4-dinitrophenol uncouples by making the membrane leaky to protons by acting as a membrane-permeable proton carrier.

The oxidation of one mole of glucose by the glycolytic pathway and the citric acid cycle generates 38 moles of ATP, or the ATP equivalent, GTP.

The electron transport chain has been simplified in this diagram.

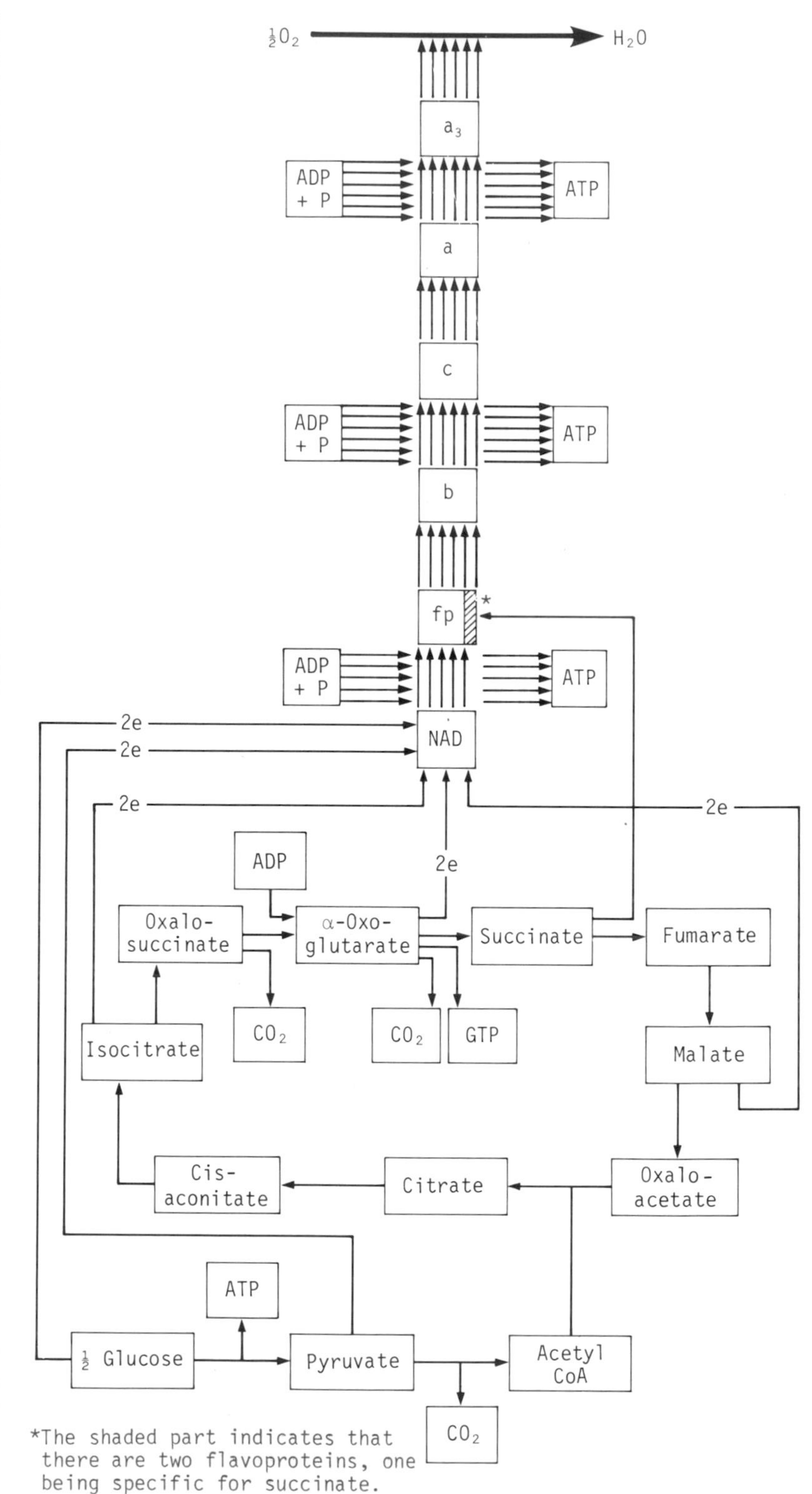

*The shaded part indicates that there are two flavoproteins, one being specific for succinate.

A

Translocases in the inner mitochondrial membrane

A number of translocases exist in the inner mitochondrial membrane which transport a molecule into or out of the mitochondrion whilst simultaneously carrying another molecule in the opposite direction (*anti-port*; see p. 255).

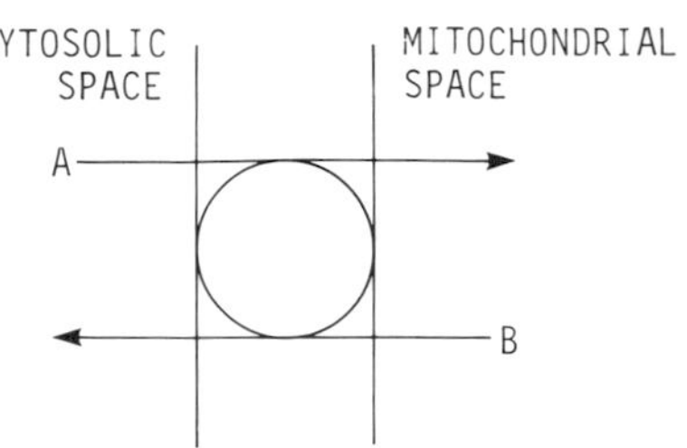

Pairs of compounds undergoing antiport translocation

A	B
Pyruvate	OH^-
Phosphate	Malate
Citrate	Malate
Phosphate	OH^-
ADP	ATP
Aspartate	Glutamate
Malate	2-oxoglutarate

B

Transport of hydrogen from the cytosol into the mitochondrion can be mediated by glycerol 3-phosphate interacting with a dehydrogenase in the inner mitochondrial membrane.

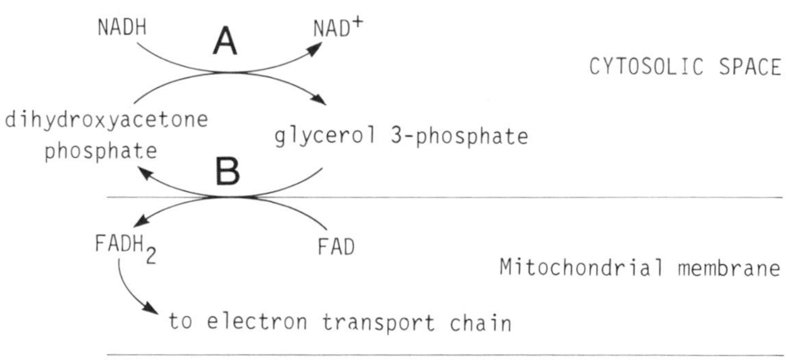

A = Soluble glycerol 3-phosphate dehydrogenase
B = membrane-bound glycerol 3-phosphate dehydrogenase

Hydrogen is first transferred from NADH in the cytosol to dihydroxyacetone phosphate forming glycerol 3-phosphate by the soluble cytosolic glycerol 3-phosphate dehydrogenase. When the glycerol 3-phosphate is oxidized back to dihydroxyacetone by the mitochondrial membrane-bound enzyme, electrons enter the electron transport chain at cytochrome b via coenzyme Q, yielding only 2 molecules of ATP for each electron.

The overall reaction is thus the passage of hydrogens from the cytosolic NADH to the intra-mitochondrial electron transport chain.

Mitochondrial regulation of cell function

Over short time intervals, the components of each of the following coenzyme systems sum to a constant total

1. ATP + ADP + AMP
2. NADH + NAD^+
3. NADPH + $NADP^+$

If one component of any of these three systems increases, it does so at the expense of the other component(s) of the system, i.e. if NADH increases, NAD^+ decreases, or if ATP increases, ADP or AMP (or both) will decrease. If this happens, the ratio NADH/NAD^+ or ATP/ADP will change.*

The rate of many different enzymes can be regulated by the change in the ratios ATP/ADP, NADH/NAD^+ and NADPH/$NADP^+$. Allosteric effects, and other effects on enzyme mechanisms, are involved in this regulation. The effects often originate in the mitochondrion, where respiratory control determines the relative concentrations of these coenzyme systems, and this can regulate citric acid cycle activity.

On the opposite page, there are some illustrations of the way that control of the citric acid cycle, in different situations in various types of tissue, influences the direction of metabolic flux.

* Note that the total amount of these nucleotides in the cell can change, over the course of hours, by changes in the rate that they are synthesized, or the rate at which they are broken down or utilized, for example for DNA synthesis. The interconversions referred to above occur on a much more rapid time scale, over a period of minutes.

Situation A: Liver during fasting
Citrate synthase activity is reduced so that citrate is not formed from oxaloacetate and acetyl CoA. This is important, as it is necessary to prevent the oxidation of these compounds. The oxaloacetate is destined for glucose synthesis, and the acetyl CoA for ketone body synthesis.

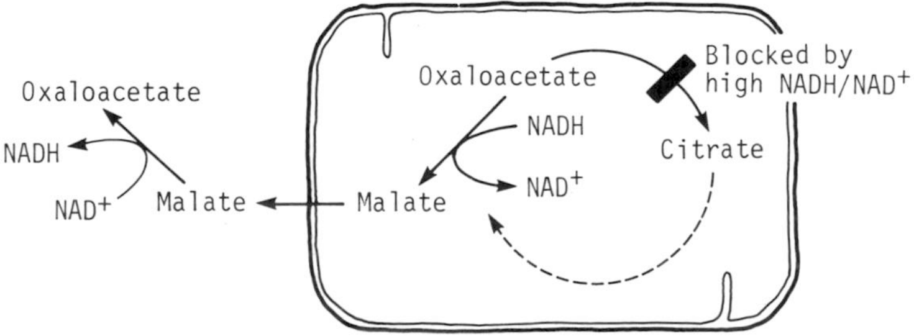

NOTE : Within mitochondrion $\frac{NADH}{NAD^+}$ is in range 0.1 → 0.2

In cytosol $\frac{NADH}{NAD^+}$ is approx. 0.001

Thus malate dehydrogenase can work in opposite directions in these two compartments.

Situation B: Liver after a meal
Citrate synthesis is active, but excess citrate is exported into the cytosol for fat synthesis.

Situation C: Brown adipose tissue
This tissue is important in thermoregulation. Rapid oxidation that does not lead to work or other forms of useful energy leads to the production of heat. If there is some means of removing respiratory control (such as by the constant hydrolysis of ATP) the effect will be similar to the uncoupling of oxidative phosphorylation; there will be unrestricted flow of electrons and oxidation of NADH, and a high rate of oxidative metabolism generating heat.

Situation D: Heart muscle
In cardiac muscle, constant contraction utilizes ATP, converting it to ADP. There are large numbers of mitochondria in heart muscle, ensuring that oxidation by the heart is essentially always aerobic.

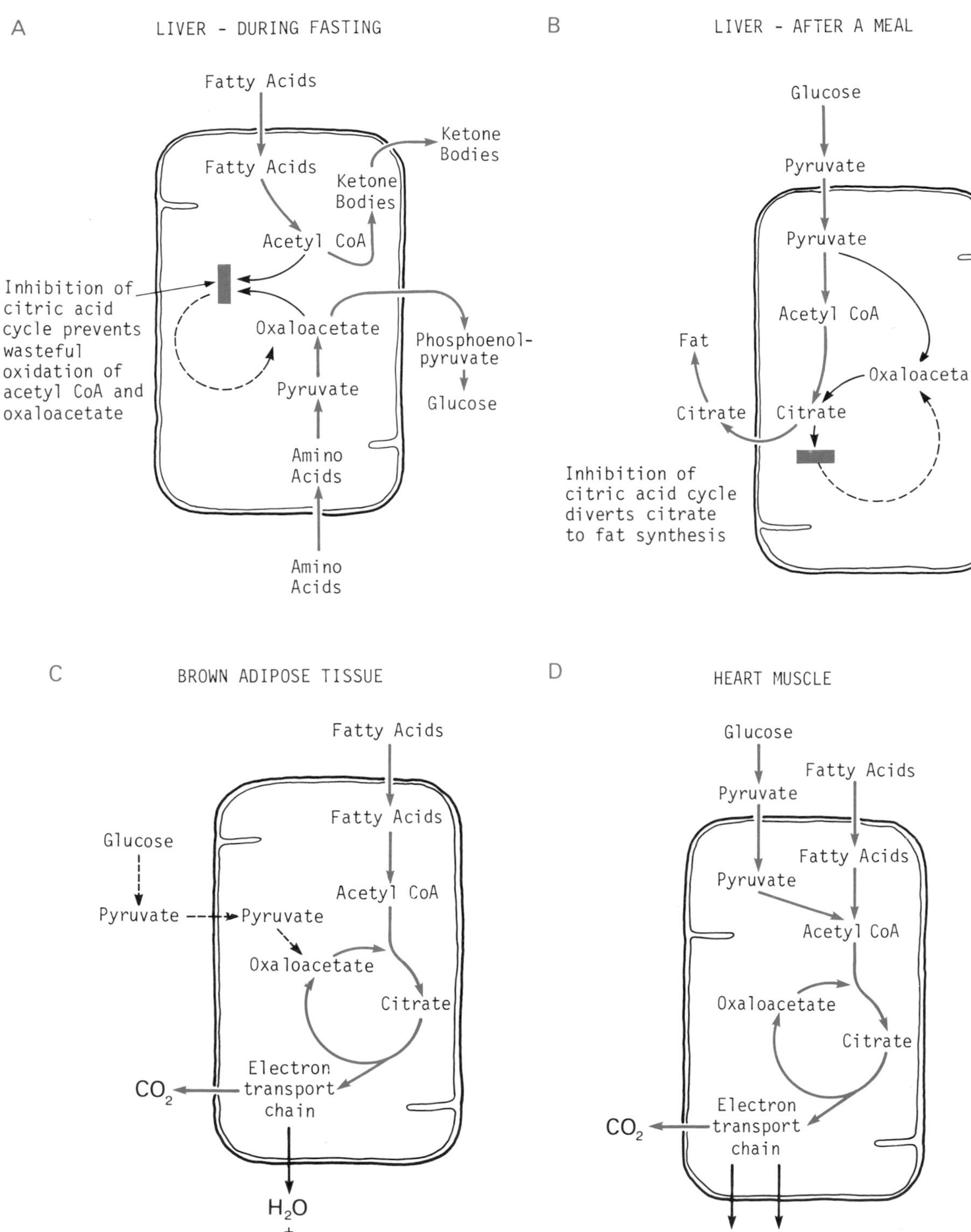
A
LIVER - DURING FASTING
Fatty Acids
Fatty Acids
Ketone Bodies
Ketone Bodies
Acetyl CoA
Inhibition of citric acid cycle prevents wasteful oxidation of acetyl CoA and oxaloacetate
Oxaloacetate
Phosphoenol-pyruvate
Pyruvate
Glucose
Amino Acids
Amino Acids
B
LIVER - AFTER A MEAL
Glucose
Pyruvate
Pyruvate
Acetyl CoA
Fat
Oxaloacetate
Citrate
Citrate
Inhibition of citric acid cycle diverts citrate to fat synthesis
C
BROWN ADIPOSE TISSUE
Fatty Acids
Fatty Acids
Glucose
Acetyl CoA
Pyruvate
Pyruvate
Oxaloacetate
Citrate
Electron transport chain
CO_2
H_2O
+
HEAT
D
HEART MUSCLE
Glucose
Fatty Acids
Pyruvate
Fatty Acids
Pyruvate
Acetyl CoA
Oxaloacetate
Citrate
Electron transport chain
CO_2
H_2O ATP

A

The glucose/fatty acid cycle.

A reciprocal relationship exists between the rates of oxidation of glucose and fatty acids by muscle. This has important consequences when liver glycogen stores are depleted, and fatty acid oxidation can spare the use of glucose.

The mechanism by which fatty acids inhibit utilization of glucose involves a number of enzymes, in particular pyruvate dehydrogenase and phosphofructokinase.

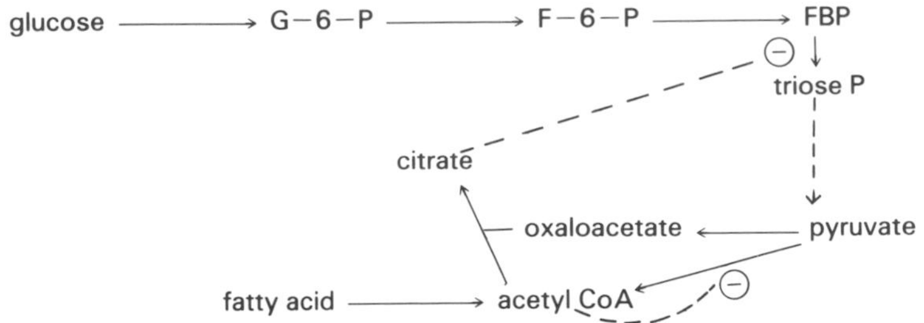

The most direct action of fatty acid oxidation is that it increases the level of acetyl CoA, and in the process utilizes CoA. An increase in the acetyl CoA/CoA ratio strongly inhibits pyruvate dehydrogenase. More indirectly, levels of citrate are elevated, leading to inhibition of phosphofructokinase.

B

Pyruvate dehydrogenase is regulated by the ratios acetyl CoA/CoA, NADH/NAD^+ and ATP/ADP. The control is exerted through a kinase/phosphatase enzyme system (see p. 193).
CoA, NAD^+ and ADP activate the enzyme, ATP, NADH and acetyl CoA inhibit

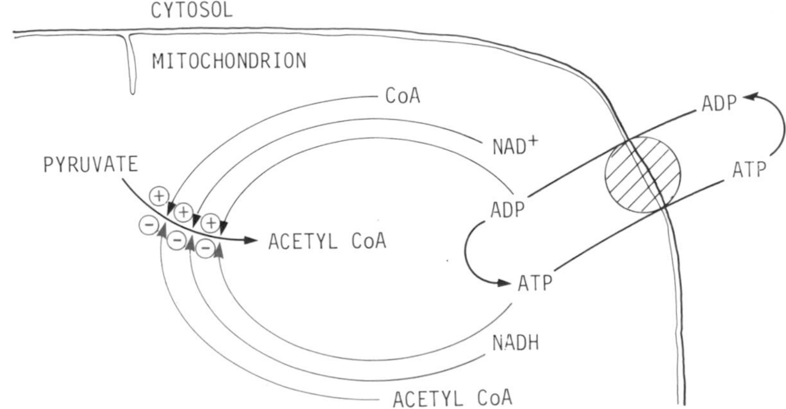

A translocase for ATP and ADP is involved in regulating the ATP/ADP ratio.

A

Regulation of the citric acid cycle

Citrate synthase, isocitrate dehydrogenase and oxoglutarate dehydrogenase are non-equilibrium reactions.

Both isocitrate dehydrogenase and oxoglutarate dehydrogenase are activated by Ca^{2+}, which may be of importance in muscle in relation to contraction stimulated by Ca^{2+}.

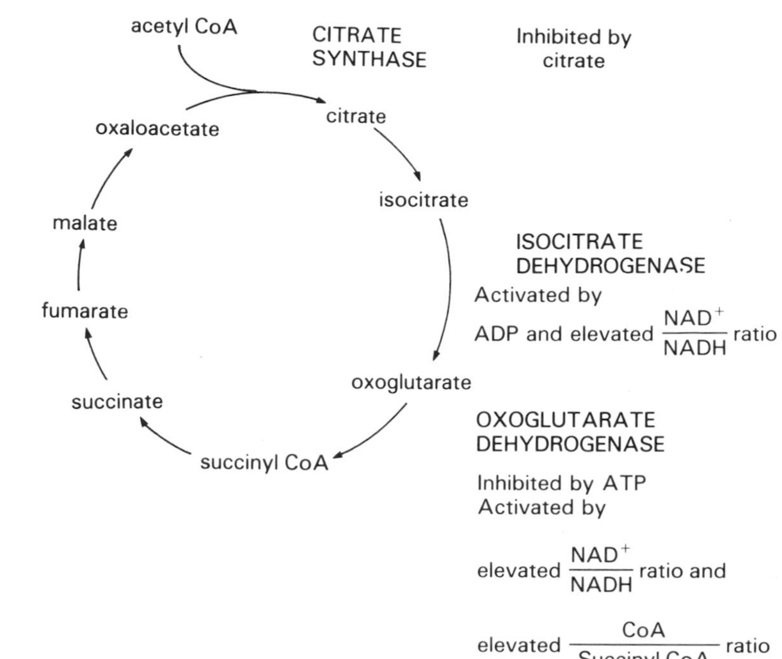

A

Oxidation in the deficiency or absence of electron transport can continue as a result of the action of lactate dehydrogenase.

The red blood cell lacks mitochondria and derives its energy entirely from glycolysis. This requires the continual oxidation of NADH back to NAD^+, otherwise the enzyme glyceraldehyde phosphate dehydrogenase would cease to function (see p. 166). This can be achieved by the enzyme lactate dehydrogenase, which catalyses the reaction:

$$\text{lactate} + NAD^+ \rightleftarrows \text{pyruvate} + NADH + H^+$$

Thus the pyruvate formed by the glycolytic enzymes is converted to lactate which is secreted into the blood.

A similar sequence of reactions occurs in muscle when the provision of oxygen is insufficient to oxidize all of the NADH produced by glycolysis, and in other tissues when they are deprived of oxygen. This is the cause of elevated blood levels of lactate during severe exercise,* or as a result of tissue anoxia of pathological origin, e.g. after cardiac infarction.

The reactions involved can be summarized as follows:

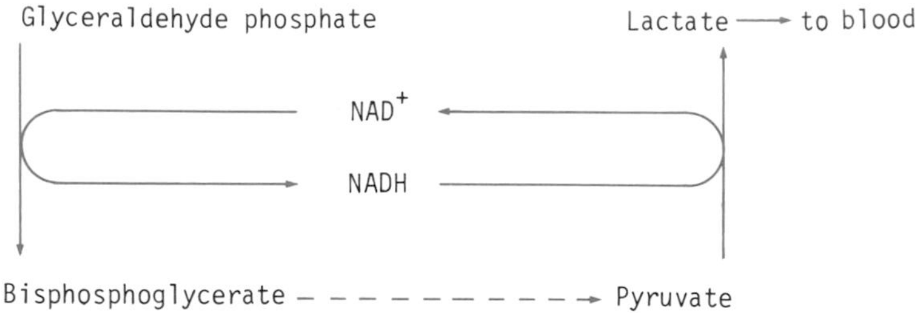

* Note that a sprinter is actually using about 10 times as much oxygen during a sprint as he does whilst at rest. The lactate is produced as a result of the need to metabolize yet more substrate than even this amount of oxygen can provide for.

B

Specialization of metabolism in muscle

The fibres of a muscle can be classified into types according to their content of different enzymes, and this is related to their function.

The muscle fibres can be classed as Type I or Type II fibres according to their content of certain enzymes, such as the enzymes of the glycolytic pathway. Fibres containing high levels of these enzymes are classed as Type II fibres, or fast twitch fibres. These fibres contain relatively few mitochondria. Other fibres that contain lower levels of glycolytic enzymes, but higher levels of citric acid cycle enzymes and cytochromes are classed as slow twitch or Type I fibres. Formerly, muscles containing large numbers of Type I or Type II fibres were referred to as red and white muscles respectively (red muscles containing high amounts of myoglobin and cytochromes). Type II fibres can contract very rapidly, but for short periods of time only (lobster abdominal muscle, responsible for the tail flick escape reaction, is an example). Type I fibres on the other hand are capable of more sustained, slower contraction.

8

CARBOHYDRATE AND FAT METABOLISM

B. Regulation in the whole body and in the cell

A

Regulation of metabolism at the whole-body level (the fed and fasting states)

After a meal, food which is surplus to immediate energy requirements is converted to glycogen and fat.

In this book we call this state *the fed state* of metabolism.

The fed state of metabolism operates whilst food is being absorbed from the intestine, and is more correctly termed the *absorptive state*.

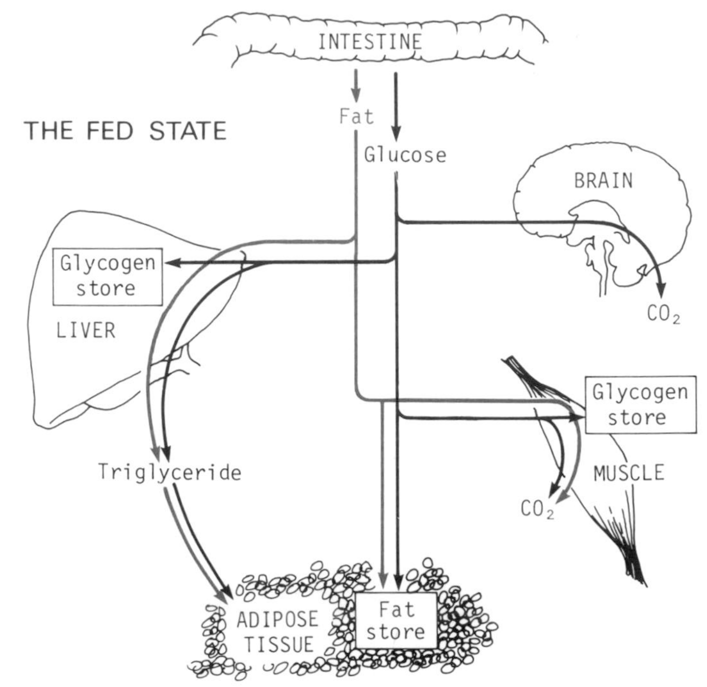

Carbohydrate and fat are oxidized to CO_2 and H_2O in peripheral tissues to drive synthetic reactions, and sustain cell function. Nutrients surplus to immediate requirements are laid down as fat.

B

During a fast, fat is mobilized and is the main substrate for oxidation in peripheral tissues, either as fatty acid or after conversion in liver to ketone bodies. This means the body needs to synthesize only minimum quantities of glucose from precious muscle amino acids, which in this state are the only long-term source of precursors for blood glucose synthesis.

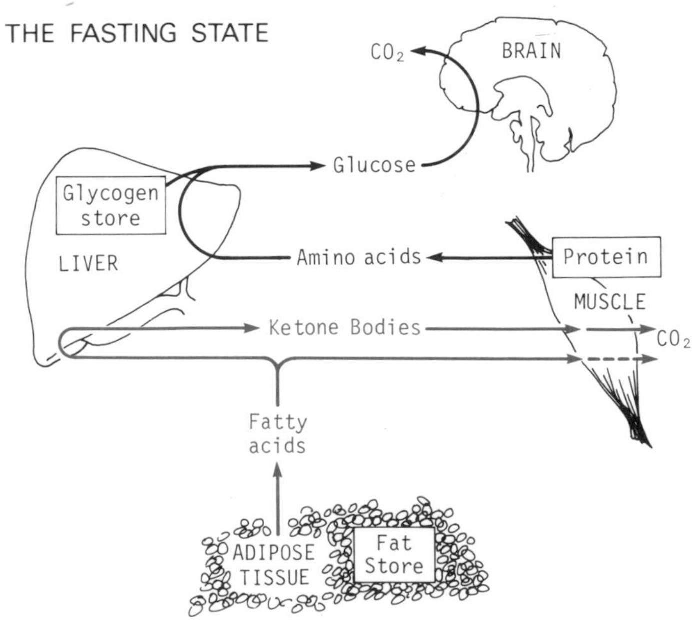

A

Hormones that regulate carbohydrate and fat metabolism

In the fed state, insulin is secreted, and promotes the synthesis of glycogen and triacylglycerol from glucose. It inhibits the release of fatty acids from adipose tissue triglyceride.

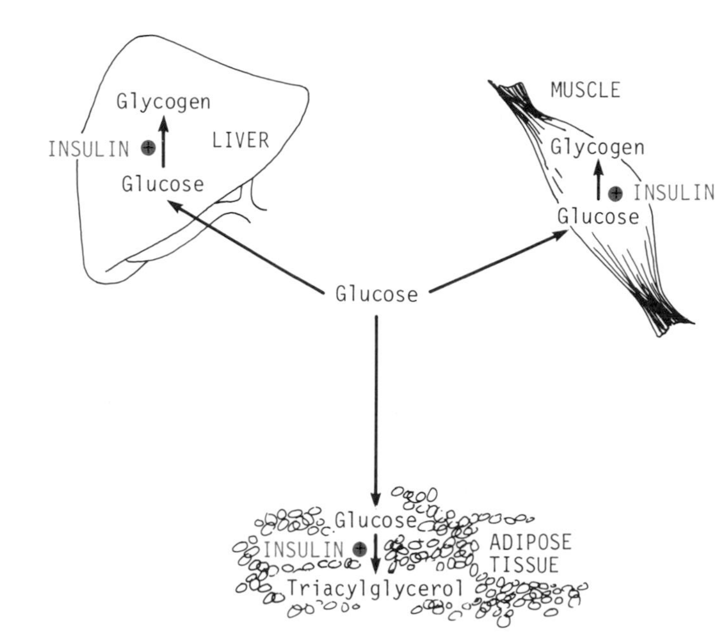

B

In the fasting state:

1. Glucocorticoids release amino acids from muscle. In liver, gluconeogenic enzymes act on these amino acids to convert them to glucose;
2. Glucagon and adrenaline activate enzymes which release glucose from glycogen;
3. Fat mobilization is the result of the action of *hormone-sensitive lipase* on triacylglycerol in adipose tissue to bring about the release of fatty acids, which are converted to ketone bodies in liver. The lipase is activated by a cAMP-dependent protein kinase system (see for example, p. 191).

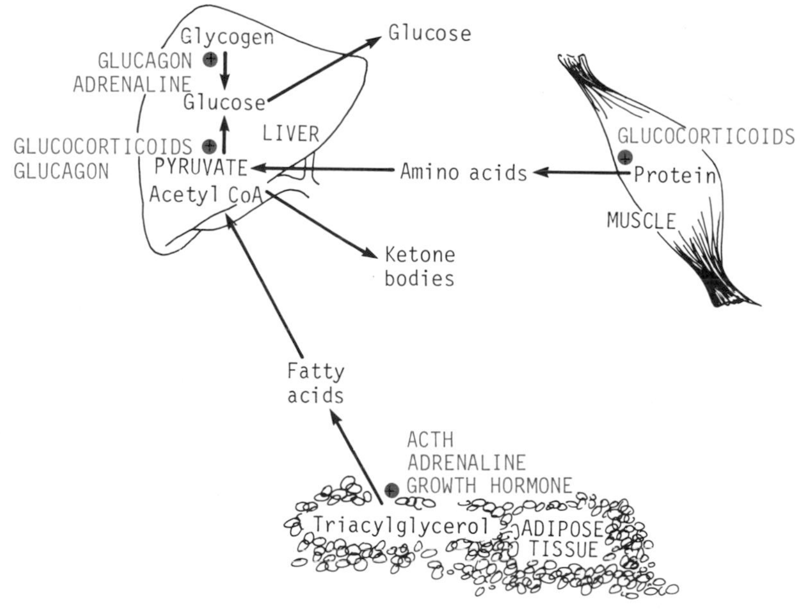

A

Many hormones act by stimulating an enzyme named adenylate cyclase.

Hormone	**Tissues**
Adrenaline	Muscle, liver, adipose tissue
Glucagon	Liver, adipose tissue
TSH	Thyroid
ACTH	Adrenals, adipose tissue

Appropriate receptors for the hormone must be present on the surface of the cell. In the absence of a specific receptor the hormone will not activate adenylate cyclase. TSH is thyroid stimulating hormone.

B

Cyclic AMP

Regulation of metabolism is also brought about by a fundamentally important and generally widespread system involving a compound called *cyclic AMP*, and a system of enzymes called *protein kinases*.

The enzyme, *adenylate cyclase*, converts ATP to adenosine 3′5′ cyclic monophosphate (cyclic AMP or cAMP) see page 259. *Phosphodiesterase* hydrolyses the bond between the phosphate and the 3′ hydroxyl of the ribose. Between them, adenylate cyclase and phosphodiesterase thus regulate the concentration of cyclic AMP.

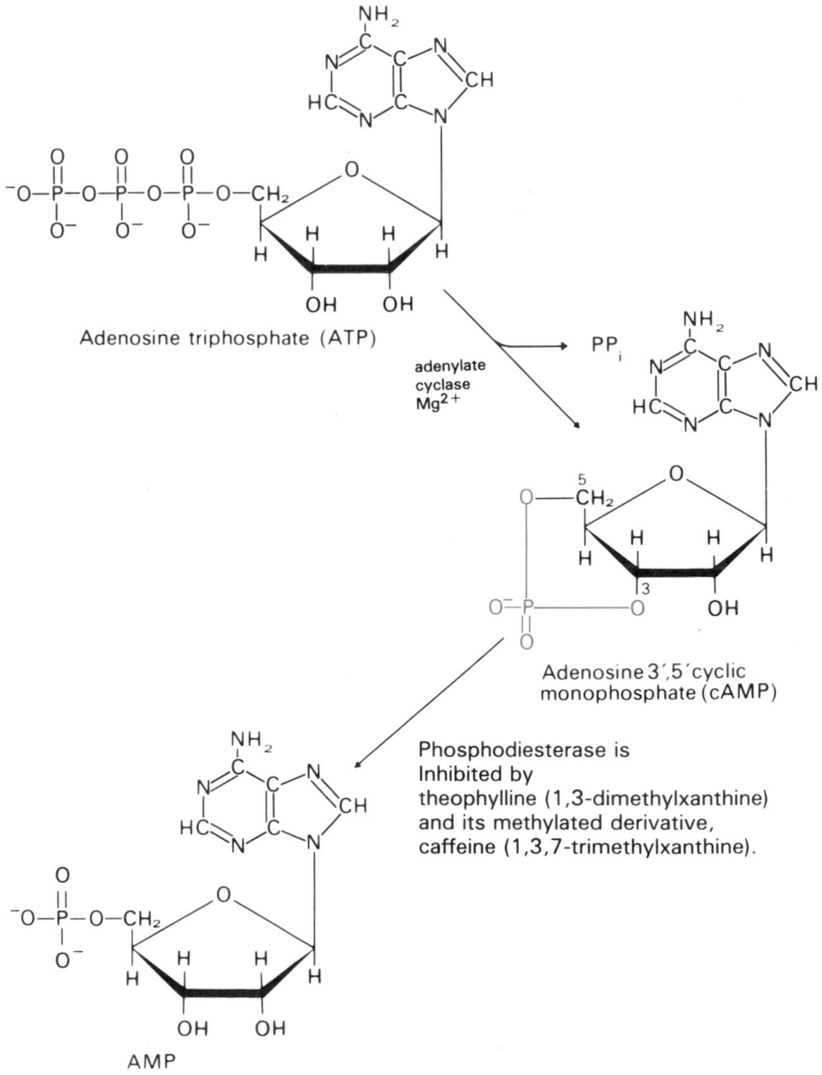

Adenosine triphosphate (ATP)

adenylate cyclase Mg^{2+}

PP_i

Adenosine 3′,5′cyclic monophosphate (cAMP)

Phosphodiesterase is Inhibited by theophylline (1,3-dimethylxanthine) and its methylated derivative, caffeine (1,3,7-trimethylxanthine).

AMP

A

Protein kinases

Protein kinases are mediators of cAMP action

Protein kinases bring about the phosphorylation of proteins. Thus the protein undergoes modification through a covalent bond addition. This modification activates some enzyme proteins. In the case of other proteins, however, such a phosphorylation reduces the enzyme activity.

Two kinases amplify the action of cyclic AMP manyfold. *Phosphorylase phosphatase* removes the phosphates and inactivates phosphorylase. Thus the balance of activity between *phosphorylase b kinase* and phosphorylase phosphatase regulates the amount of active phosphorylase.

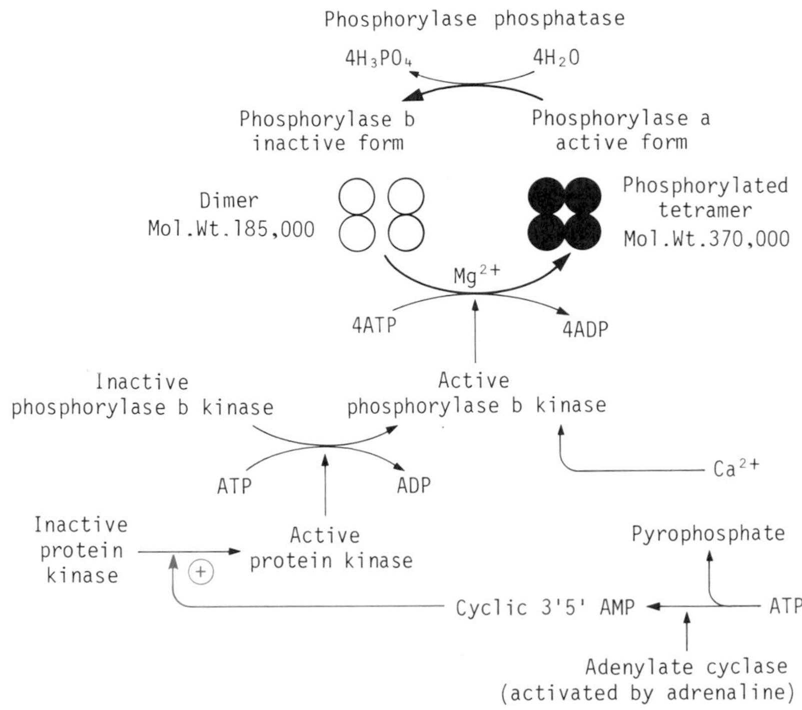

The system of protein kinase activity most thoroughly studied is that involving phosphorylase, the enzyme that degrades glycogen. The inactive form of phosphorylase is a dimer, termed phosphorylase b. Phosphorylase b is activated by the addition of four phosphate groups, donated by ATP, a reaction brought about by phosphorylase b kinase. The addition of the phosphates causes the dimers to associate to form a tetramer. Phosphorylase b kinase itself exists in an inactive, non-phosphorylated form, and an *active*, phosphorylated form, a conversion brought about by another protein kinase that is allosterically activated by cyclic AMP.

B

Serine and tyrosine kinases.

In the case of kinases involved in regulation of pathways of intermediary metabolism the phosphate is esterified to hydroxyls of serine or threonine residues. Other protein kinases, involved in growth regulation, add phosphate to the phenolic group of tyrosine.

OH
|
CH₂
|
−HN−CH−CO−
Seryl residue

→ Protein kinase →

O⁻
|
⁻O−P=O
|
O
|
CH₂
|
−HN−CH−CO−
Serylphosphate residue

OH
|
(benzene ring)
|
CH₂
|
−HN−CH−CO−
Tyrosyl residue

→ Protein kinase →

O⁻
|
⁻O−P=O
|
O
|
(benzene ring)
|
CH₂
|
−HN−CH−CO−
Tyrosylphosphate residue

A

Simultaneous control of glycogen synthesis and degradation by cAMP.

Cyclic AMP activates kinases that phosphorylate enzymes which may be in pathways that have opposing actions. However, phosphorylation may activate some enzymes, and inactivate others.

Note: The D form of *glycogen synthase* is dependent on the presence of glucose 6-phosphate for full activity, whilst the I form is not (i.e. it is the independent form).

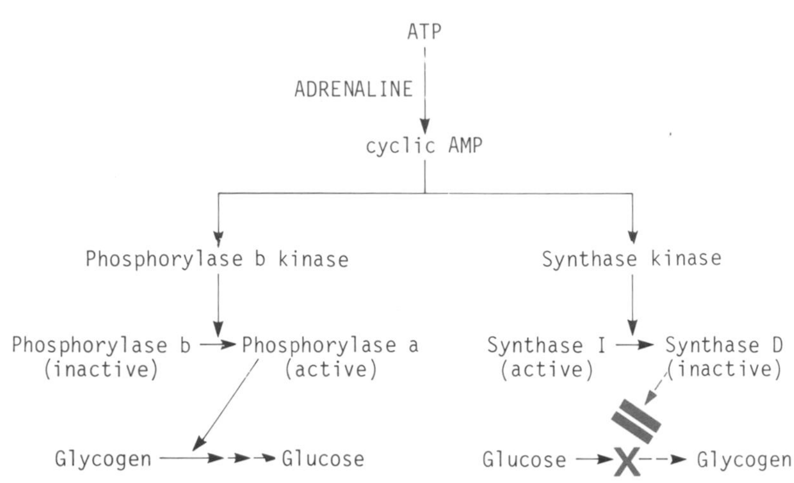

Glycogen synthase, the enzyme involved in the regulation of glycogen synthesis, is also phosphorylated by a kinase. In this case the phosphorylated enzyme is less inactive. Thus, by activating both phosphorylase b kinase and glycogen synthase kinase, cyclic AMP activates glycogen breakdown and inhibits glycogen synthesis.

B

Activities of the enzymes of glycolysis can be assayed in the homogenates of the disrupted cells of various tissues.

Phosphofructokinase is rate-limiting in all tissues.

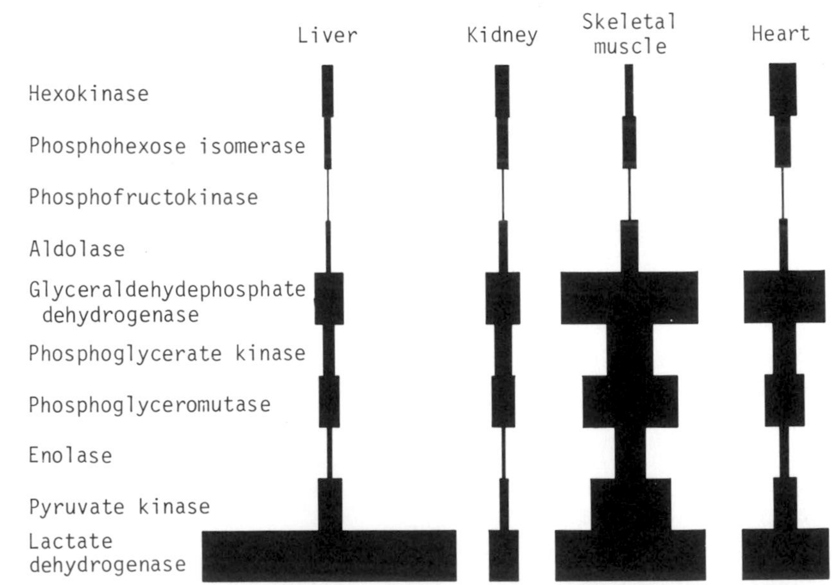

Enzyme activity is proportional to the width of the solid bars. For reference the activity of kidney aldolase was 2μmol/min per g fresh weight of tissue.

A

Pyruvate dehydrogenase is regulated by the action of a protein kinase and a phosphatase.

The protein kinase is an example of the class of protein kinases that are not cyclic-AMP-dependent.

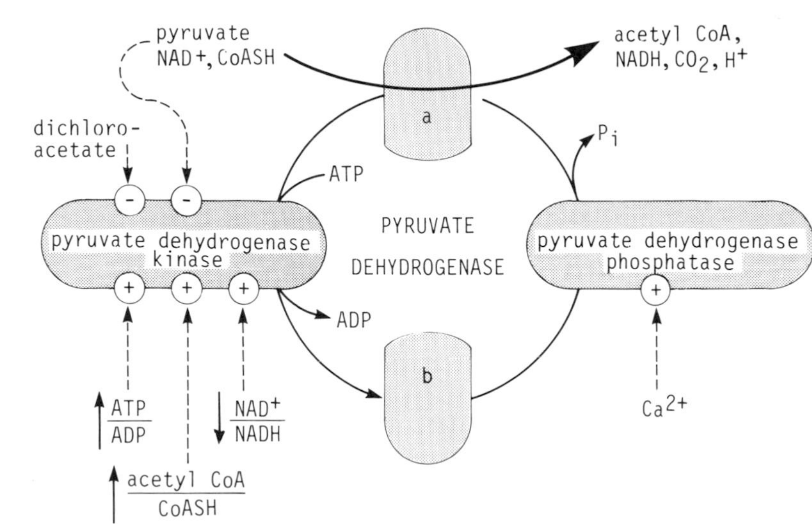

Pyruvate dehydrogenase is regulated by a kinase, that phosphorylates the enzyme, thereby inactivating it, and a phosphatase, that hydrolyses the phosphates and converts the enzyme to the non-phosphorylated, active form.

The kinase is activated by high ratios of ATP/ADP and acetyl CoA/CoA, and a decreased NAD^+/NADH ratio. Pyruvate inhibits the kinase. Activation of this kinase leads to inhibition of pyruvate dehydrogenase (see p. 184).

The compound dichloroacetate has been used as a drug to alleviate lactic acidosis. Its mechanism of action appears to include inhibition of the kinase which phosphorylates pyruvate dehydrogenase. As phosphorylation inhibits pyruvate dehydrogenase, the action of dichloroacetate is thus to accelerate conversion of pyruvate to acetyl CoA, thereby reducing formation of lactate.

A

Hormone receptors.

The site of action of many hormones is at the cell surface, but hormones may also penetrate into the cell and bind to receptors in the cytosol.

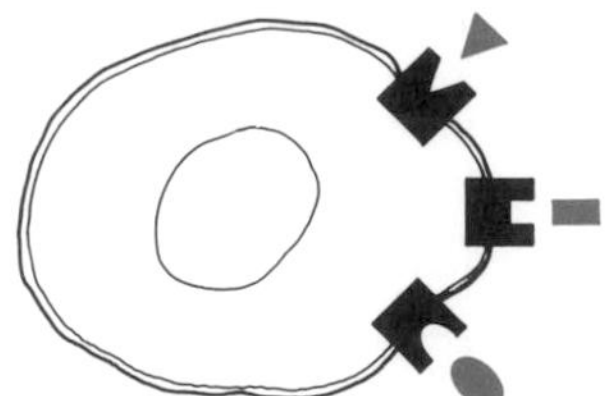

Specific receptors for peptide hormones, and adrenaline, are present on the cell surface.

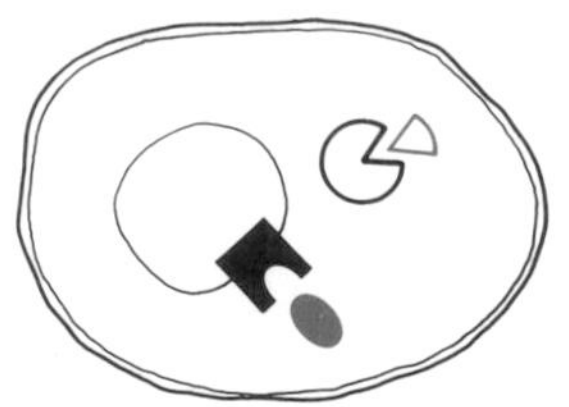

Steroid hormones have no receptors on the cell surface, but there are receptors in the cytosol.

Some peptide hormones are found within the cell. Many peptide hormones act by stimulating adenylate cyclase, but for others there is no evidence of this and they may be found in the cytosol. These and the steroid hormones are thought to interact with the nucleus after combination between hormone and receptor. Receptors which bind insulin have been found, for example, in the nuclear membrane as well as the plasma membrane.

B

Hormones may exercise control by regulating the cytosolic Ca^{2+} concentration.

Some hormonal effects are thought to be mediated by changes in the cytosolic Ca^{2+} concentration resulting from the hormone action. Adrenaline promotes glycogenolysis in liver in the presence of agents that block its action on β-receptors, as also do specific agonists that act via the α-adrenergic receptors, with no effect on the adenylate cyclase system. Vasopressin and oxytocin also stimulate glycogenolysis without an accompanying effect on adenylate cyclase. These effects have given rise to the hypothesis of an alternative mechanism involving Ca^{2+} (see p. 266).

8

CARBOHYDRATE AND FAT METABOLISM

C. Utilization of energy stores

A

Details of metabolism in the fasting state

The pathways of major importance in the fasting state are *glycogenolysis, gluconeogenesis, ketogenesis* and ketone body utilization.

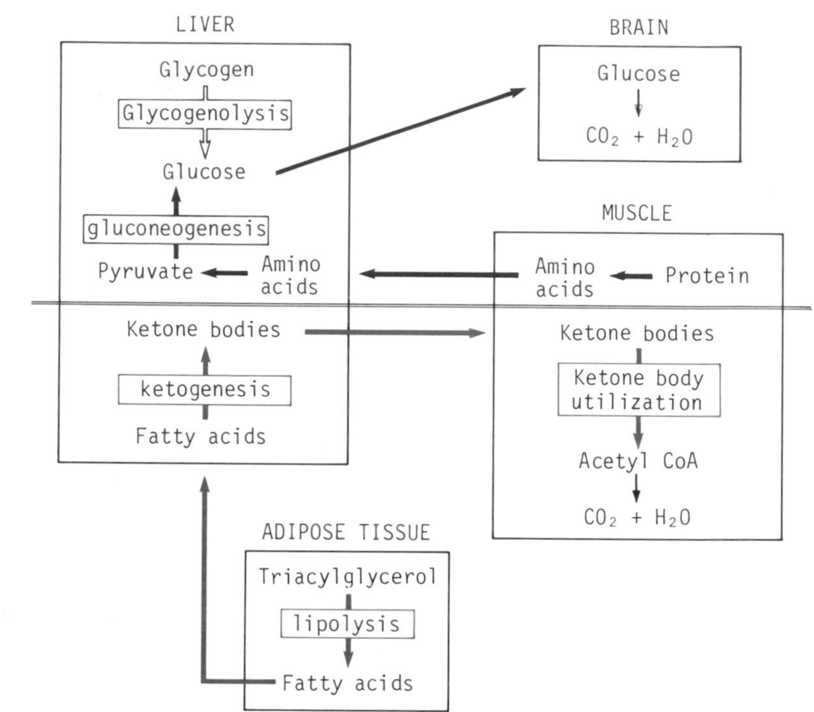

The system can be understood more easily by segregating it into the two main systems involved. Thus, the black arrows relate to the system for converting muscle protein to blood glucose, for use mainly by nervous tissue. The red arrows relate to the system for converting adipose tissue triglyceride to ketone bodies, used in place of glucose by tissues other than nervous tissue.

B

Glycogen and its degradation

Glycogen is a branched polymer of glucose. The main chain consists of α 1,4 links between the glucose units. The branches are formed by α 1,6 links, at approximately every 12th glucose unit.

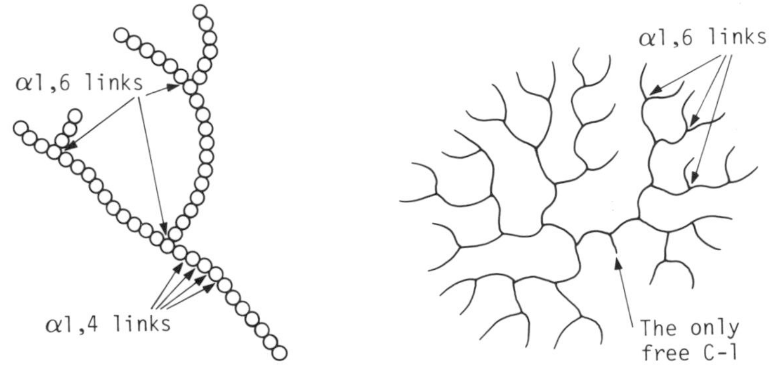

The glycogen chain is synthesized (see p. 211) by linking an additional glucose by its C-1 to the C-4 of an existing chain of glucose units, in other words the chain grows away from a single glucose, which is the only glucose with a free C-1 group. Thus glycogen gives a poor colour reaction with reagents that react only with free C-1 reducing groups.

A

The formation of glucose from glycogen (glycogenolysis)

The diagram shows the detail of the *α* 1,4 and *α* 1,6 links in the glycogen molecule.

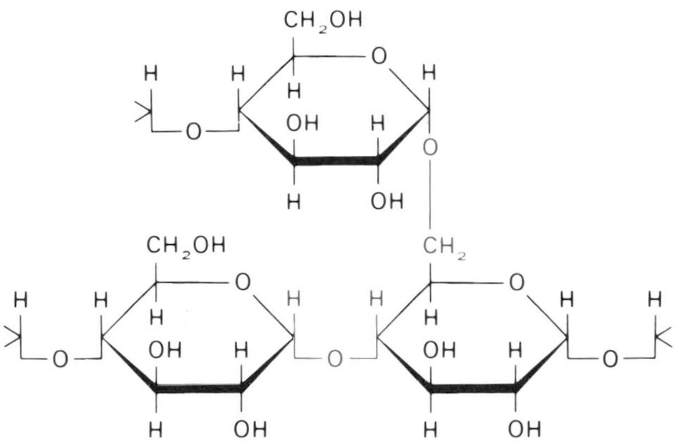

B

When glycogen is degraded, terminal *α* 1,4 links are broken by the enzyme phosphorylase.

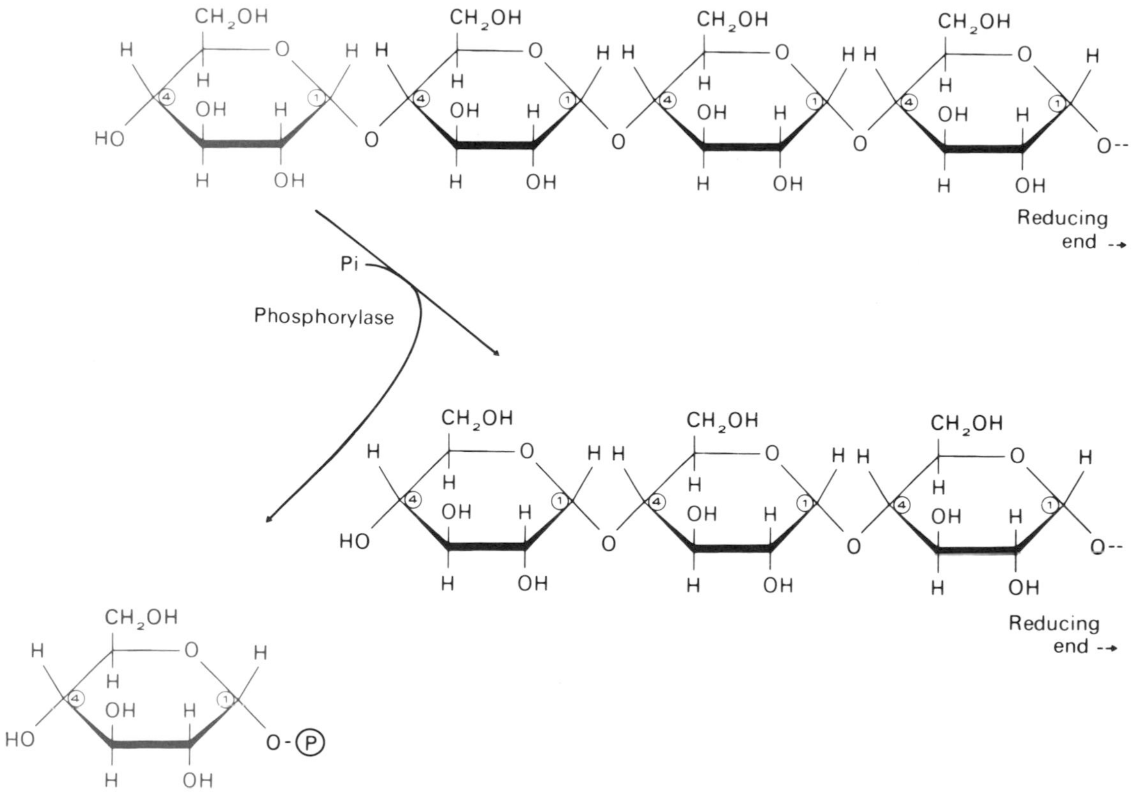

Pi

Phosphorylase

Reducing end →

Reducing end →

Note: (1) The product is glucose 1-phosphate
(2) The phosphate is derived from inorganic phosphate, *not* ATP.

α1,6 links must be broken by other enzymes before phosphorylase can pass the branch point.

A

Degradation of glycogen to free glucose involves four enzymes:

1. Phosphorylase
2. 'Debranching enzyme'
3. Phosphoglucomutase
4. Glucose 6-phosphatase

'Debranching enzyme' releases free glucose.

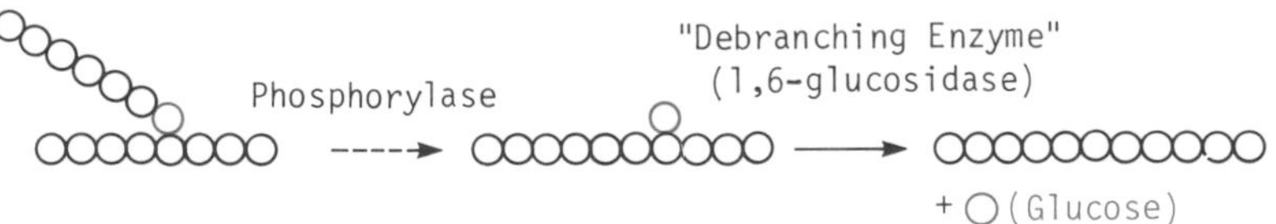

Phosphorylase releases G-1-P. This must be converted to G-6-P which is then hydrolysed to glucose.

Glucose 1-phosphate → Glucose 6-phosphate → Glucose

Phosphoglucomutase   Glucose 6-phosphatase

Glucose 6-phosphatase is present in liver. It is not present in muscle. Thus liver glycogen can act as a source of blood glucose, but muscle glycogen does not directly provide blood glucose (but see page 201A).

B

The de novo formation of glucose (gluconeogenesis)

The first step in the conversion of muscle amino acids to glucose is the removal of the amino group.

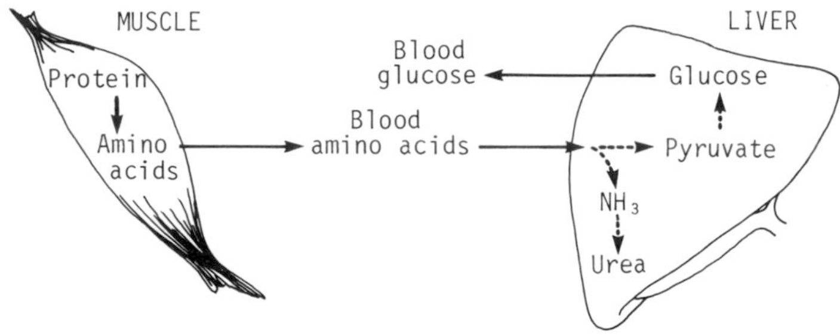

Phosphorylase can be activated within seconds by hormones such as adrenaline and glucagon, and possibly by nervous impulses, to meet the needs of the body for the immediate supply of glucose to the blood. However, the liver glycogen stores are sufficient to provide glucose for a few hours only. Thereafter, blood glucose can only be replenished by the conversion of muscle protein to glucose, after transport to the liver as amino acid.

A

The amino group is converted to urea by the *urea cycle*, which has increased activity during gluconeogenesis, giving an elevated urinary concentration of urea indicative of the protein breakdown that is occurring.

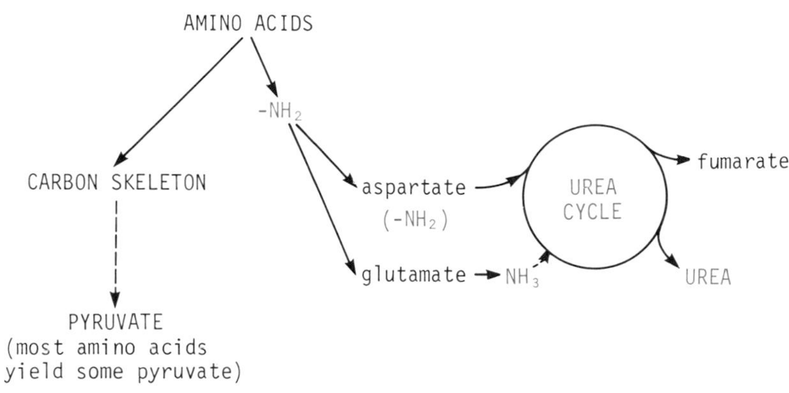

B

Substances other than amino acids may be used as substrates for gluconeogenesis.

GP = glycerol 3-phosphate (see p. 211)

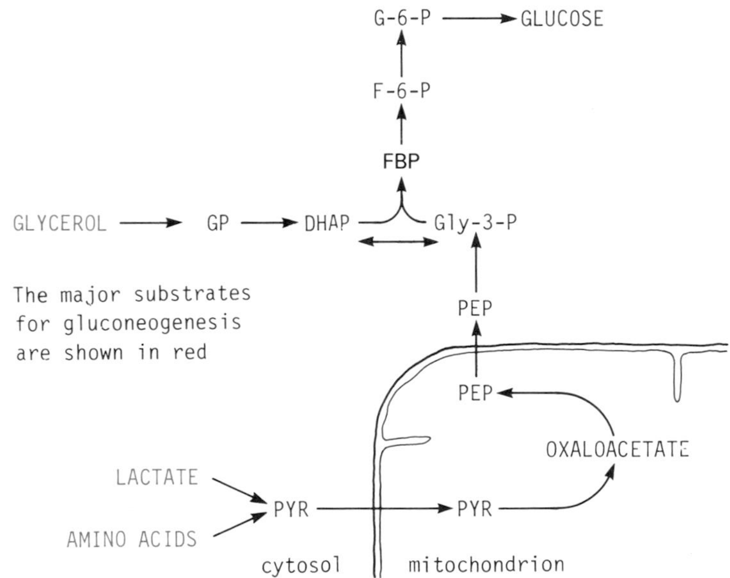

The major substrates for gluconeogenesis are shown in red.

Only alanine directly provides pyruvate after removal of the amino group. The carbon skeletons of other amino acids must undergo a series of metabolic conversions to yield pyruvate. There is some evidence that this may be partly carried out in muscle, after which the pyruvate is converted to alanine by transamination, and delivered to the liver where it readily yields pyruvate after another transamination step.

Lactate provides pyruvate directly after oxidation by lactate dehydrogenase. Glycerol can be converted to glucose if it is first phosphorylated (by glycerol kinase) to glycerol phosphate, which can be oxidized to dihydroxyacetone phosphate by glycerol phosphate dehydrogenase.

A

Key enzymes in gluconeogenesis

The enzymes of particular importance in gluconeogenesis are located at the beginning and at the end of the pathway. Two ATP equivalents are required by the first two enzymes, that convert pyruvate to phosphoenolpyruvate.

Control of the gluconeogenic pathway is primarily exercised at the steps involved in the formation of oxaloacetate and PEP. Elevated levels of acetyl CoA stimulate pyruvate carboxylase. The levels of pyruvate carboxylase and PEP carboxykinase are raised in situations requiring elevated glucose synthesis.

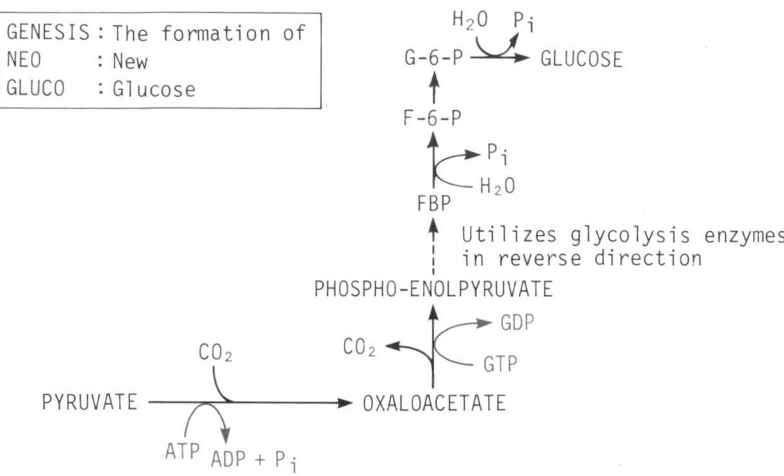

Four enzymes are of particular importance in gluconeogenesis:

1. *Pyruvate carboxylase.* This is the same enzyme as that which provides oxaloacetate for citric acid cycle activity. It is located within the mitochondrion.
2. *Phosphoenolpyruvate* (PEP) carboxykinase converts oxaloacetate to phosphoenolpyruvate. This enzyme is partially located in the mitochondrion, partially in the cytosol, and the location varies with the species of animal. It is most likely that oxaloacetate is converted to PEP in the cytosol in man.
3. *Fructose bisphosphatase* is a hydrolytic enzyme in the cytosol.
4. *Glucose 6-phosphatase* is a hydrolytic enzyme in the membrane of the endoplasmic reticulum.

B

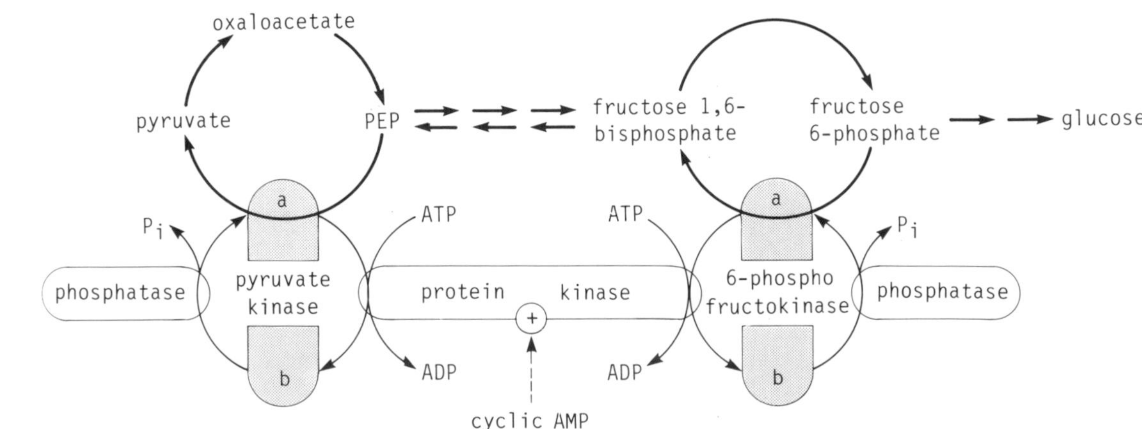

Pyruvate kinase and 6-phosphofructokinase are both subject to covalent control by protein kinases. In each case, action of the protein kinase converts the enzyme from active form (a) to an inactive form (b), and these protein kinases are stimulated by cyclic AMP, which thus inhibits glycolysis.

A

The Cori Cycle

The lack of glucose 6-phosphatase in muscle prevents the formation of free glucose in that tissue. However, muscle metabolism can contribute to blood glucose indirectly. This has been termed the Cori cycle.

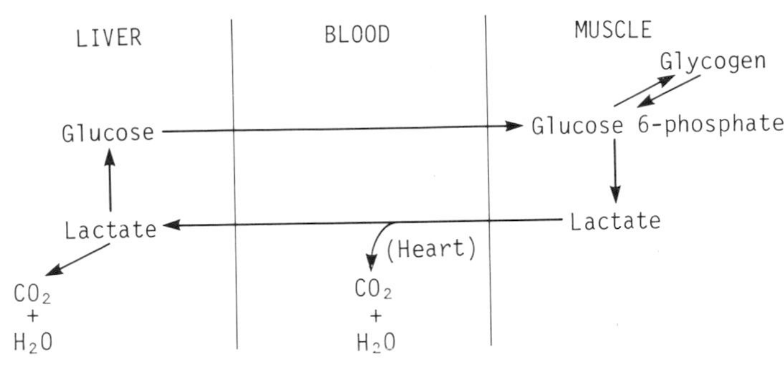

B

The formation of ketone bodies (ketogenesis)

Gluconeogenesis is usually accompanied by ketogenesis.

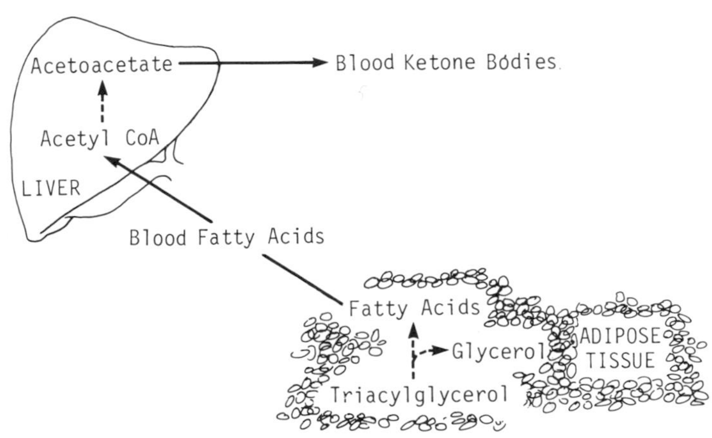

Ketogenesis is the term used to describe the formation of the "ketone bodies".

C

Three molecules of acetyl CoA are involved in the synthesis of acetoacetate. Acetoacetate (*NOT* acetoacetyl CoA) is the form secreted by liver into the blood.

The synthetic pathway for ketone bodies is found only in liver mitochondria but HMG CoA can also be synthesized in the cytosol.

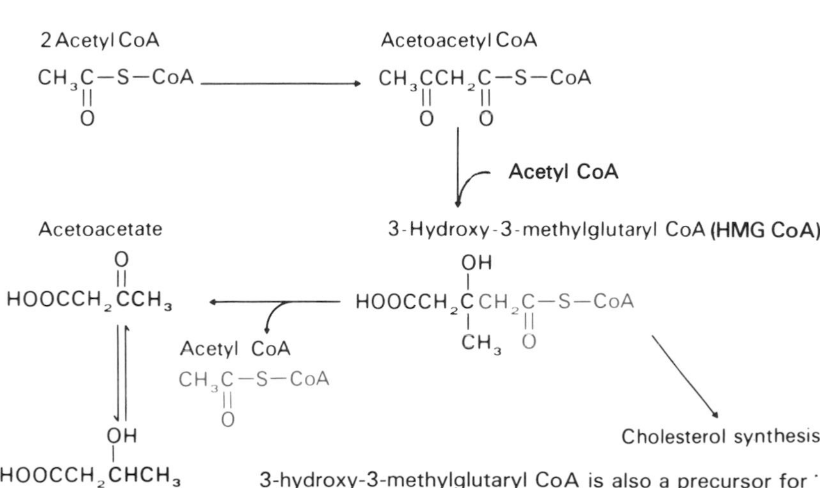

3-hydroxy-3-methylglutaryl CoA is also a precursor for cholesterol synthesis. A cytosolic enzyme produces this HMGCoA.

A

During a fast ketogenesis and gluconeogenesis occur simultaneously in liver.

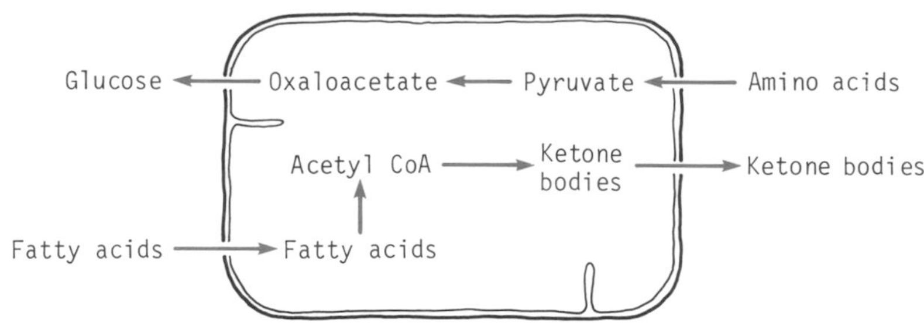

Both oxaloacetate and acetyl CoA are being formed in liver mitochondria at rapid rates. If they were oxidized via the citric acid cycle, the entire purpose of their formation would be defeated, and in particular, the valuable muscle protein would be oxidized to CO_2 and H_2O instead of being converted to blood glucose. Such a situation seems to indicate that synthesis of citrate must be inhibited under these conditions. This is one of the reasons that pyruvate and lactate accumulate in pathological conditions such as lactic acidosis, glycogen storage disease and other diseases. In these states, there is mobilization of muscle amino acid, with consequent production of large amounts of pyruvate in the liver, which cannot be oxidized because of the inhibition of the citric acid cycle. If the gluconeogenic pathway is then also blocked, as in glycogen storage disease, lactic acid accumulates (see p. 205).

B

The utilization of ketone bodies

Ketone bodies are produced in the liver, circulate in the blood and are utilized in muscle.

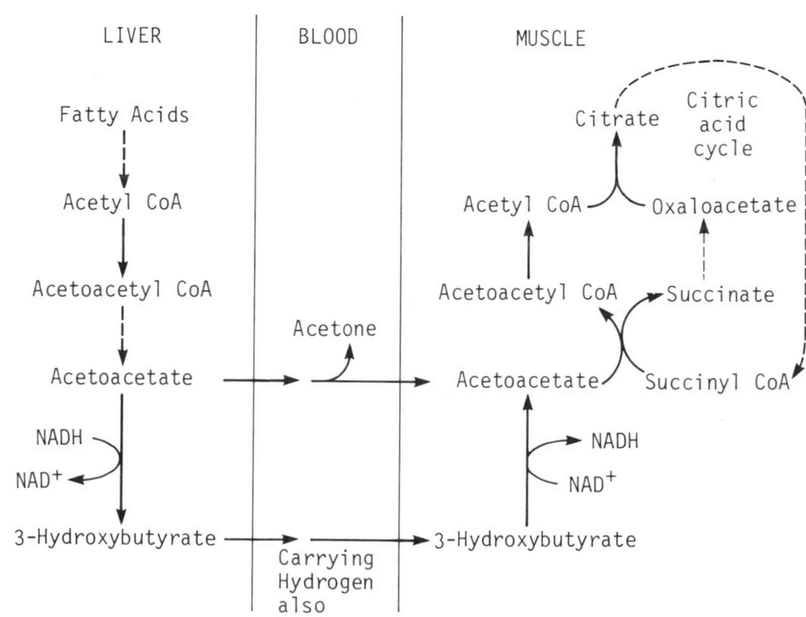

In the blood, some acetoacetate spontaneously decomposes to form acetone, which is expired through the lungs.

The availability of ketone bodies reduces to the minimum the amount of glucose which needs to be synthesized from muscle protein. The body need synthesize only that amount of glucose required by the brain and other cells of the central nervous system. Even these, during a prolonged fast, can utilize ketone bodies as a respiratory substrate, but it is still essential that the blood glucose level be maintained above about 2 or 3 mmol/litre.

A

Regulation of fatty acid oxidation and ketogenesis

Fatty acid oxidation takes place within the mitochondrion and the rate of oxidation is regulated by modulation of the rate at which fatty acids are transported across the mitochondrial membrane.

Fatty acids are transported across the mitochondrial membrane as esters of a quaternary ammonium hydroxyacid, *carnitine*. The carnitine acyl esters are formed in a reversible reaction catalysed by the enzyme *carnitine acyltransferase.*

$$CH_3-\overset{+}{N}(CH_3)_2-CH_2-CH(OH)-CH_2-C(=O)O^- \quad + \quad \text{palmitoyl CoA} \longrightarrow CH_3-\overset{+}{N}(CH_3)_2-CH_2-CH(O-C(=O)-CH_2-(CH_2)_{14}-CH_3)-CH_2-C(=O)O^- \quad + \quad \text{CoA}$$

carnitine + palmitoyl CoA ⟶ palmitoyl carnitine + CoA

B

The role of carnitine

Two distinct isoenzymes of carnitine acyltransferase (CAT) perform different roles in fatty acyl group transport. Only CAT I is subject to regulation.

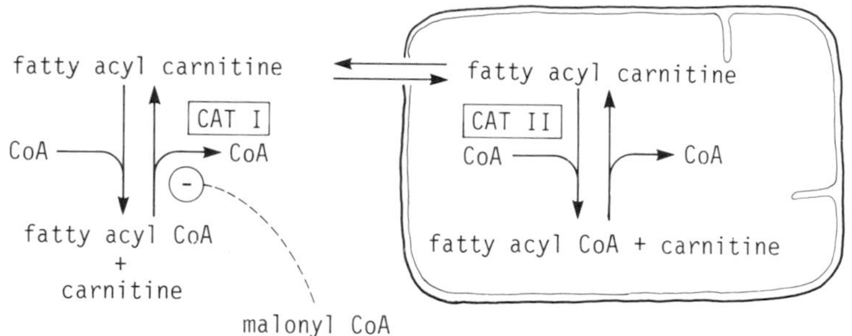

Carnitine acyltransferase exists as 2 isoenzymes. Carnitine acyltransferase I (CAT I) is bound to the cytosolic surface of the inner mitochondrial membrane, whilst carnitine acyltransferase II (CAT II) is inside the inner membrane. CAT I is subject to inhibition by malonyl CoA, a precursor in fatty acid synthesis (see p. 215). As the level of malonyl CoA rises during activation of fatty acid synthesis, the resulting inhibition of CAT I reduces transport of fatty acid for oxidation, thereby preventing oxidation of newly synthesized fatty acid.

A

Control of the blood sugar level

The glucose tolerance test is used to assist diagnosis in diseases that involve abnormal blood glucose levels. If insulin and free fatty acid levels are measured during the course of the test, they reflect the change in metabolic states from the fasting state, which is operating at the start of the test, to the fed state, to which the tissues have adjusted after about one hour.

The glucose tolerance test is carried out after an overnight fast. Glucose is then given orally, and blood samples are taken at intervals thereafter. The concentration of glucose and other substances can be measured in these blood samples.

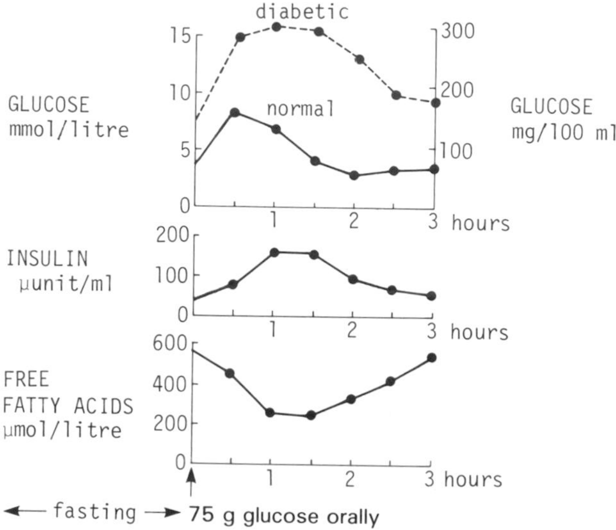

B

The control of the blood glucose level in mammals

During a fast, the liver can regulate with very great precision and rapidity the amount of glucose secreted into the blood, to match exactly that amount being removed. This ensures a constant blood sugar level during fasting.

The constancy of the fasting blood sugar level is to be contrasted with the way in which the blood sugar level rises on feeding.

The fasting blood sugar is maintained within precise limits. A fasting person can leap suddenly out of a chair, and run rapidly, with only minor effects on the blood glucose level, because of the speed with which the liver can increase or decrease glucose output.

In contrast, glucose entry from the gut, and glucose utilization by tissues, are not regulated with the same degree of precision. Thus, during ingestion of carbohydrates, the blood sugar rises. It cannot fall below the fasting level, because any tendency for uptake by peripheral tissues to exceed the rate of entry from the gut calls into play the mechanisms for the control of the secretion of glucose by the liver.

The above applies only in health. Variation of the fasting blood sugar level, either above or below its normal range, indicates a pathological situation.

A few hours in the history of John Smith's blood sugar.

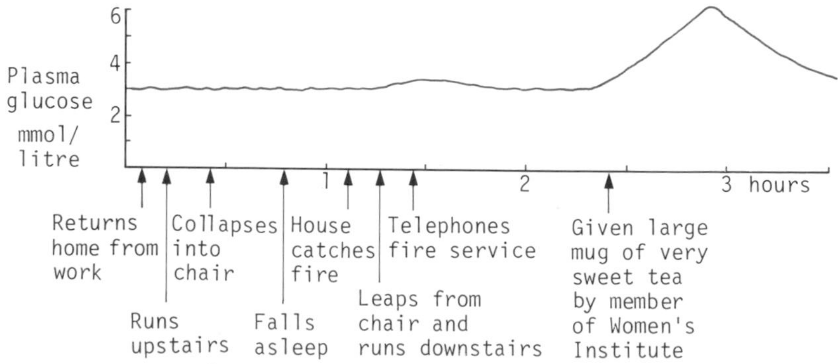

A

Pathological aspects of carbohydrate metabolism

Hypoglycaemia resulting in acidosis

The rare inborn error, *Type I glycogen storage disease* (Von Gierke's disease), is an excellent model in which to study interrelationships between carbohydrate and fat metabolism, because the metabolic lesion is precisely known.

The specific defect is a deficiency of liver glucose 6-phosphatase.
Over a period of years large deposits of glycogen may form in the liver.
On fasting for only a short period, there is severe hypoglycaemia, mild ketosis and mild *lactic acidosis*.

In this condition, the pancreas functions normally. Excessive ketosis does not result because a *basal* level of insulin is secreted that keeps fatty acid mobilization within reasonable limits.

Glucose 6-phosphatase deficiency

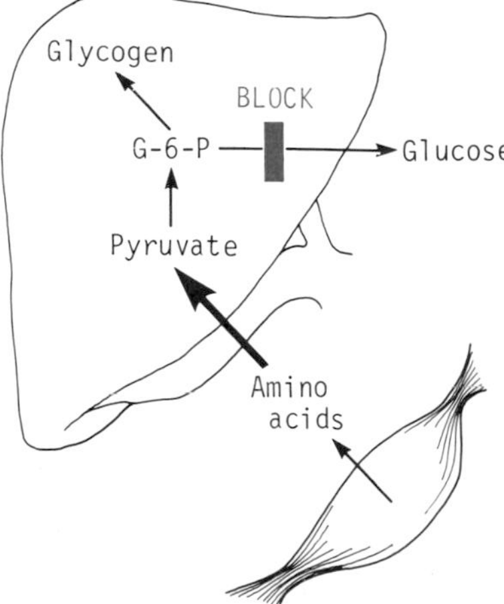

In the FASTING STATE

1. Glucose formation by liver cannot occur, because
   (a) Glycogen breakdown to glucose is blocked
   (b) Gluconeogenesis is blocked.

2. Hypoglycaemia results. Hormones are secreted in response to the hypoglycaemia e.g. glucocorticoids, which cause excessive mobilization of muscle amino acids, which enter the liver.

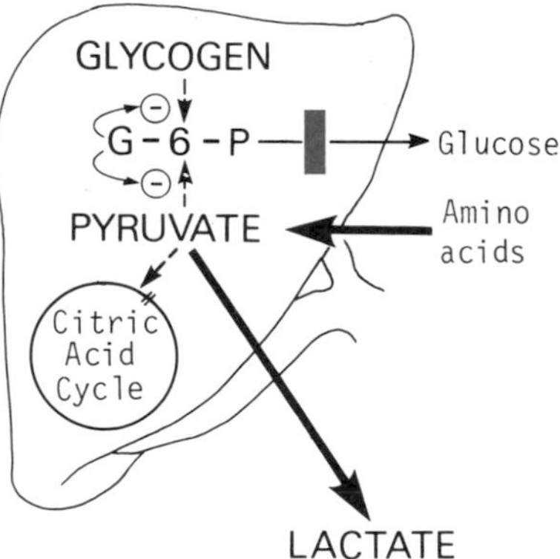

3. G-6-P levels build up and feedback inhibition blocks glycogen breakdown, and also the route from pyruvate to G-6-P.

4. Because of the hypoglycaemia and the fasting state of the body, citric acid cycle activity is reduced.

5. Thus all normal routes for disposal of pyruvate are blocked.

6. This forces it towards lactate, which spills over into the blood (lactic acidosis).

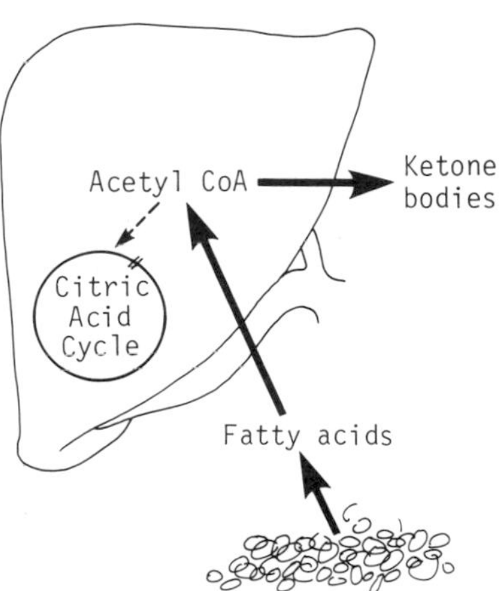

7. Hormones secreted in response to the hypoglycaemia cause release of fatty acid from adipose tissue.

8. Formation of ketone bodies results.

A

Diabetic ketosis

Diabetic ketosis can be thought of as an extreme of the type of metabolism seen during fasting.

In the absence of insulin there is no regulator to keep the action of hormones such as the glucocorticoids, glucagon, ACTH, and growth hormone within limits. Excessive production of glucose and ketone bodies occurs. The continual excretion of these by the kidney causes dehydration and loss of cations.

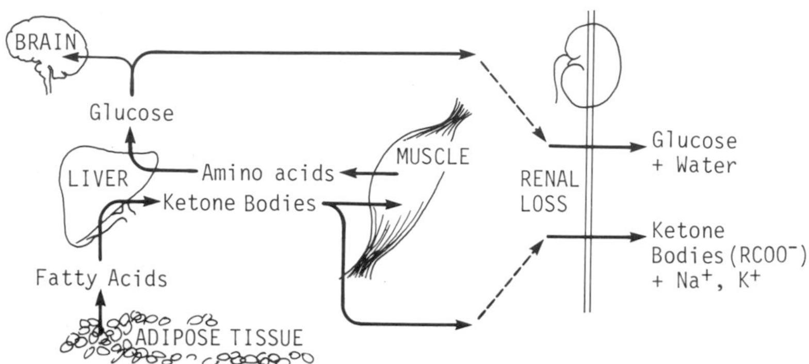

In diabetic ketosis, large amounts of oxaloacetate are continually being formed in the liver, due to the exaggerated gluconeogenic response that occurs in the absence of insulin. At the same time large amounts of acetyl CoA are being formed due to the exaggerated ketogenic response that is also occurring. However, these do not condense to form citrate to be oxidized by the citric acid cycle. For many years this situation appeared paradoxical, but it can reasonably be explained in terms of the changes in enzyme activities found in the extreme fasting state, discussed in previous sections.

B

Hyperinsulinaemia; excessive secretion of insulin

Excessive secretion of insulin occurs in cases of *insulinoma*, a tumour of the pancreas. Severe hypoglycaemia may result on fasting, due to the excess insulin blocking the action of fasting state hormones.

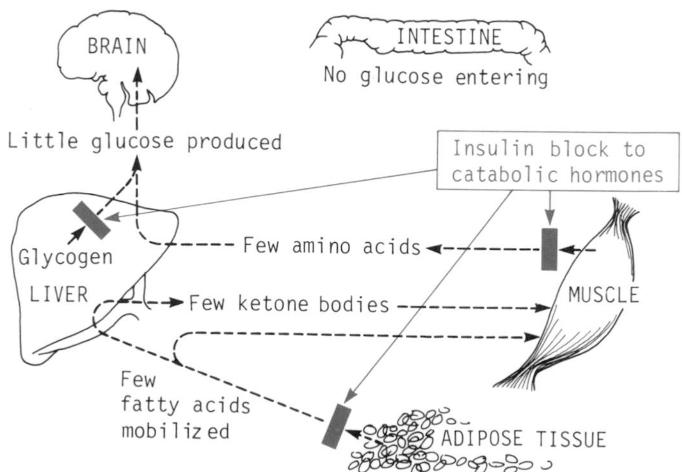

Note: The block of hormonal responses will probably not be total; the dotted lines indicate very low amounts of the relevant materials.

8

CARBOHYDRATE AND FAT METABOLISM

D. The absorptive state—synthesis of energy storage compounds

A

Digestive processes during the absorptive state

Before food can be absorbed from the digestive tract it must be processed to convert it to a form suitable for absorption.

Enzymes released during digestion hydrolyse food constituents.

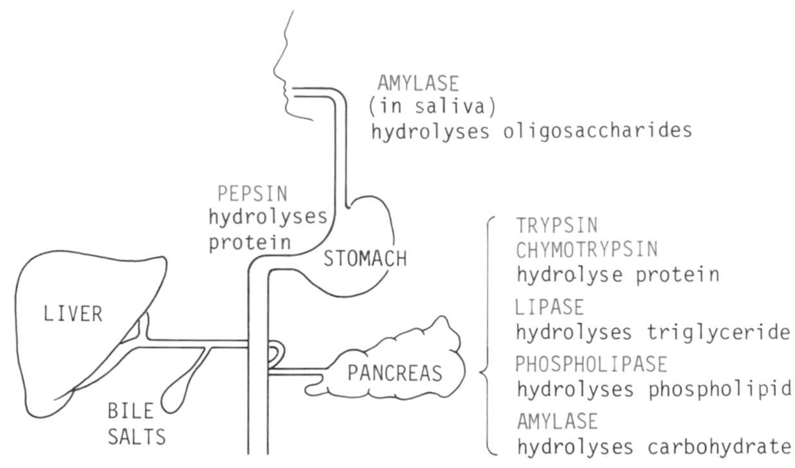

B

Protein digestion

Proteins are hydrolysed to smaller peptides by proteases such as pepsin, trypsin and chymotrypsin. These peptides may then be hydrolysed to amino acids by peptidase enzymes or absorbed as peptides.

Enzyme	Secreted by	Precursor	Converted to active enzyme by
Pepsin	Stomach	Pepsinogen	HCl
Trypsin	Pancreas	Trypsinogen	Enterokinase
Chymotrypsin	Pancreas	Chymotrypsinogen	Trypsin

	Leu		Phe		Lys		Ser		Tyr		Gly		Gly		Phe
Enzyme acting on following bond		Leucine amino-peptidase		Pepsin		Trypsin				Chymotrypsin				Carboxypeptidase A	
Alternative residues that permit enzyme action			Leu, Tyr		Arg				Phe						Neutral amino acids

$$H_2N-CH(CH_2CH(CH_3)_2)-CO\ominus NH-CH(CH_2C_6H_5)-CO\ominus NH-CH(CH_2CH_2CH_2CH_2NH_2)-CO\ominus NH-CH(CH_2OH)-CO-NH-CH(CH_2C_6H_4OH)-CO\ominus NH-CH_2-CO-NH-CH_2-CO\ominus NH-CH(CH_2C_6H_5)-COOH$$

The proteolytic enzymes act on the peptide bond on the carboxyl side of certain R groups, as shown (carboxypeptidase acts on the amino side of appropriate C-terminal residues). The red arrows show which bond is hydrolysed.

A

Carbohydrate digestion
Amylases in saliva and pancreatic juice break α1,4 bonds and hydrolyse amylose to free glucose, and maltose. Because α-amylases do not break α1,6 bonds, amylopectin is degraded also to oligosaccharides known as dextrins.

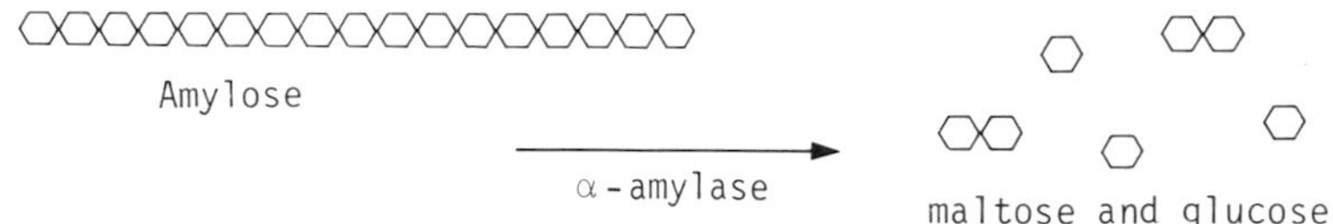

Other enzymes are present in the gut, especially disaccharidases.

$$\text{sucrose} \xrightarrow{\text{sucrase}} \text{glucose} + \text{fructose}$$

$$\text{maltose} \xrightarrow{\text{maltase}} \text{glucose} + \text{glucose}$$

$$\text{lactose} \xrightarrow{\text{lactase}} \text{galactose} + \text{glucose}$$

Absence of disaccharidases occurs, either as an hereditary trait (when it may manifest itself in infancy) or as a result of infection. Such a deficiency may be the cause of chronic diarrhoea. This is especially so in the case of lactase. A deficiency of this enzyme gives rise to the condition known as lactose intolerance.

Cellulose cannot be degraded by mammalian gut enzymes. This and other plant polysaccharides (see p. 137) which pass through to the lower bowel may have an important function in bowel action. They contribute to dietary fibre and are a component of 'roughage'.

B

Fat digestion

Fats are hydrolysed to monoacylglycerol and fatty acid.

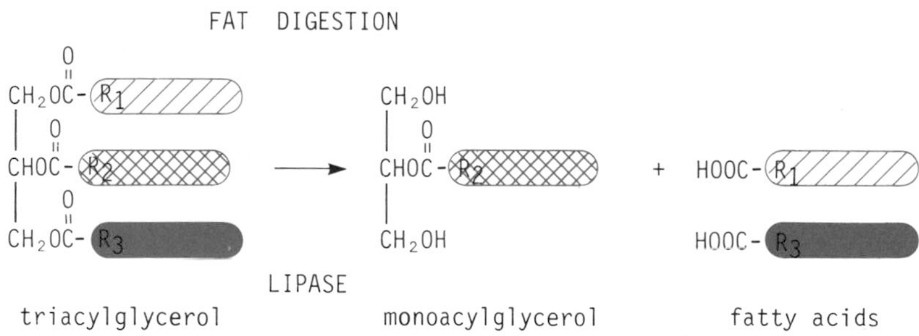

The coloured symbols represent the alkyl chains of the acyl residues.

Lipase acts on micelles of triacylglycerol and bile salts; i.e. the bile salts emulsify the fat droplets containing the triacylglycerol

$$\text{Large fat droplets} \xrightarrow{\text{bile salts}} \text{micelles of triacylglycerol and bile salts}$$

C

Nomenclature of enzymes

Enzymes that are involved in the degradation of macromolecules may be of two general types: (1) Enzymes that attack bonds in the interior of the macromolecule have the prefix endo-. Thus α-amylase is an *endo-amylase*, and trypsin is an *endo-peptidase*. (ii) Enzymes that only break bonds at the terminals of the chains of the macromolecule bear the prefix exo-. Thus β-amylase, which is an enzyme that removes only the terminal maltose of amylose, is an *exo-amylase*, and carboxypeptidase is an *exo-peptidase*.

A

Detailed metabolism in the fed state

The fed (or absorptive) state involves the synthesis of glycogen and fat from blood glucose.

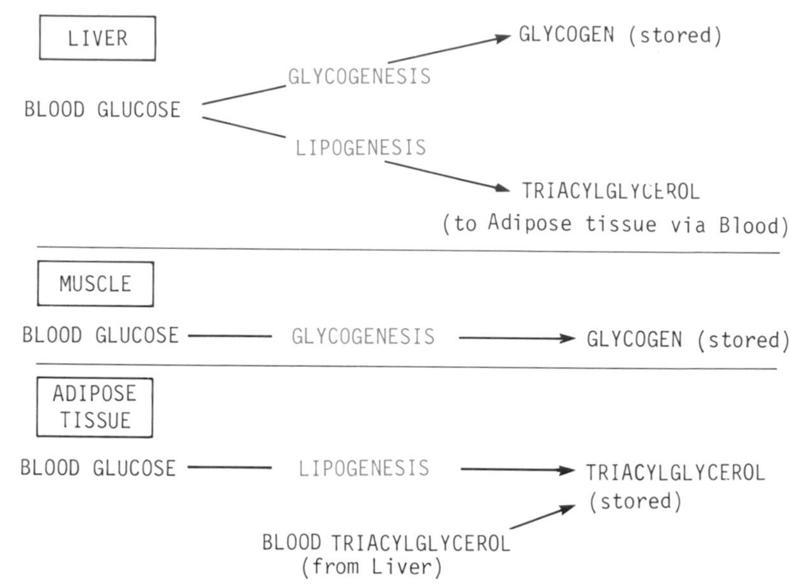

B

The pathways involved in the synthesis of glycogen and fat in liver can be divided into three groups.

Note that glycerol phosphate, (glycerol 3-phosphate) derived from dihydroxyacetone phosphate, is an important link between fat and carbohydrate metabolism. It provides the glycerol backbone of many lipid molecules.

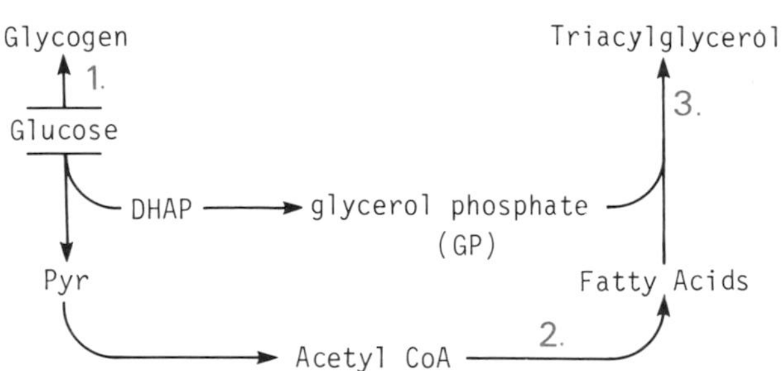

1. The formation of glycogen from glucose (GLYCOGENESIS)
2. The synthesis of fatty acid from acetyl CoA
3. The synthesis of triacylglycerol from glycerol phosphate and fatty acid

(2 and 3: LIPOGENESIS)

C

Glycerol 3-phosphate dehydrogenase interconverts dihydroxyacetone phosphate and glycerol 3-phosphate

Glycerol 3-phosphate dehydrogenase

$$\text{Glycerol 3-phosphate} + NAD^+ \rightleftarrows \text{dihydroxyacetone phosphate} + NADH + H^+$$

A

There is a close structural relationship between glycerol 3-phosphate, dihydroxyacetone phosphate and glyceraldehyde 3-phosphate, compounds that are related to the glycerol backbone of lipid molecules.

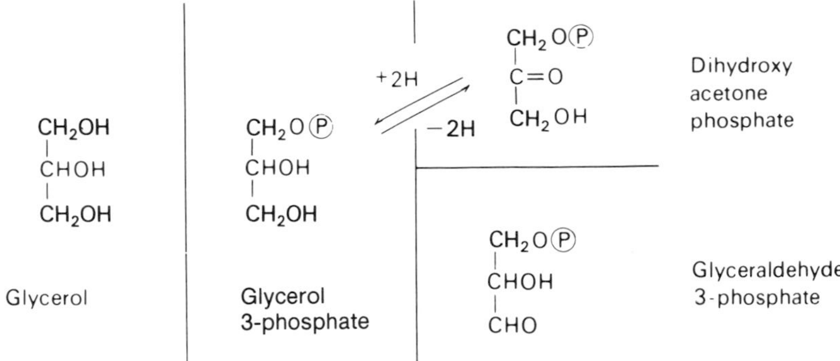

B

Biosynthesis of glycogen

Glycogen is synthesized by addition of glucose units to the (non-reducing) C-4 of the terminal glucose of a pre-existing α1,4 chain primer.

This is catalysed by glycogen synthase, and the glucose to be added reacts in the form of UDP-glucose.

UDP-glucose

Glycogen synthase

UDP

A

Branches in the glycogen molecule are formed by breaking the growing α1,4 chain and transferring the residues to form an α1,6 link.

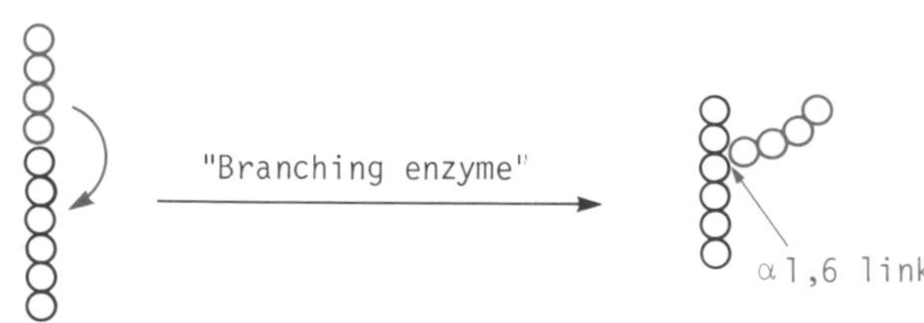

B

Initiation of glycogen synthesis requires the presence of an *initiator protein* to which glucose residues are attached by *glycogen initiator synthase.* Glycogen synthase and branching enzyme then act on the glycosylated initiator protein, Glycogenin.

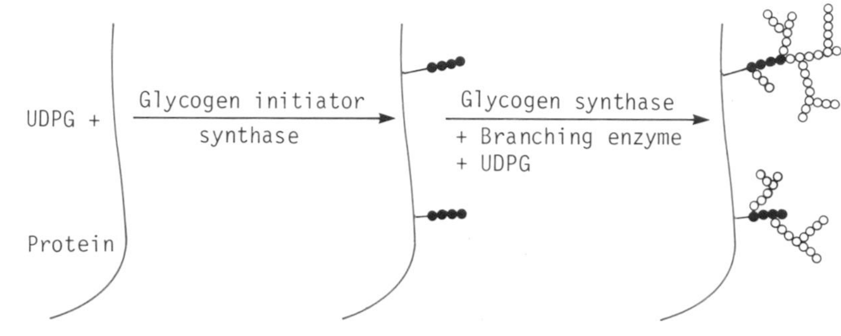

The linkage between the growing glycogen chain and the initiator protein is through tyrosyl residues

C

The combination of all of these reactions results in the branched macromolecule, glycogen.

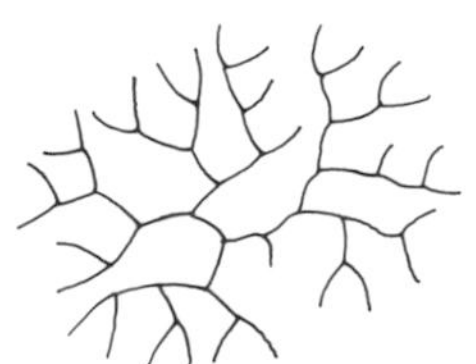

The distance between branch points is on average 12–14 glucose units, but near the reducing end may be less than this.

D

Glycogen storage diseases

An inherited deficiency of one or other of the enzymes involved in glycogen metabolism can lead to glycogen storage disease, some types of which are listed.

Type	Name	Enzyme affected	In which tissue	Remarks
I	Von Gierke's disease (see p. 205)	Glucose 6-phosphatase	Liver Kidney	Low fasting blood glucose, enlarged liver with elevated glycogen levels.
III	Limit dextrinosis	Debranching enzyme	Liver	Phosphorylase acts on α1,4 bonds until an α1,6 bond is reached. The resulting molecule is called a limit dextrin.
V	McArdle's disease	Phosphorylase	Muscle	Muscular pain, weakness and stiffness after only mild exercise.
VI		Phosphorylase	Liver	Has some features similar to type I, but as gluconeogenesis is not blocked hypoglycaemia may not be so severe.

A

Lipogenesis

The synthesis of triglycerides from glucose
Lipogenesis is the term which means the synthesis of fat from non-fat materials, especially carbohydrate, and includes synthesis of fatty acids and triacylglycerol

Triacylglycerol (TG) synthesized in liver is exported, whilst that synthesized in fat cells is stored.

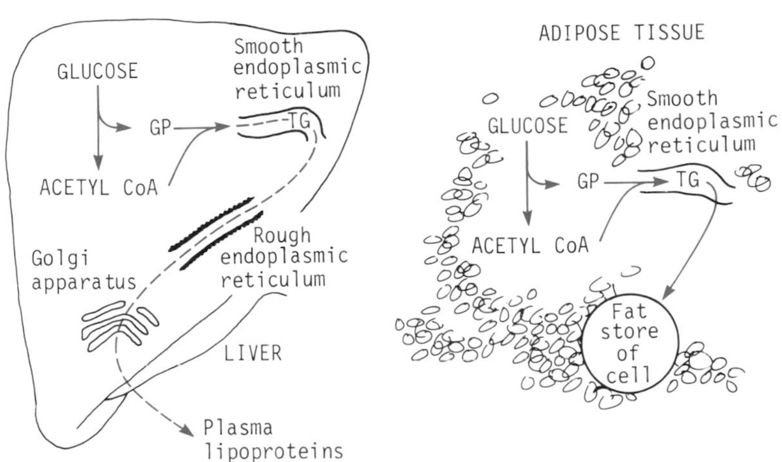

Lipogenesis from glucose requires not only the formation of acetyl CoA by the glycolytic pathway and pyruvate dehydrogenase, but also conversion of dihydroxyacetone phosphate to glycerol 3-phosphate (GP). Triacylglycerol (TG) is synthesized in the smooth endoplasmic reticulum. It then combines with protein synthesized in the rough endoplasmic reticulum and is packaged and secreted as plasma lipoprotein (see p. 218) through the Golgi apparatus.

B

Pentose phosphate pathway ("pentose shunt")

The enzymes of the pentose phosphate pathway are important in providing NADPH for fatty acid synthesis. They also yield pentoses for nucleotide and nucleic acid synthesis.

Two dehydrogenases are involved in the production of this NADPH:

1. Glucose 6-phosphate dehydrogenase;
2. 6-phosphogluconate dehydrogenase.

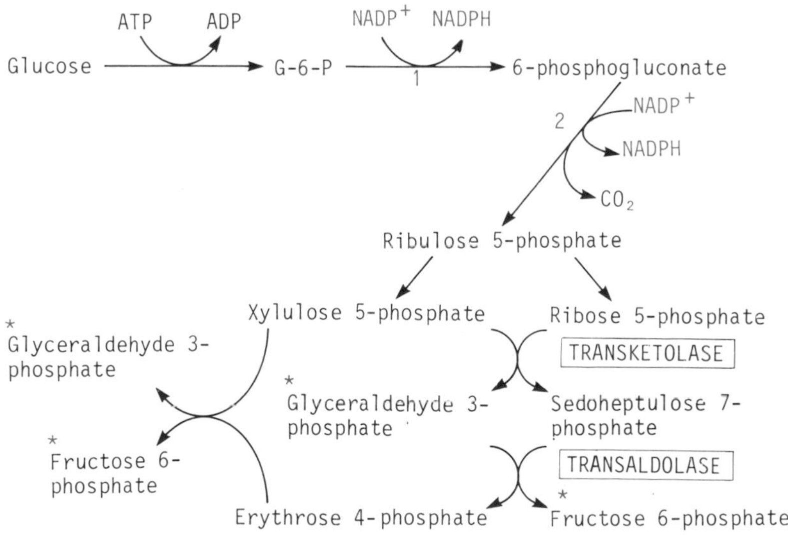

* Enter other pathways such as the glycolytic pathway

A mnemonic for the pentose shunt is that after formation of ribulose 5-phosphate 3 pairs of reactants successively sum to 10 carbons. Then a 4-C compound and 5-C compound (total of 9-C) combine to give fructose 6-phosphate and glyceraldehyde 3-phosphate, which enter other pathways of metabolism.

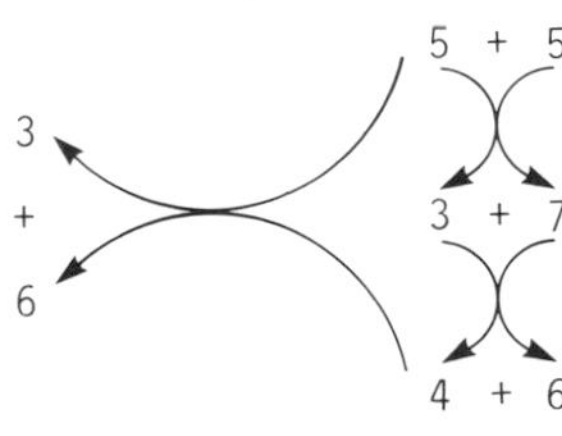

Steps in lipogenesis These involve

1. Conversion of glucose to acetyl CoA (glycolytic pathway)
2. formation of glycerol phosphate (via glycolytic pathway)
3. Synthesis of fatty acid from acetyl CoA
4. Synthesis of triacylglycerol from fatty acids and glycerol phosphate

Further NADPH is formed by the malic enzyme; cytosolic malate dehydrogenase together with the malic enzyme act as a transhydrogenase reaction in the cytosol, transferring hydrogen from NADH to NADPH.

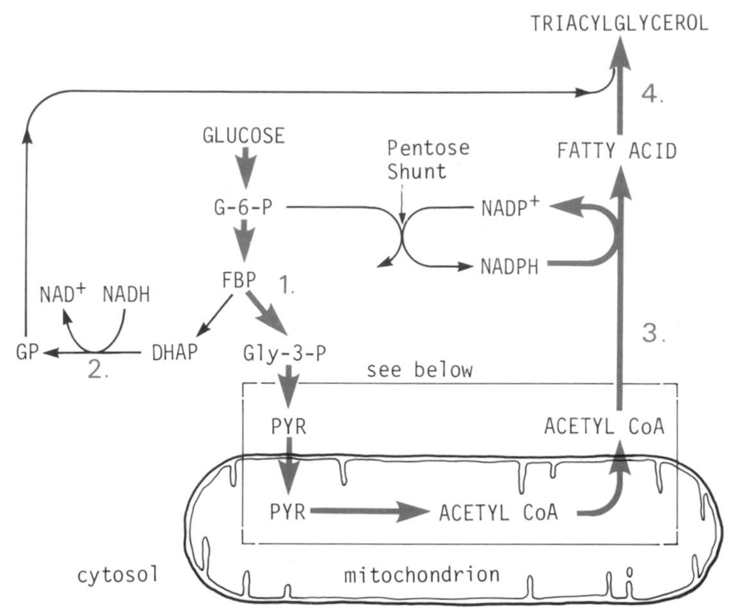

detail of reactions occurring in inset above

Acetyl CoA
Oxaloacetate
NADH
Malate dehydrogenase
NAD^+
Citrate cleavage enzyme
$NADP^+$
Malate
NADPH
Malic enzyme
Citrate
Pyruvate
CO_2
Citrate
Pyruvate
Acetyl CoA
Oxaloacetate

Acetyl CoA and oxaloacetate are not transported readily through the mitochondrial membrane. They are converted to citrate, which is cleaved in the cytosol to acetyl CoA and oxaloacetate.
An alternative mechanism for transporting acetyl groups across the mitochondrial membrane involves the synthesis of acetylcarnitine.
Acetyl CoA + carnitine ⇄ acetylcarnitine + CoA.

Acetyl carnitine is transported from the mitochondrion to the cytosol, where a similar enzyme catalyses the formation of acetyl CoA and carnitine (compare the reactions for fatty acyl carnitine on p. 203).

A

Biosynthesis of fatty acids

Fatty acid synthesis proceeds by addition of 2-carbon units to acetyl CoA.

Malonyl CoA is the intermediate by which the 2-carbon units are added.

NADPH is the coenzyme which carries out the reductive steps.

* Details of the reductive steps:

- CO - CH_2 -
+ 2H
- CH - CH_2 - (OH)
- H_2O
- CH = CH -
+ 2H
- CH_2 - CH_2 -

Chemically analogous but reverse of the steps of β-oxidation.

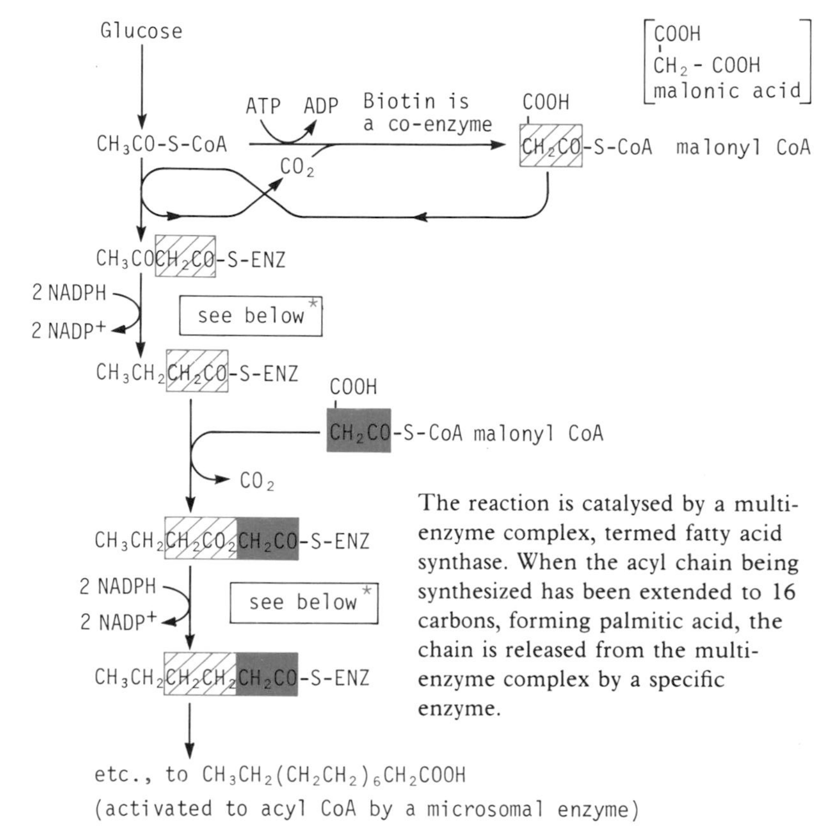

The reaction is catalysed by a multi-enzyme complex, termed fatty acid synthase. When the acyl chain being synthesized has been extended to 16 carbons, forming palmitic acid, the chain is released from the multi-enzyme complex by a specific enzyme.

B Palmitic acid synthesized by fatty acid synthase is converted to palmitoyl CoA in the endoplasmic reticulum.

Palmitic acid + ATP + CoA → Palmitoyl CoA + ADP + P_i

C

Regulation of acetyl CoA carboxylase.

Regulation of fatty acid synthesis can be exerted at the point of synthesis of malonyl CoA by *acetyl CoA carboxylase.*
The effect of malonyl CoA on carnitine acyl transferase helps to prevent oxidation of newly synthesized fatty acid (see p. 203).

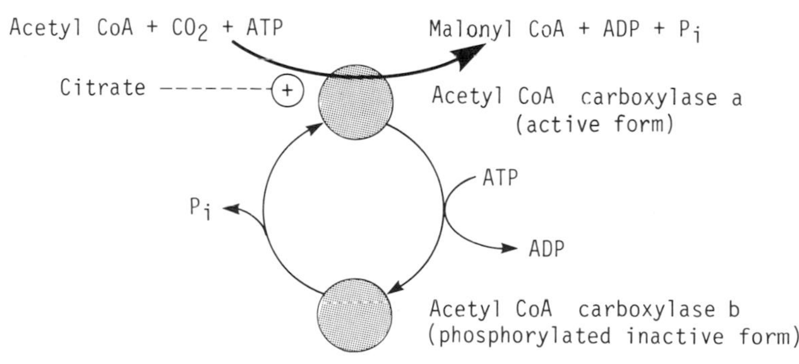

Acetyl CoA carboxylase is strongly activated by citrate, and is subject to control by phosphorylation to an inactive form by a protein kinase and conversion to the active form by a phophatase. Biotin is required as coenzyme, CO_2 being introduced via carboxybiotin (see p. 128B).

Fatty acid structures

Palmitic acid synthesized by fatty acid synthase is only one of a number of different fatty acids required by the body. *Palmitoyl CoA* can be elongated by microsomal enzyme systems, which add carbons as 2-carbon units, utilizing acetyl CoA, to the carboxyl end of the molecule. Different desaturase enzymes can insert double bonds at specific carbons between C-1 and C-9. Thus there is *Δ4 desaturase, Δ5 desaturase* and *Δ9 desaturase.* There are no mammalian desaturases which insert double bonds between C-9 and the methyl group. Thus, as the body needs fatty acids with double bonds in this region, precursors known as essential fatty acids must be taken in from the diet. Linoleic and *α*-linolenic acids are examples.

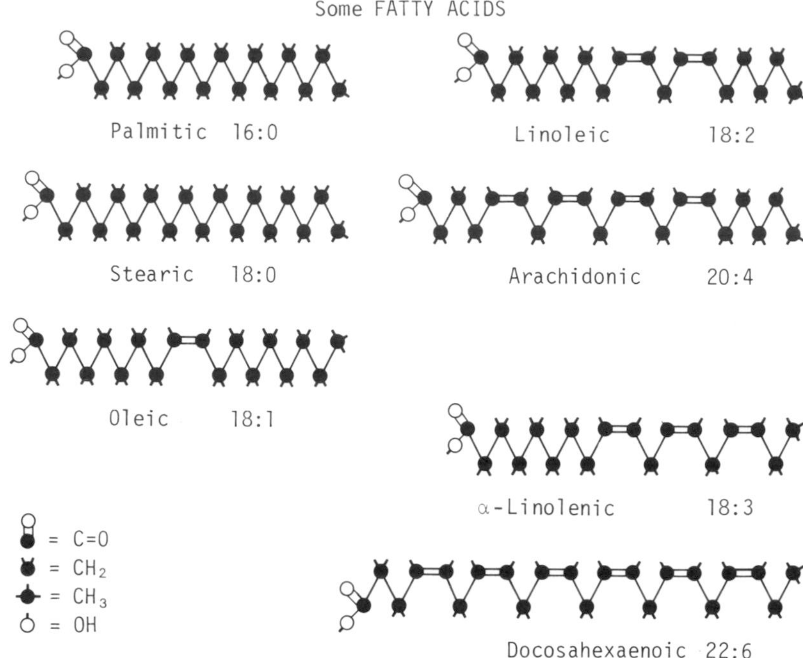

Linoleic acid cannot be synthesized by the body and must be provided in the diet and is hence termed an essential fatty acid. Linoleic acid serves as a precursor for arachidonic acid. Oleic acid and stearic acid can be synthesized from palmitic acid and thus need not be provided in the diet.

As indicated above, a fatty acid can be designated by the convention—no. of carbons:no. of double bonds. Thus linoleic acid, with 18 carbons and 2 double bonds, is 18:2. The double bond position can be designated in two ways:

1. From the carboxyl group, when the symbol Δ is used. Thus linoleic acid is Δ9, 12;

2. From the methyl end, when the symbol n- (or *ω*) is used. Thus linoleic acid is n-6,9 (or *ω* 6,9)

The advantage of the second convention is that it emphasizes a metabolic relationship between the acids that depends on the fact that in mammals, double bonds cannot be introduced enzymically on fatty acid carbons nearer to the methyl end of the molecule than the *ω* 9 carbon.

HOOC CH_2 CH_2 CH_2 CH_2 CH_2 CH_2 CH_2 CH_2 CH_2 CH_2 CH_2 CH_2 CH_2 CH_2 CH_2 CH_2 CH_3

[bonds cannot be inserted in this region]

This means that there are three 'families' of fatty acids, arising from precursor acids taken in the diet; these are shown on the next page.

A

There are 3 major families of fatty acids, derived from precursors with ω 6 or ω 3 double bonds or from palmitic acid.

ω 6 family

Linoleic acid (18 : 2.ω 6,9) $\xrightarrow{\text{elongation \& desaturation}}$ γ-linolenic acid (18 : 3 ω 6,9,12) $\longrightarrow$ arachidonic acid (20 : 4 ω 6,9,12,15)

ω 3 family

α-linolenic acid (18 : 3 ω 3,6,9) $\xrightarrow{\text{elongation \& desaturation}}$ docosahexaenoic acid (22 : 6 ω 3,6,9,12,15,18)

ω 9 family

As double bonds can be inserted at ω 9–10, and points nearer to the carboxyl group than this, these acids can be synthesized from the saturated fatty acid, palmitate, which in turn can be synthesized from acetyl CoA.

palmitic acid $\longrightarrow$ stearic acid $\longrightarrow$ oleic acid $\longrightarrow$ eicosatrienoic acid (20 : 3 ω 9,12,15)

Eicosatrienoic acid is synthesized in animals when no ω 6 acids are contained in the diet, and it substitutes, to some extent, for the long chain unsaturated fatty acids such as arachidonic acid, that are then much reduced in the membrane lipids. Absence of the ω 6 acids in the diet is known as essential fatty acid deficiency and may be corrected by the addition to the diet of ω 6 acids such as linoleic acid.

B

Formation of triacylglycerols

Triglyceride synthesis involves sequential addition of fatty acyl CoA molecules to glycerol 3-phosphate to form phosphatidic acid, which is hydrolysed to diacylglycerol. This is then acylated to triacylglycerol.

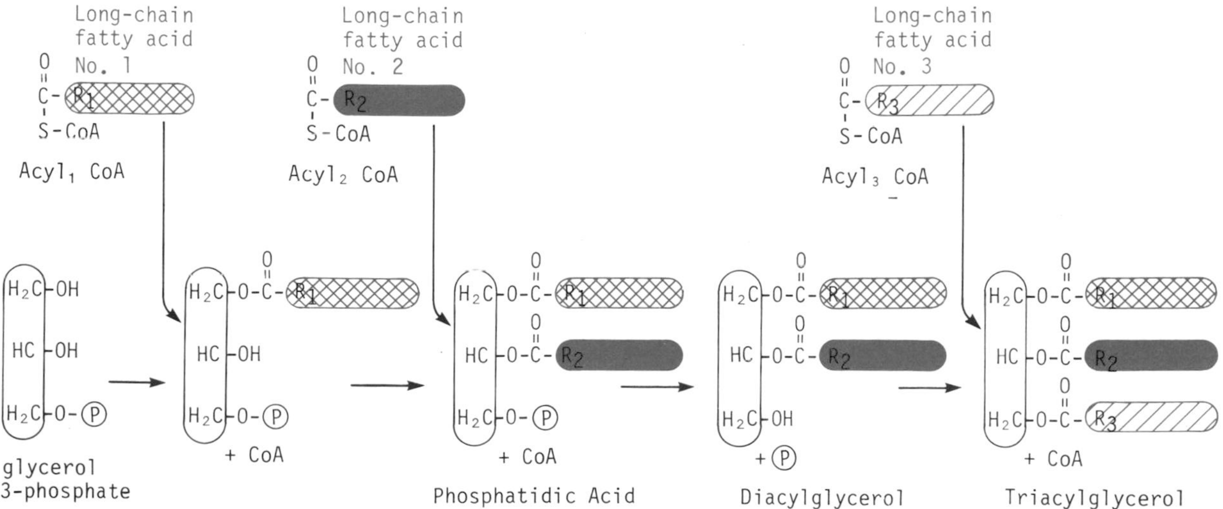

A

Chemistry and metabolism of plasma lipoproteins

Lipids circulate in the blood as complex molecules called *lipoproteins*

Lipoproteins are large assemblies of lipid and protein. As the different classes of lipoprotein vary in their relative content of lipid, they have a wide range of density. They are also of widely different size. They can be separated either by electrophoresis, or by centrifugation in solutions of appropriate density. This has led to their classification as shown below.

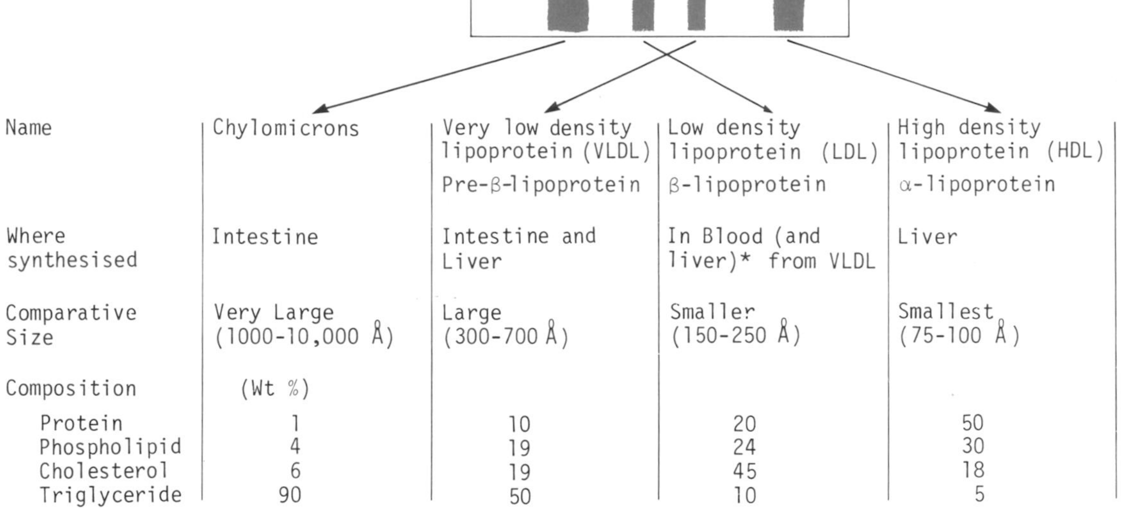

Name	Chylomicrons	Very low density lipoprotein (VLDL) Pre-β-lipoprotein	Low density lipoprotein (LDL) β-lipoprotein	High density lipoprotein (HDL) α-lipoprotein
Where synthesised	Intestine	Intestine and Liver	In Blood (and liver)* from VLDL	Liver
Comparative Size	Very Large (1000-10,000 Å)	Large (300-700 Å)	Smaller (150-250 Å)	Smallest (75-100 Å)
Composition	(Wt %)			
Protein	1	10	20	50
Phospholipid	4	19	24	30
Cholesterol	6	19	45	18
Triglyceride	90	50	10	5

* It is not certain whether all β-lipoprotein is formed from VLDL in blood or whether some may be directly synthesised in liver.

B

The composition of the lipoproteins can be represented in diagrammatic form.

The proteins associated with the lipoproteins have been purified and characterized and are termed apolipoproteins (or simply Apo B etc.).

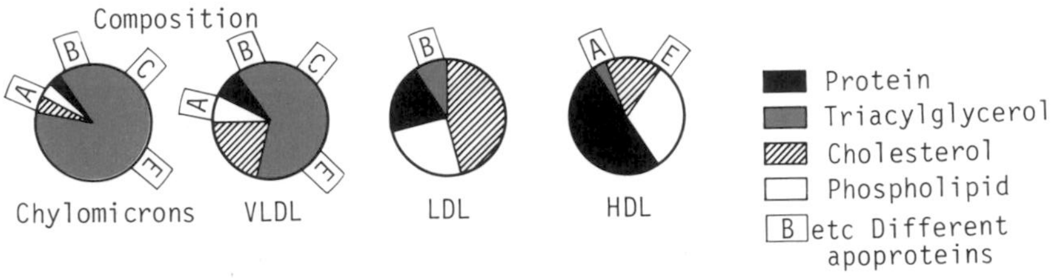

The different apoproteins are designated A_I, A_{II}, B, C_I, C_{II}, C_{III}, D and E. Each lipoprotein has a characteristic pattern of apoproteins.

Chylomicrons & VLDL	LDL	HDL
A_I, B, C_I, C_{II}, C_{III}E	B	A_I, A_{II}, E

Apo B synthesized by liver has M_r 100 000 and is designated B–100 to distinguish it from that synthesized by intestine (B–48, M_r 48 000)

A

An important enzyme in plasma is lipoprotein lipase, which hydrolyses triacylglycerols and has the effect of reducing *chylomicrons* and VLDL to smaller fragments. This enzyme used to be called *'clearing factor'*, as after a fatty meal it cleared the milky appearance of plasma that was due to the large quantities of chylomicrons being absorbed from the intestine.

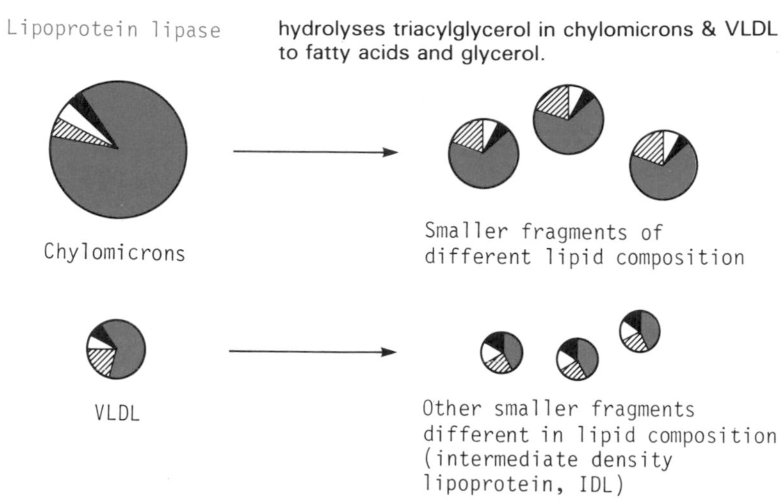

Lipoprotein lipase is activated by apoprotein C_{II}

B

In order for fatty acids of triacylglycerols to enter cells, the triacylglycerol must be hydrolysed at the plasma membrane. The fatty acids can then be transported through the membrane into the cell. Within the cell, triacylglycerol may be resynthesized from the fatty acids, but glucose is essential for the provision of glycerol 3-phosphate by means of the glycolytic pathway. The glycerol released outside the membrane can be oxidized to CO_2 and H_2O in liver, or can be converted to glucose, after phosphorylation to glycerol 3-phosphate by glycerol kinase.

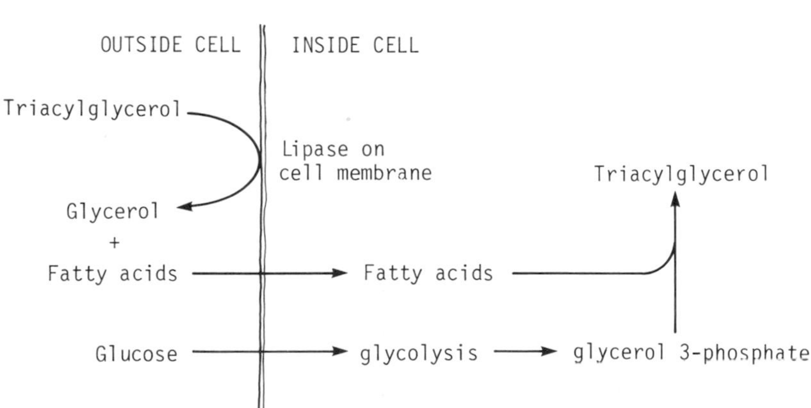

In the intestine, lipase action results in the formation of monoacylglycerol and fatty acids. The interstitial cells of the intestine contain enzymes that can catalyse the reaction:

monoacylglycerol + 2 fatty acyl CoA ⟶ triacylglycerol + 2 CoA

This forms an alternative pathway for triacylglycerol synthesis, independent of a direct supply of glycerol 3-phosphate. However, an oxidizable substrate must be available, as energy is required for the synthesis of the fatty acyl CoA derivatives from the fatty acids released by the lipase.

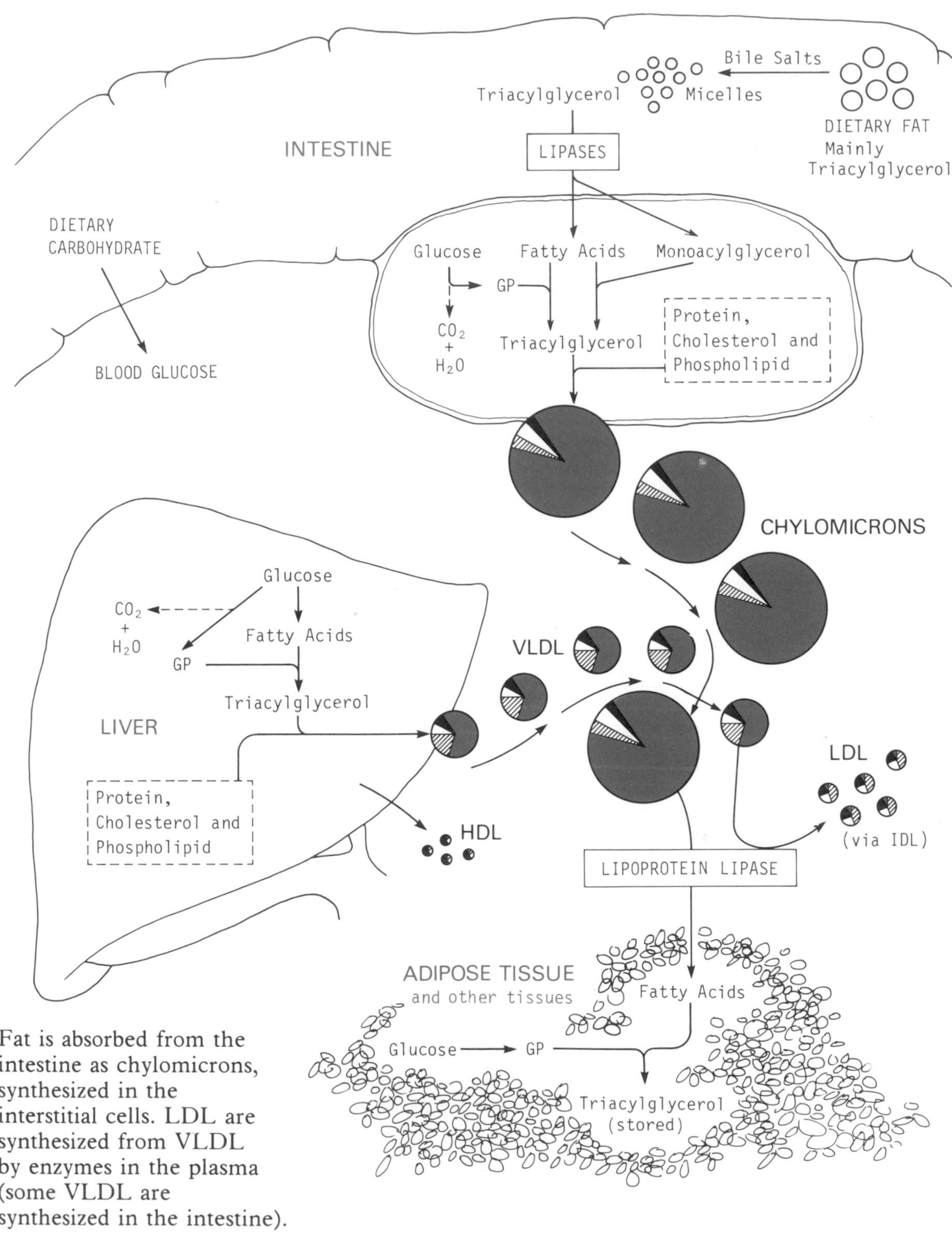

Fat is absorbed from the intestine as chylomicrons, synthesized in the interstitial cells. LDL are synthesized from VLDL by enzymes in the plasma (some VLDL are synthesized in the intestine).

A

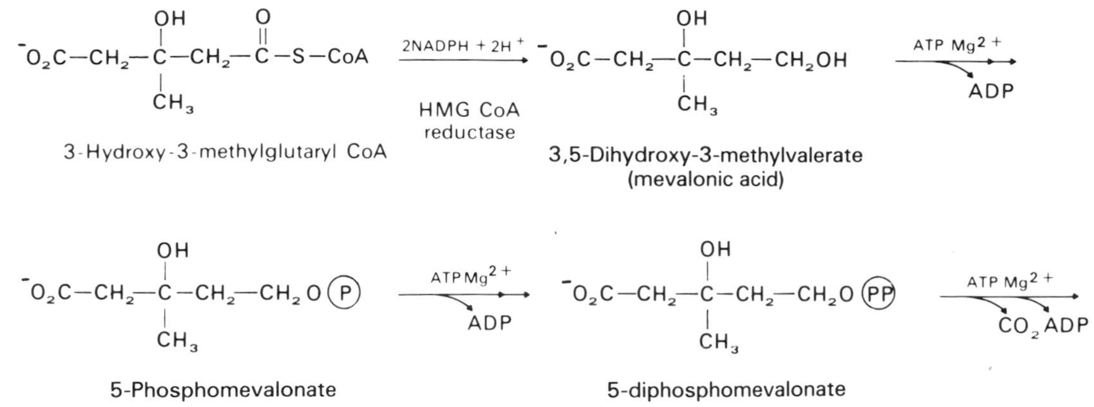

$^{-}O_2C-CH_2-C(OH)(CH_3)-CH_2-C(=O)-S-CoA$ —(2NADPH + 2H$^+$; HMG CoA reductase)→ $^{-}O_2C-CH_2-C(OH)(CH_3)-CH_2-CH_2OH$ —(ATP Mg^{2+} → ADP)→

3-Hydroxy-3-methylglutaryl CoA

3,5-Dihydroxy-3-methylvalerate (mevalonic acid)

$^{-}O_2C-CH_2-C(OH)(CH_3)-CH_2-CH_2O$ (P) —(ATP Mg^{2+} → ADP)→ $^{-}O_2C-CH_2-C(OH)(CH_3)-CH_2-CH_2O$ (PP) —(ATP Mg^{2+} → CO_2 ADP)→

5-Phosphomevalonate

5-diphosphomevalonate

Biosynthesis of cholesterol.

Cholesterol synthesis in liver is the subject of much interest because of the evidence that high blood cholesterol levels may be associated with increased risk of *atherosclerosis*. *HMG CoA reductase* is a key enzyme that regulates the activity of the pathway for cholesterol synthesis.

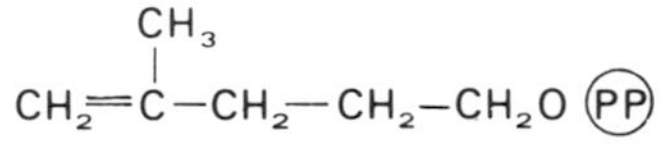

$CH_2=C(CH_3)-CH_2-CH_2-CH_2O$ (PP)

Isopentenyl diphosphate

HMG CoA reductase = 3-hydroxy-3-methylglutaryl CoA reductase.

This is an important key enzyme that regulates cholesterol synthesis.

B

Further metabolism of *isopentenyl diphosphate* (a 5-carbon compound) results in polymerization of six 5-carbon compounds to squalene (30 carbons) which is converted, in a further series of reactions, to cholesterol.

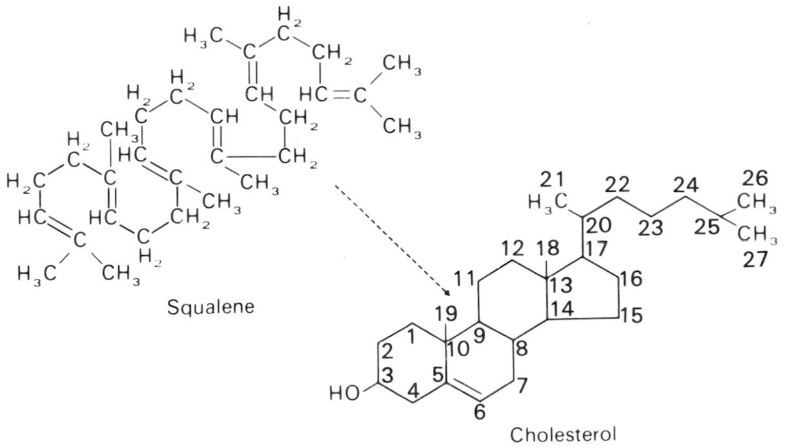

Squalene

Cholesterol

Cholesterol is an important precursor of the sex hormones, the steroid hormones of the adrenal cortex, and of the bile acids. All cells may take up cholesterol from the blood as it is a major constituent of the plasma lipoproteins. The major site of cholesterol synthesis is the liver, from which it is secreted as a lipoprotein constituent. Many tissues can, however, synthesize cholesterol, including the adrenals and reproductive organs, and also skin fibroblasts, which can be readily cultured and used to investigate cholesterol metabolism in humans.

A

Cholesterol metabolism in blood lipoproteins

HDL is thought to be synthesized as small discs, rich in unesterified cholesterol and lecithin. These discs are converted to HDL as a result of the action of LCAT, which enriches the lipoprotein in esterified cholesterol.

Action of lipoprotein lipase on VLDL causes formation of a molecule less rich in triacylglycerol (IDL). IDL accepts cholesterol esters from HDL, to form LDL.

Lecithin–cholesterol acyltransferase (LCAT)

Phosphatidylcholine + cholesterol $\rightleftarrows$ lysophosphatidylcholine + cholesterol ester

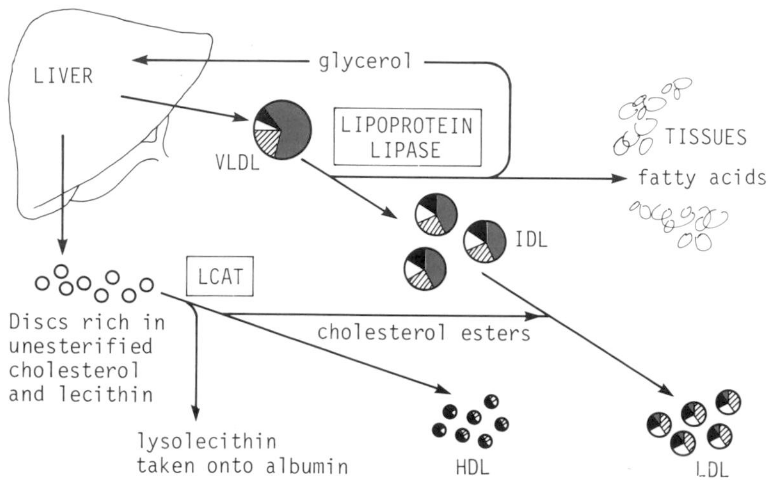

The cholesterol of lipoproteins is predominantly in the esterified form. These esters are synthesized from cholesterol and lecithin by the enzyme, found in plasma, lecithin-cholesterol acyltransferase (LCAT, pronounced El Cat!). This enzyme is activated by apoprotein A_I

lysolecithin = lysophosphatidylcholine

B

Apolipoproteins

The apoprotein moieties of the plasma lipoproteins have a variety of functions.

Apoprotein	*Function*
A_I	This apoprotein is an activator of lecithin—cholesterol acyl transferase.
B-100	Apoprotein B-100 is recognized by receptors on liver cells and other cells within the peripheral circulatory system, and plays an important role in the uptake by these cells of lipoproteins that carry this apoprotein.
C_{II}	Apoprotein C_{II} activates lipoprotein lipase. A deficiency of this apoprotein has in some cases been associated with elevated plasma triacylglycerol levels.
E	Liver cells carry receptors for apoprotein E, which is important for efficient uptake by liver of lipoproteins in which it occurs.

A **Hyperlipidaemias**

Electrophoresis of plasma proteins, using a stain for lipid, may be used to aid diagnosis of *hyperlipoproteinaemias*.

By comparison of these electrophoresis patterns with the composition of individual lipoproteins on page 218, it can be seen how the elevated plasma level of cholesterol or triacylglycerol arises in each type of hyperlipidaemia.

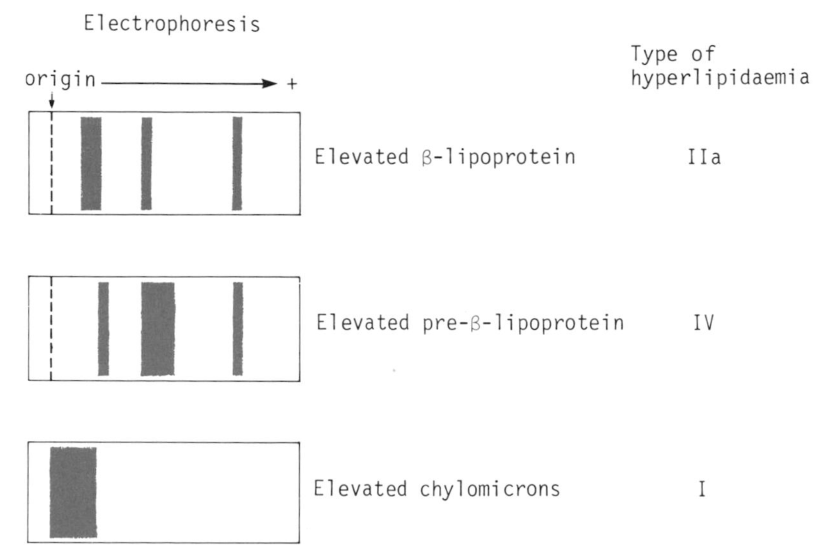

B

A more extensive analysis of the lipoprotein pattern is revealed by the analytical ultracentrifuge.

A further classification of hyperlipidaemias is given on page 228.

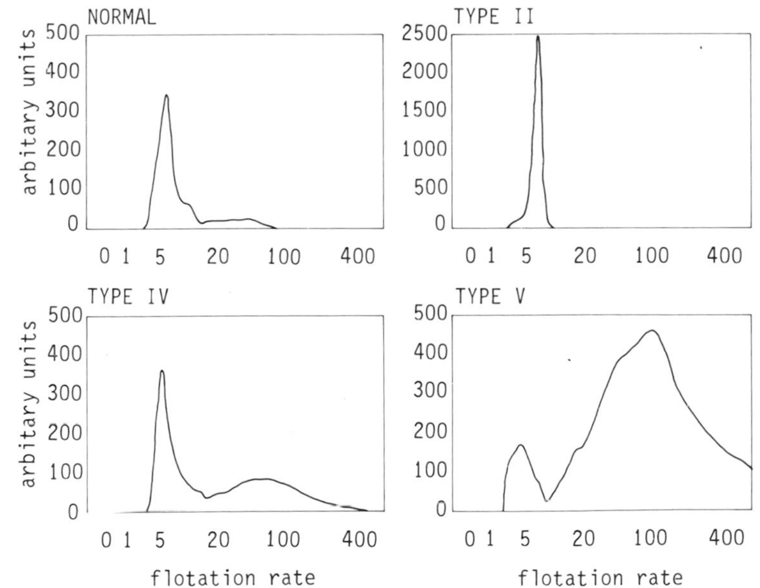

In the pattern obtained from the ultracentrifuge, the scale on the abscissa relates to the density of the lipoprotein, and is marked in Svedberg flotation units (S_f is the usual abbreviation). The low density (*β*) lipoprotein is delineated approximately by the range of 0–20 S_f units, and the very low density (pre-*β*) lipoprotein by the range 20–400 S_f units. High density lipoprotein is centrifuged at a different density, and does not appear in the patterns shown above.

These curves were drawn by computer. The ordinate shows arbitrary units. Note the different scale on the ordinate of the Type II hyperlipoproteinaemia.

LDL receptor traffic

In normal individuals LDL binds through its B-100 apoprotein to a receptor in a *clathrin coated pit*. It is internalized as a *coated vesicle*, and separated from its receptor at the endosome stage (see p.263). The receptor is recycled to the plasma membrane, whilst the vesicle containing the LDL fuses with a lysosome. The cholesterol esters of the LDL are hydrolysed to cholesterol and fatty acid, and the phospholipids, triacylglycerols and apoprotein are degraded. Free cholesterol that is internalized inhibits the synthesis of the Apo B receptor, and also inhibits HMG CoA reductase, reducing synthesis of cholesterol *de novo*. Elevated cholesterol levels also activate the synthesis of storage cholesterol esters by acyl CoA cholesterol acyl transferase (ACAT).

NOTE: ApoE receptors are also found on cell membranes, especially of liver and are thought to regulate the uptake of apo E-bearing lipoproteins.

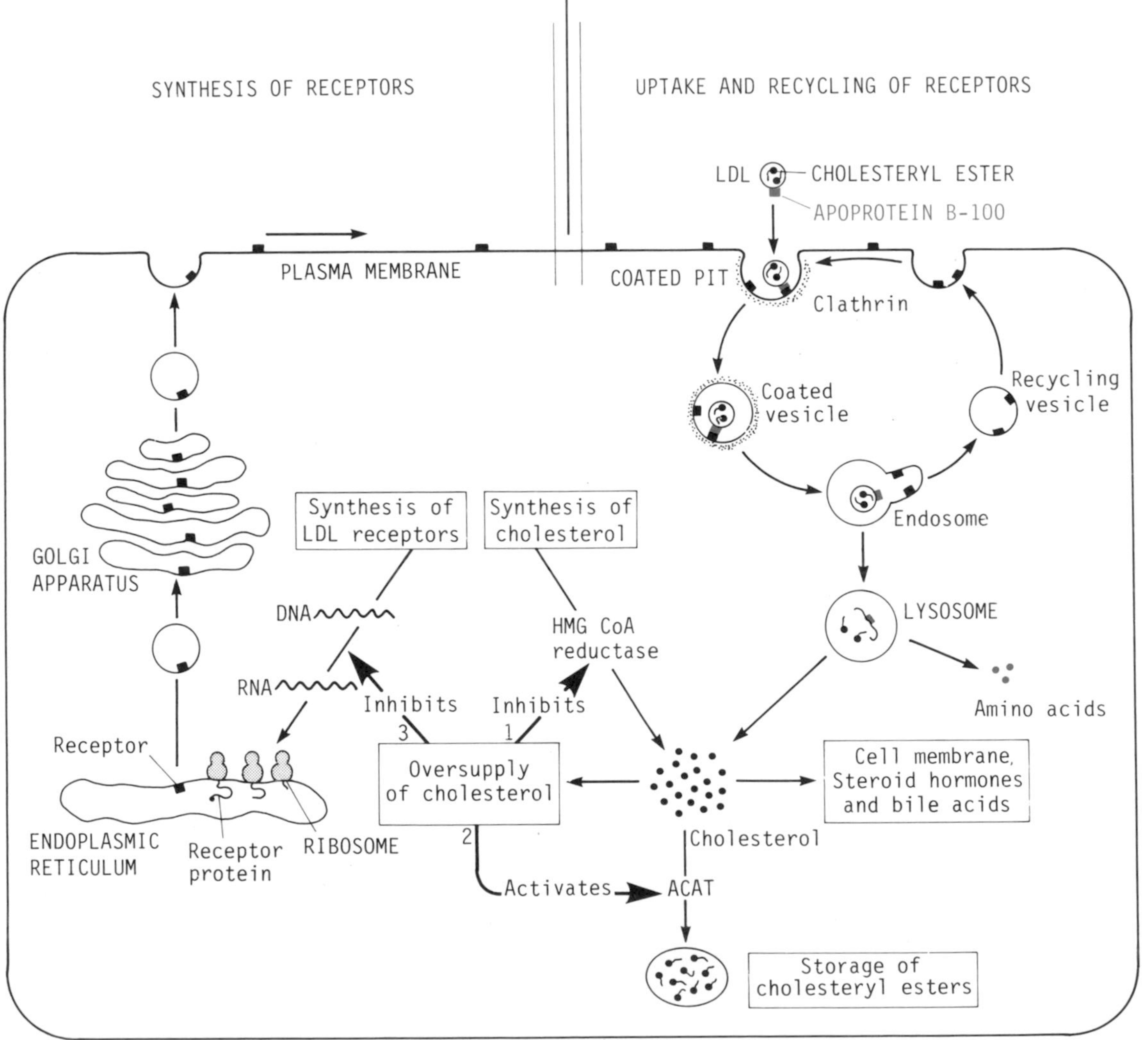

In the heterozygote in Type II familial hyperlipoproteinaemia (see next page), the strategy of treatment is to increase the number of receptors by lowering the cholesterol level in the liver. This can be achieved by preventing the recycling of bile acids, which thus activates 7α-hydroxylase, leading to accelerated conversion of cholesterol to bile acids. It is also useful to block cholesterol synthesis using the drug *compactin*, a potent inhibitor of HMG CoA reductase. Thus, less VLDL is synthesized and LDL levels are reduced.

A

The control of the level of plasma LDL

LDL (carrying the B-apoprotein) and the lipoprotein from which it is formed (IDL carrying B- and E- apoproteins) are removed from the plasma after binding to apoprotein receptors on the surface of liver and other cells.

Much of the IDL is removed from the circulation before it is converted into LDL. The E apoprotein is lost before conversion of IDL to LDL. The rate of removal of IDL from the blood regulates to some extent the rate of its conversion to LDL.

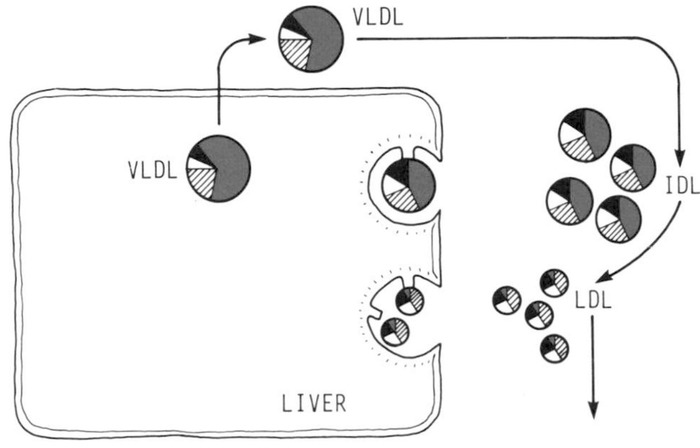

B

Type II familial hyperlipoproteinaemia.

In the homozygote, the gene for the synthesis of the Apo B receptor is totally lacking. This leads to exceptionally high levels of LDL (and thus also of cholesterol) in the blood. In the heterozygote the gene of only one chromosome is functional, leading to the synthesis of reduced concentrations of receptor, and to increased risk of early death due to atheromatous plaques in the arterial wall. In the homozygote, death may occur in childhood, in heterozygotes often before the age of 40.

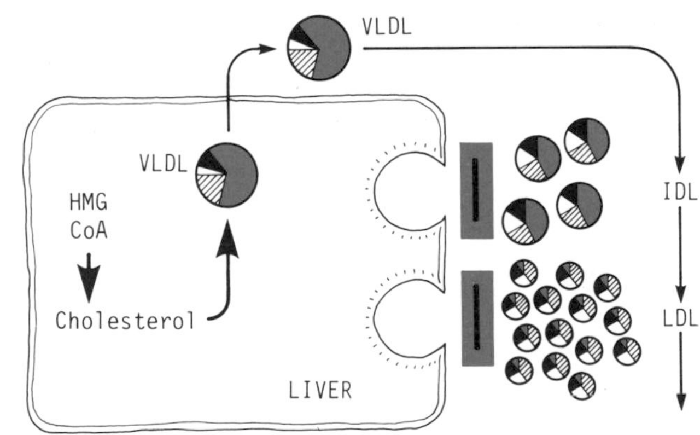

C

A two-pronged approach to therapy is required. An anion-exchange resin (cholestyramine), taken by mouth, binds bile acids in the gut and prevents their re-absorption.

This can be combined with an inhibitor of cholesterol synthesis, compactin or mevinolin, which inhibits HMG CoA reductase. (see previous page).

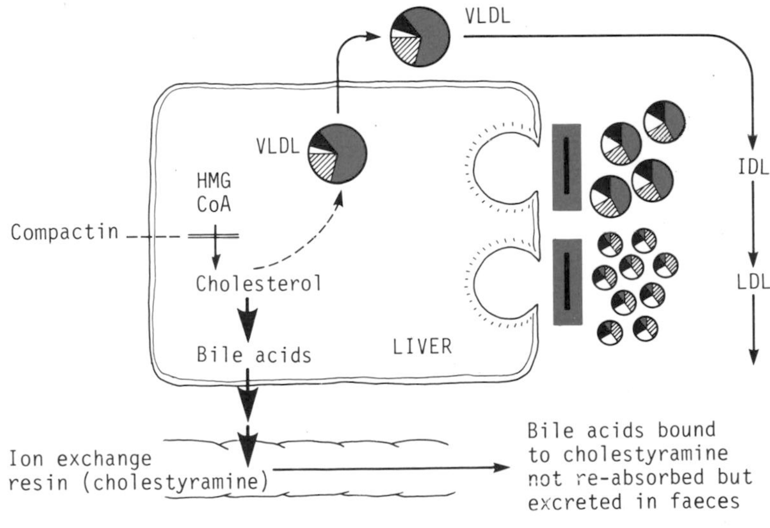

Formation of bile acids

The bile acids are synthesized from cholesterol. Two distinct series of acids are synthesized, one yielding *cholic acid*, and the other *chenodeoxycholic acid*. From these, *deoxycholic acid* and *lithocholic acid* can be synthesized in the gut by bacteria. The acids are secreted into the bile largely as *glycocholic* and *taurocholic acids*.

In cholic acid, all 3 hydroxyl groups project from the same face of the molecule (see p. 227). The synthetic reactions are thus directed to achieving this, including inversion of the 3*a* hydroxyl group in cholesterol to a 3 *β* hydroxyl group in cholic acid.

Taurocholic acid is a compound in which taurine is linked to cholic acid by an amide linkage between the carboxyl group of cholic acid and the amino group of taurine. *Taurine* ($H_2NCH_2CH_2SO_3H$) is formed from a derivative of cysteine, in which the sulphydryl group has been oxidized to a sulphonic acid group ($—SH \rightarrow —SO_3H$).
Glycocholic acid has glycine linked to cholic acid in a similar manner.

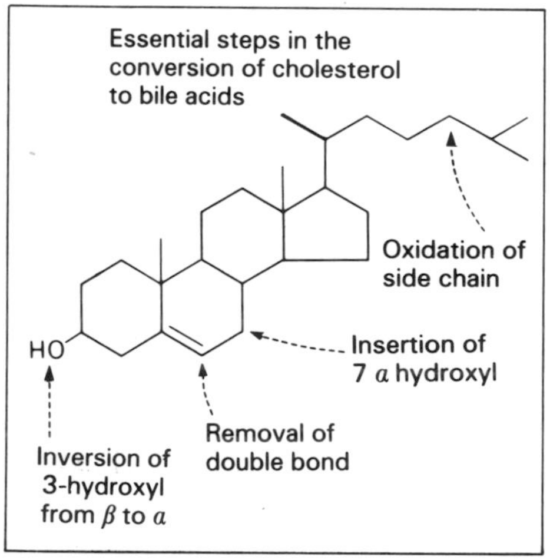

The enzyme that hydroxylates the 7 position (7α hydroxylase) regulates the rate of synthesis of the acids

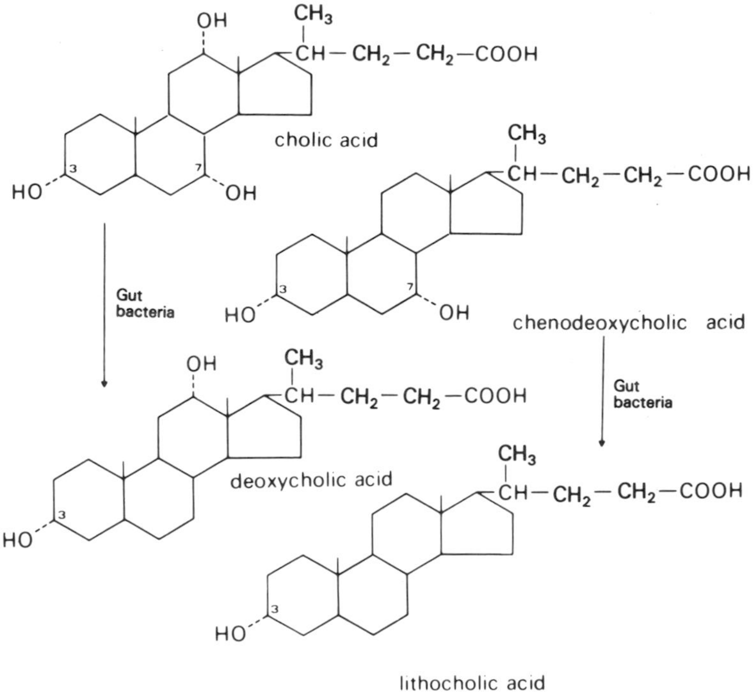

Essential steps in the conversion of cholesterol to bile acids

Oxidation of side chain

Insertion of 7 *a* hydroxyl

Removal of double bond

Inversion of 3-hydroxyl from *β* to *a*

HO

OH

CH_3

$CH—CH_2—CH_2—COOH$

cholic acid

chenodeoxycholic acid

Gut bacteria

deoxycholic acid

lithocholic acid

A

Bile

Bile is formed in the liver and secreted down the *bile duct*, into the *gall bladder*, where it passes to the intestine. Here the bile salts facilitate the degradation of ingested fats (see p. 209).

Bile is a complex solution of salts and protein, and contains micelles of cholesterol, phospholipids and bile salts. The composition of these micelles is critical: imbalance may result in crystallization of cholesterol in the gall bladder, leading to the formation of gallstones. This may result from small differences in composition.

	Normal Bile	**Abnormal bile (taken from a patient with cholesterol gallstones)**
	%	%
Lecithin	74	71
Bile salts	20	13
Cholesterol	6	16

B

Bile micelles include bilayers of phospholipid containing cholesterol

These form a disc with an ionic flat surface above and below, and hydrophobic sides. These are coated with bile salts, which present their hydrophobic face to the hydrophobic region of the disc, and a hydrophilic face directed towards the aqueous medium.

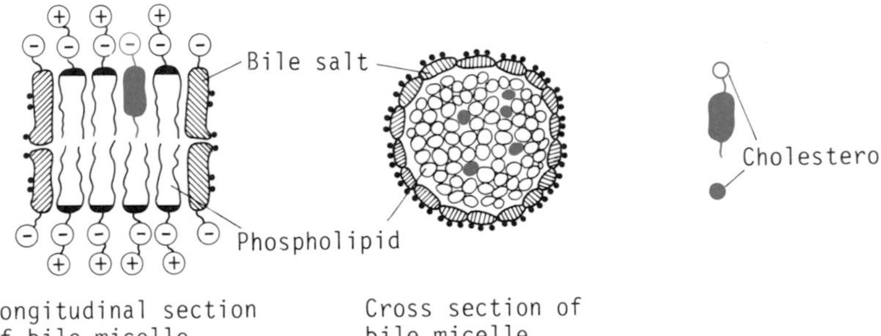

The bilayers of phospholipid and cholesterol are essentially similar formations to those found in cell membranes.

C

Bile acids are a highly specialized form of detergent

However, rather than consisting of a molecule with a hydrophobic and hydrophilic *end*, as in a normal detergent, bile acids have a hydrophilic and a hydrophobic *face*.

COOH
Hydrophobic face
OH
OH
CHOLIC ACID
Hydrophilic face
OH

The complex series of reactions converting cholesterol into the bile acids is directed to converting configurations of the hydroxyl groups and the ring conformations to give this structure highly specialized for its function, and particularly that all hydroxyls project from the same face of the molecule.

A

Whole body regulation of cholesterol metabolism

Control of the cholesterol level in the plasma is largely dependent on regulation of cholesterol synthesis in the liver. Cholesterol excreted into the intestine in the bile mixes with dietary cholesterol, and is absorbed back into the blood, but routes of excretion and re-adsorption are not closely regulated.

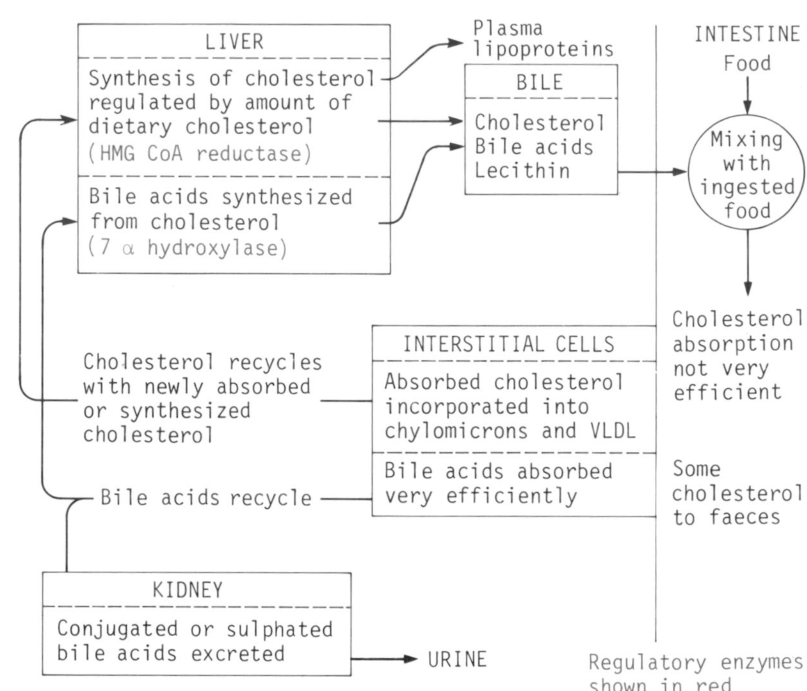

The bile contains mixed micelles of cholesterol, bile acids and lecithin. If these micelles contain too much cholesterol this may crystallize out and form gallstones. Excretion of cholesterol in the bile is thus concerned mainly with maintaining the composition of the bile and is not a route for controlling cholesterol excretion, although, as the absorption of cholesterol is not very efficient, some cholesterol is lost by this route. Conversion of bile acids in the gut to forms that are not re-absorbed, and urinary excretion of bile acids that have been conjugated or sulphated in the kidney, are other routes of excretion of cholesterol.

B

Hyperlipidaemias (Hyperlipoproteinaemias)

(see also page 223).

Patients in whom high plasma lipoprotein levels are found, may often be classified into types, first described by Fredrickson. In some cases (comparatively rarely) there is a familial trait that gives rise to the abnormality.

Types of hyperlipidaemia

Fredrickson Type	*Lipoprotein elevated*	*Cholesterol level*	*Triacylglycerol level*
I	Chylomicrons (also possibly VLDL)	+	+ + +
IIa	LDL	+ + +	±
IIb	LDL & VLDL	+ +	+ +
III	'floating' LDL	+ + +	+ + +
IV	VLDL	±	+ +
V	VLDL and chylomicrons	+	+ + +

± normal to slightly increased. + + moderately increased. + + + greatly increased.

8

CARBOHYDRATE AND FAT METABOLISM

E. Phospholipids, other lipid substances and complex carbohydrates

Structure of phospholipids

Membrane lipids consist of
1. Phospholipids
2. Sphingolipids
3. Cholesterol
4. Glycolipids (containing carbohydrate)

Phospholipids have a glycerol backbone. A fatty acid is esterified to two of the OH groups of the glycerol. The third is esterified by a phosphate group and a nitrogenous compound (choline, ethanolamine or serine) except that one family of phospholipid contains inositol, a 6-C sugar alcohol. The nitrogenous moiety is often referred to as a 'base', and the phosphate and 'base' together comprise the 'headgroup'.

Phospholipids isolated from natural sources vary in their fatty acids esterified at positions C-1 and C-2. Thus many isomers (often called species) are possible for each phospholipid, i.e. 1-palmitoyl-2-linoleoyl-*sn*-glycero-3-phosphocholine and 1-stearoyl-2-arachidonoyl-*sn*-glycero-3-phosphocholine are species of phosphatidylcholine. Diacyl-*sn*-glycero-3-phosphate has the trivial name phosphatidic acid. Thus diacyl-*sn*-glycero-3-phosphocholine is often alternatively termed phosphatidylcholine.

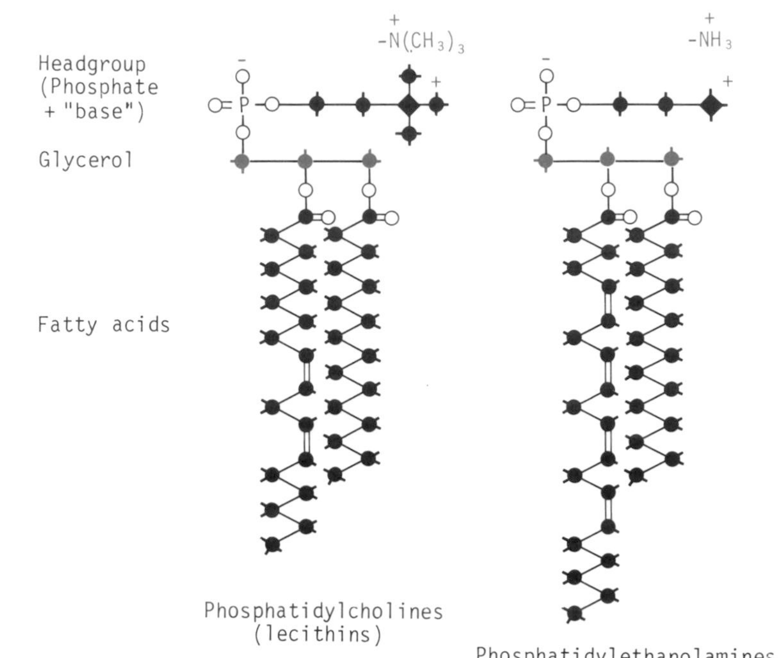

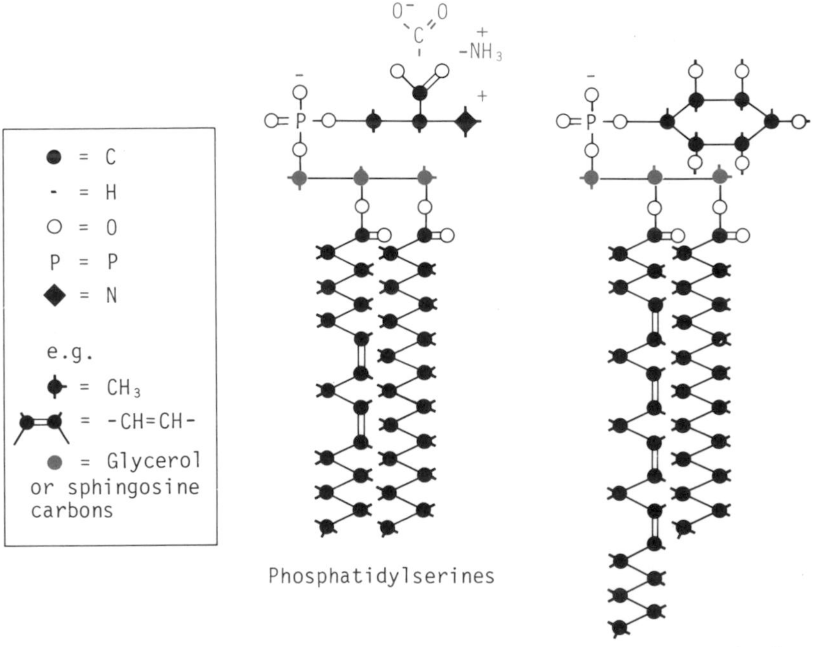

The drawings are diagrammatic and do not attempt to represent molecular orientations in space.

A

Phospholipids are derived from *sn*-glycerol 3-phosphate.

$$\begin{array}{ll} CH_2OH & \text{C-1} \\ HO\text{—}C\text{—}H & \text{C-2} \\ CH_2OPO_3H_2 & \text{C-3} \end{array}$$

NOTE: *Sn* indicates *S*tereospecific *n*umbering. It derives from an arbitrary convention which numbers the glycerol carbons as in the diagram, with the chiral carbon oriented as shown.

B

Synthesis of phospholipids

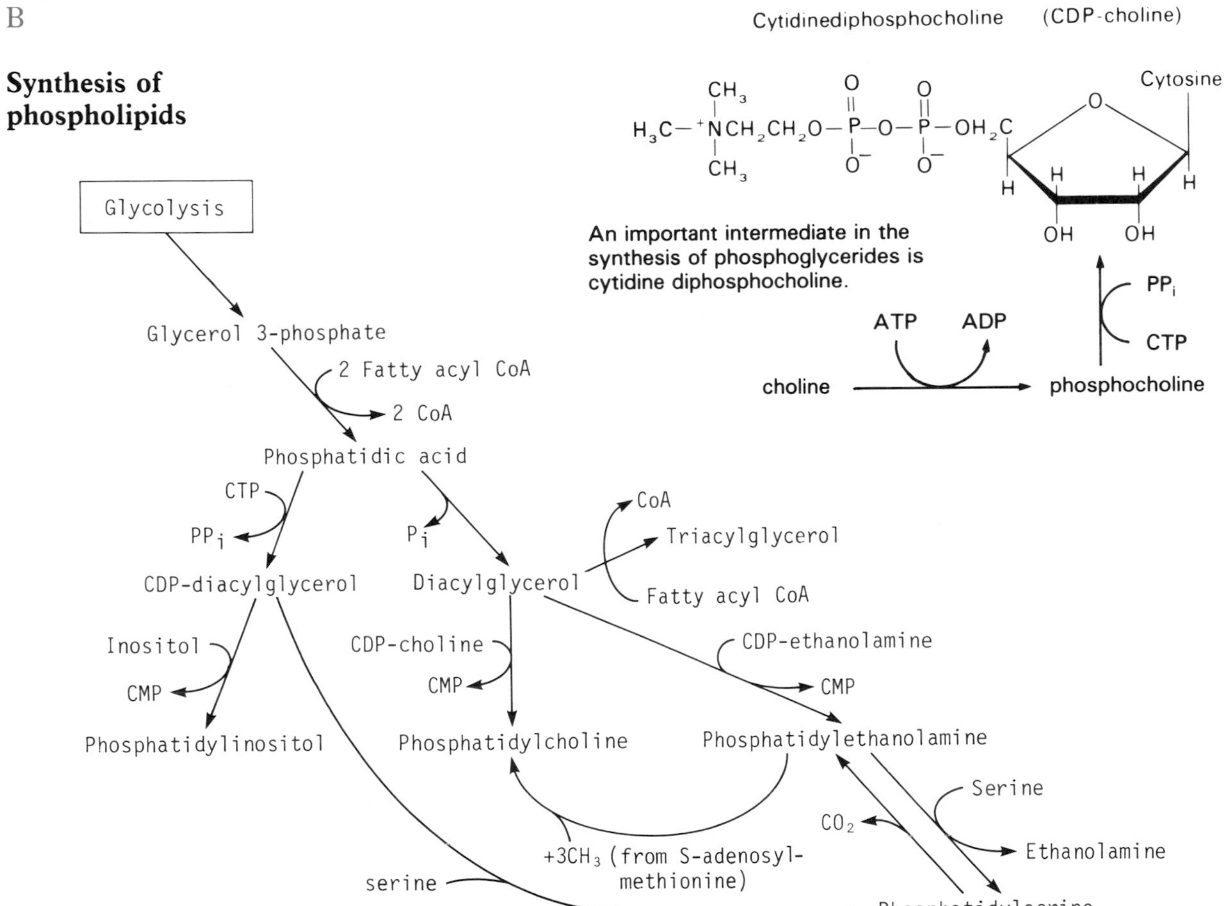

The synthesis of phospholipids and sphingolipids involves many complex reactions and only an outline is presented here. Synthesis of phosphatidic acid involves the acylation of glycerol phosphate as previously described for the synthesis of triacylglycerol. The compounds CDP-diacylglycerol, CDP-choline and CDP-ethanolamine involve an ester linkage between the hydroxyl group of diacylglycerol, choline or ethanolamine and the terminal phosphate of CDP.

A **Phospholipases**

Phospholipids play an important role in cell stimulation as substrates for phospholipases.

Phospholipase A_2 removes unsaturated fatty acids from the *sn*-2 position of phospholipids. If arachidonic acid is released it acts as a substrate for formation of prostaglandins.

Marked in red are the bonds attacked by different phospholipases. Thus fatty acids are removed from the *sn*-1 position by phospholipase A_1 and from the *sn*-2 position by phopholipase A_2. Phospholipase C removes the whole of the head group leaving diacylglycerol, and phospholipase D removes the terminal hydroxy compound, leaving phosphatidic acid.

B

Of particular importance in many mammalian cells is an enzyme with phospholipase C action which is specific for phosphatidylinositol and its phosphorylated derivatives. This enzyme has the name *phosphatidylinositol phosphodiesterase*. Thus it attacks diesters of phosphoric acid, i.e. phosphoric acid with two of its acidic —OH groups involved in ester linkage.

Release of the head group from PIP_2 yields inositol trisphospate (IP_3) and diacylglycerol (DG).

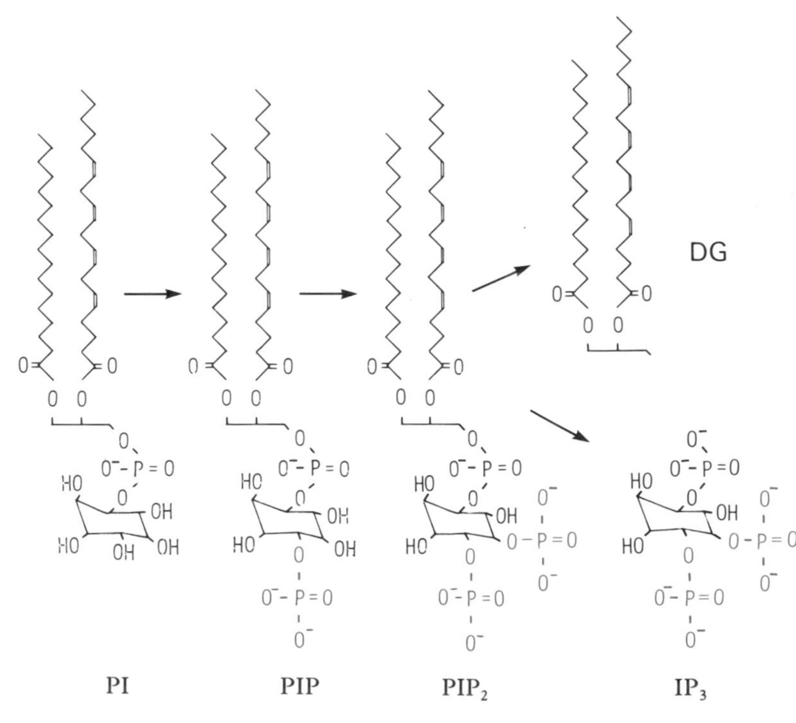

Phosphatidylinositol = PI

Phosphatidylinositol 4-monophosphate = PIP
Phosphatidylinositol 4,5-bisphosphate = PIP_2
Inositol 1,4,5-trisphosphate = $1P_3$.

A

Structure of sphingolipids

The backbone of sphingolipids is the base *sphingosine.*

$CH_2OHCHNH_2CHOHCH{=}CH(CH_2)_{12}CH_3$

A fatty acid is attached to the nitrogen in amide linkage. To the terminal hydroxyl a sugar or chain of sugars is attached, except in the case of sphingomyelin, which has a phosphocholine grouping on this hydroxyl group.

The atoms derived from sphingosine are shown in red.

phosphocholine

fatty acid

Sphingomyelins

Ceramide (Common to all Sphingolipids)

galactose

fatty acid

Galactocerebroside

The sphingolipids include sphingomyelins, cerebrosides and gangliosides. Sphingomyelin may be described as a phospholipid as it contains phosphorus. Cerebrosides and gangliosides contain no phosphorus, but do contain carbohydrate structures, and thus are both sphingolipids and glycolipids. Sphingomyelin is an important structural lipid in membrane bilayers, comprising 10–20% of the total phospholipid of the bilayer. As their name suggests, cerebrosides (and their sulphated derivatives, sulphatides) are found in appreciable quantities in brain, especially myelin, as also applies to gangliosides. Cerebrosides and gangliosides are found in much smaller quantities in most membranes.

B

Sphingosine is synthesized from serine and palmitoyl CoA

Palmitoyl CoA: $O{=}C{-}S{-}CoA$, CH_2, $(CH_2)_{13}$, CH_3 + Serine: CH_2OH, $HC{-}NH_3^+$, COO^-

$\xrightarrow{H^+ \quad CO_2}$

Dihydrosphingosine: CH_2OH, $HC{-}NH_3^+$, $HO{-}C{-}H$, CH_2, CH_2, $CH_3{-}(CH_2)_{12}$

$\xrightarrow{NADP^+ \quad NADPH}$

Sphingosine: CH_2OH, $H{-}C{-}NH_3^+$, $HO{-}C{-}H$, CH, $\|$, CH, $CH_3{-}(CH_2)_{12}$

C

Sphingomyelin is synthesized by addition of fatty acid to the hydroxy group of sphingosine to form *ceramide*, which reacts with CDP-choline to yield sphingomyelin.

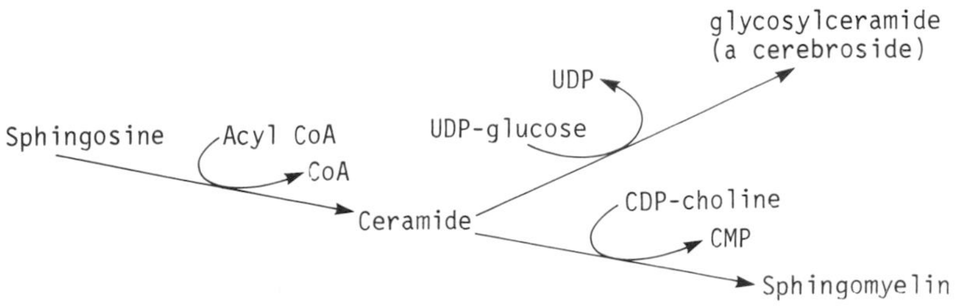

Ceramide also reacts with UDP-glucose or UDP-galactose to form glucosylceramide or galactosylceramide (the cerebrosides).

A

The steroid hormones

The three main types of steroid hormone are

1. Female sex hormones, or *oestrogens*
2. Male sex hormones, or *androgens*
3. Corticoid hormones, including *cortisol*, *corticosterone* and *aldosterone*

Progesterone, which serves as an intermediate in the synthetic pathway, also has important hormonal properties.

The main oestrogens are oestradiol and oestrone. Oestriol has a structure similar to that of oestradiol, with a third hydroxyl group at C-16. Testosterone is the main androgen, but its metabolite 5-dihydrotestosterone is also important. Cortisol has glucocorticoid and anti-inflammatory action, and aldosterone is the important mineralocorticoid. The precursor of all these steroid hormones is cholesterol. The initial step in all the synthetic pathways is the conversion of cholesterol to pregnenolone by the enzyme cholesterol desmolase. 3β Dehydrogenase converts pregnenolone to progesterone.

Cholesterol

Cholesterol desmolase

3 β dehydrogenase

Pregnenolone

Progesterone

B

Feedback regulation of steroid synthesis by the adrenal cortex results from cortisol action on the pituitary.

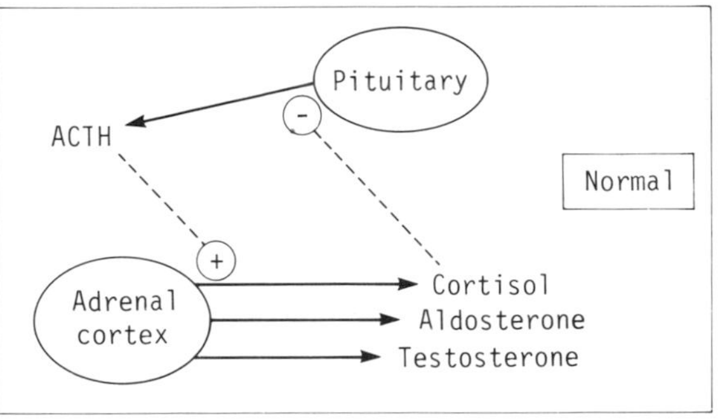

Inhibition by cortisol of ACTH secretion by the pituitary reduces the plasma ACTH level, and thus reduces the stimulus to the adrenal cortex to synthesize steroids.

Pathways of Steroid Hormone Synthesis

1 & 2 show reactions missing in the deficiency states described on page 236.

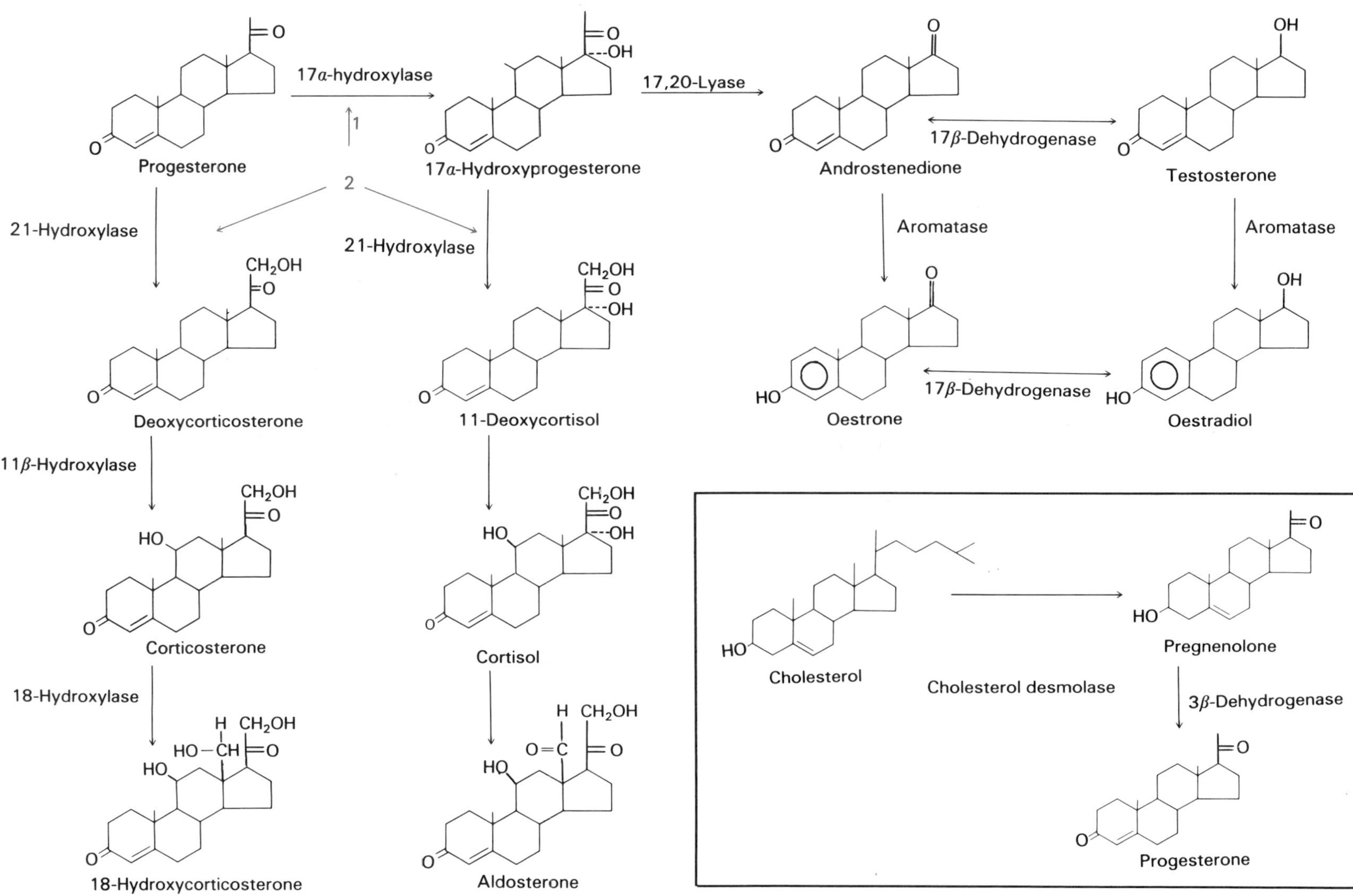

Inherited diseases are known, which involve a deficiency of one or other of the enzymes of the synthetic pathways shown on the previous page.

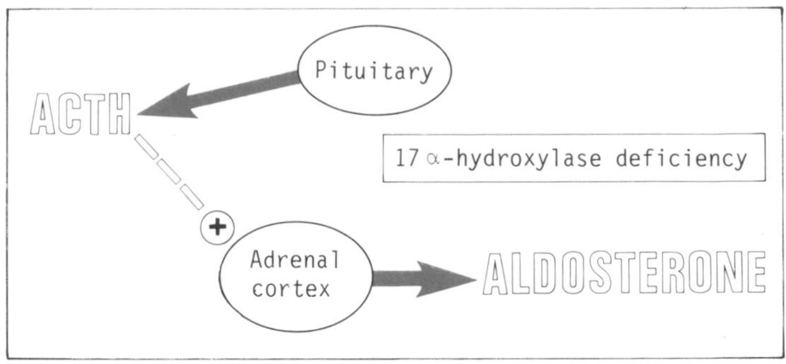

A deficiency of cholesterol desmolase or 3 *β*-dehydrogenase will result in failure to synthesize any of the steroid hormones. A deficiency of 17 *α*-hydroxylase results in failure to synthesize cortisol. This is the hormone mainly responsible for negative feedback on the release of adrenocorticotrophin (ACTH), the hormone which stimulates the synthesis of steroid hormones in the adrenal cortex (see page 234). Thus there is overproduction of ACTH, and this results in excessive production by the adrenal cortex of aldosterone, the synthesis of which does not involve 17 *α*-hydroxylase. Deficiency of 21-hydroxylase also leads to failure to synthesize cortisol with resulting overproduction of ACTH (the mechanism underlying the condition known as congenital adrenal hyperplasia). In the case of 21-hydroxylase deficiency, there is not only failure to synthesize aldosterone and cortisol, which requires this enzyme, but overproduction of testosterone due to the action of excessive concentrations of ACTH on the adrenal.

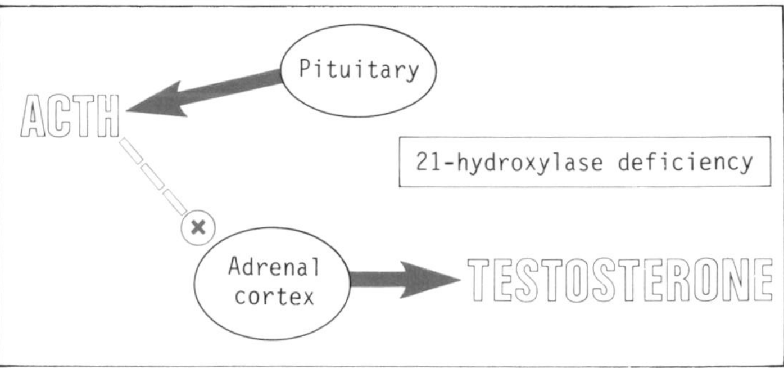

A

The lipid soluble vitamins A, E, K and D

Vitamin A

At least 3 different physiological functions are dependent on proper vitamin A nutrition.

1. *Somatic function* — including growth and differentiation e.g. of epithelial structures and bone
2. *Reproduction* — it is essential for spermatogenesis, oogenesis, placental development and embryonic growth
3. *Visual process* — for vision in the dark

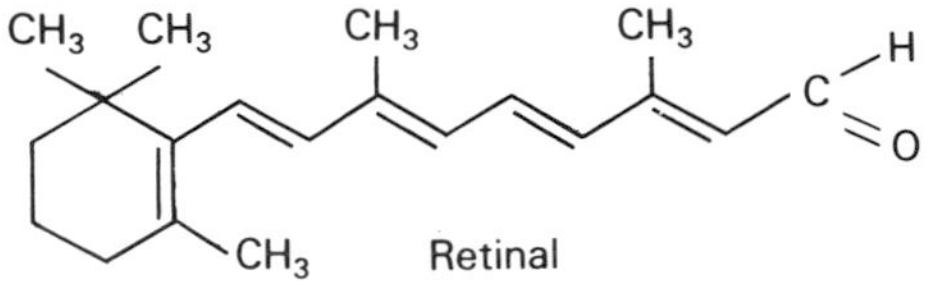

Retinal

Vitamin A (retinol) has a similar structure to retinal but with a primary alcohol in place of the aldehyde group.

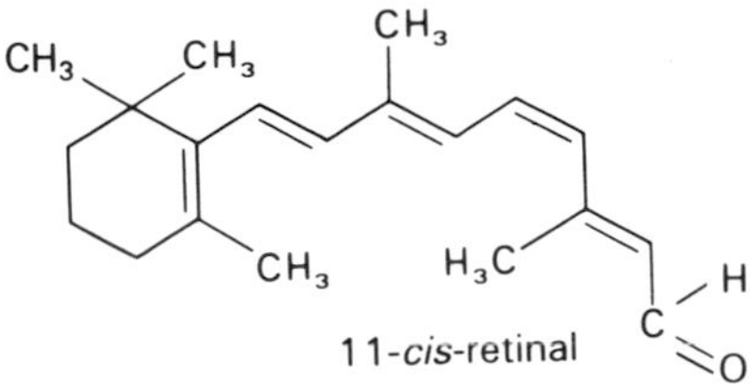

11-*cis*-retinal

Vitamin A is important for night vision, 11-*cis*-retinal interacts with the protein opsin, to form the visual pigment rhodopsin.

11-*cis*-retinal + opsin ⟶ rhodopsin

If rhodopsin is exposed to light, the double bond at C-11 of *cis*-retinal isomerizes from *cis* to *trans*, causing dissociation of rhodopsin to opsin and the *trans* form of retinal, which can then be regenerated enzymically to 11-*cis*-retinal. Vitamin A deficiency causes a number of other symptoms, but the mechanism of its action in preventing these is unknown. Excessive doses of vitamin A are highly toxic.

It is also now suspected that excessive doses of vitamin A could be teratogenic, a fact which is of considerable significance as vitamin A or derivatives are an effective treatment for acne in some young women.

B

Vitamin E

The main function of vitamin E appears to be to act as an anti-oxidant, preventing the formation of free radicals in cell membranes. [A free radical can be represented as R•]

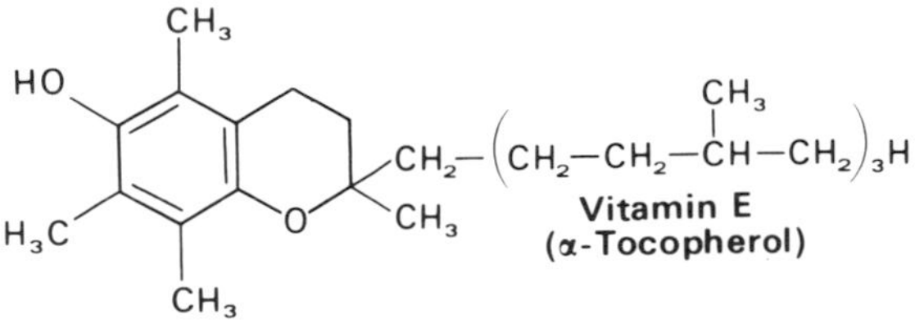

Vitamin E (α-Tocopherol)

Vitamin E is an anti-oxidant, and prevents formation of the peroxides which may otherwise be produced when polyunsaturated fatty acids are exposed to oxygen. Its physiological role is, however, not well characterized. Polyunsaturated fatty acids are very susceptible to oxidation to peroxides, through the formation of free radical intermediates. A free radical can be formed by breakage of a covalent bond in such a way than an unpaired electron is left. Thus, loss of a hydrogen atom from a methylene group forms a free radical. Then subsequent addition of O_2 can initiate formation of a peroxide.

$$-CH_2-CH=CH-CH_2-CH=CH-CH_2- \longrightarrow -CH_2-CH=CH-\dot{C}H-CH=CH-CH_2-$$

free radical

$$-CH_2-CH=CH-CH(O-O^\bullet)-CH=CH-CH_2- \longrightarrow -CH_2-CH=CH-CH(O-OH)-CH=CH-CH_2-$$

hydroperoxide

Each free radical can initiate formation of another free radical by passing on its free electron, thus starting a chain reaction. Any compound which can trap a free electron to break the chain reaction will act powerfully to prevent formation of free radicals.

A

Vitamin K

Vitamin K is a cofactor in post-translational formation of γ-carboxy glutamyl residues in certain proteins especially prothrombin.

$$-NH-CH(CH_2CH_2COO^-)-C(=O)-NH- \xrightarrow[\text{Vitamin K (reduced form)}]{CO_2 \quad O_2} -NH-CH(CH_2CH(COO^-)_2)-C(=O)-NH-$$

glutamyl residue → γ-carboxy glutamyl residue (Gla)

In vitamin K deficiency, prothrombin cannot be synthesized, and therefore the blood clotting mechanism functions imperfectly.

B

Vitamin K exists in a reduced form and as an epoxide.

Formation of the epoxide is coupled to carboxylation of the glutamyl residue by an as yet unresolved mechanism.

Reduced form Vitamin K → epoxide form of Vitamin K (Epoxidase; Glu, CO_2, O_2 → Gla, H_2O)

epoxide form of Vitamin K → Reduced form Vitamin K (Reductase)

Warfarin blocks reductase

R = [isoprenoid side chain]$_3$

Warfarin inhibits formation of the reduced form of vitamin K that is required for the γ-carboxylation reaction, and thus mimics vitamin K deficiency. It can thus be used clinically as an anticoagulant. It is also a very effective rat poison.

C

Vitamin K-dependent carboxylation occurs in other proteins of the blood clotting cascade, and in proteins in tissues.

Simplified scheme of the coagulation mechanism with the four vitamin K-dependent clotting factors (in red) PL refers to phospholipid; V, proaccelerin; VII, proconvertin; VIII, antihaemophilic factor A; IX, Christmas factor or antihaemophilic factor B; X, Stuart factor. The subscript 'a' means that the factor is in the active form.

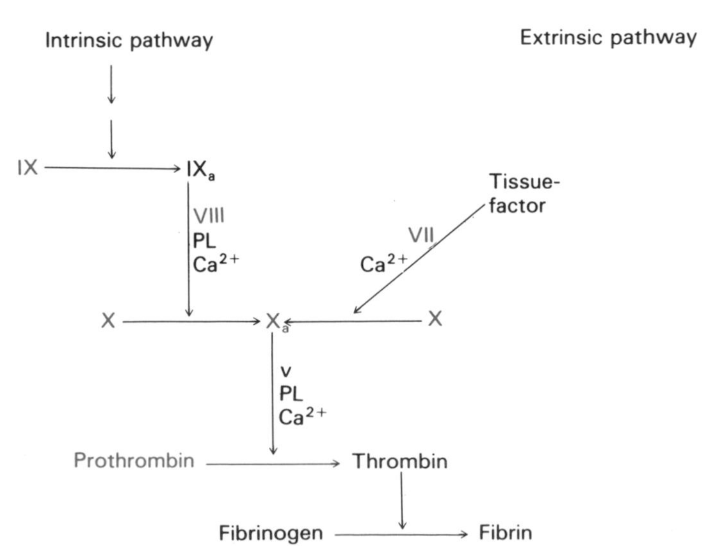

Carboxyglutamate-containing proteins are powerful Ca^{2+}– binding proteins, because of the chelating properties of the γ-carboxyglutamyl group.

A

Vitamin D

Vitamin D (cholecalciferol) may be absorbed from the intestine or derived by the action of UV light from 7-dehydrocholesterol. It is converted by the liver to 25-hydroxycholecalciferol, which is converted by kidney to the active form of the vitamin, 1,25-dihydroxycholecalciferol.

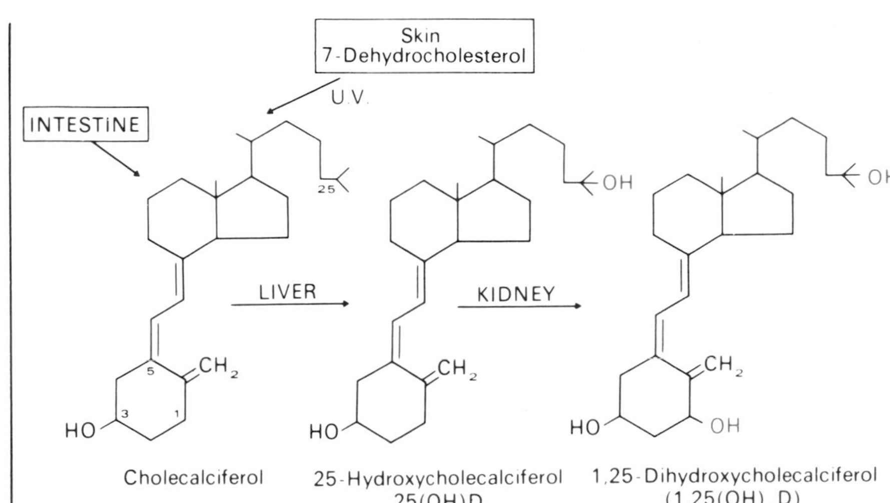

B

Calcium metabolism

The blood Ca^{2+} level is controlled by the combined action of vitamin D, *parathyroid hormone* and *calcitonin*.

Parathyroid hormone
1. is secreted by the parathyroids in response to a drop in the blood Ca^{2+} level
2. increases production of 1,25 $(OH)_2D$ by kidney
3. increases Ca^{2+} resorption from bone
4. decreases Ca^{2+} excretion in urine.

1,25-Dihydroxycholecalciferol
1. is the hormonally active form of vitamin D
2. increases Ca^{2+} absorption from gut
3. increases Ca^{2+} resorption from bone.

Calcitonin
1. is produced by the chief cells of the thyroid
2. reduces Ca^{2+} resorption from bone.

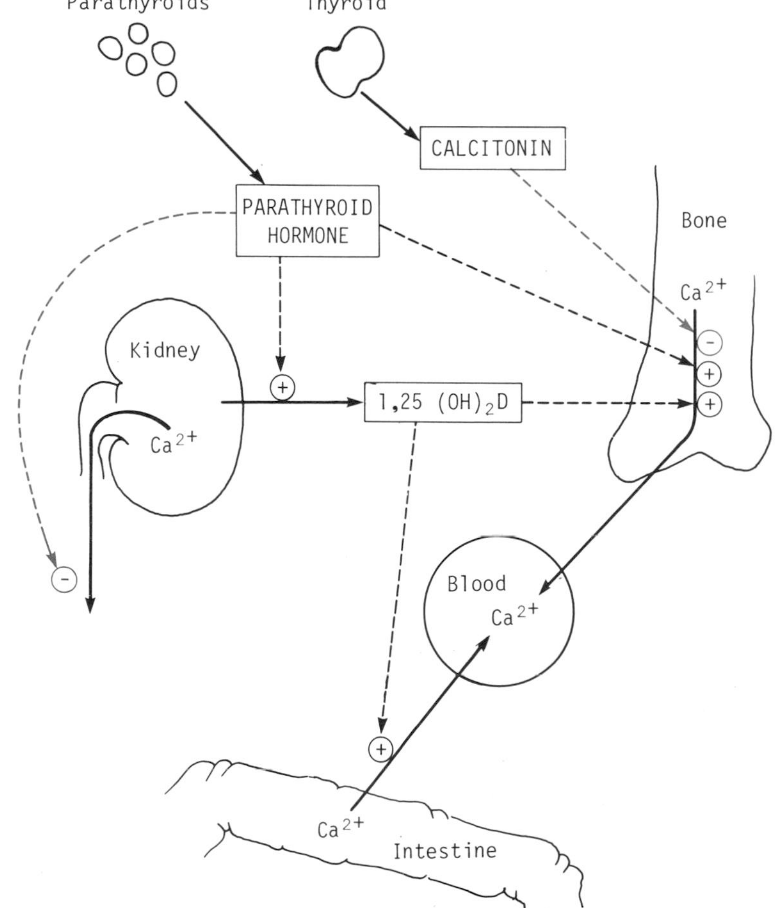

Dietary deficiency of vitamin D causes rickets, a disease in which there is defective bone formation.
Calcitonin may be administered in the treatment of Paget's disease.

Prostaglandins

The prostaglandins are synthesized in many tissues in minute amounts. The different prostaglandins have highly potent pharmacological actions that include contraction of smooth muscle, including that of the uterus, vasodilation and platelet aggregation.

Prostaglandins are synthesized from the two endoperoxides PGG_2 and PGH_2, which are the products of the action of the enzyme *cyclooxygenase* on arachidonic acid.

In recent years it has been found that another enzyme, *lipoxygenase*, acts on arachidonic acid to form a series of hydroxy fatty acids, one of which is termed 12-hydroxyeicosatetraenoic acid (HETE). A number of pharmacological actions of these acids are being identified (they are chemotactic, for example, for macrophages, and thus may be important in the inflammatory response).

The different prostaglandin classes (see opposite page) are distinguished by the substituents on the ring part of the structure, and the different series by the number of double bonds (this depends on which acid is utilized as the precursor of the prostaglandin).

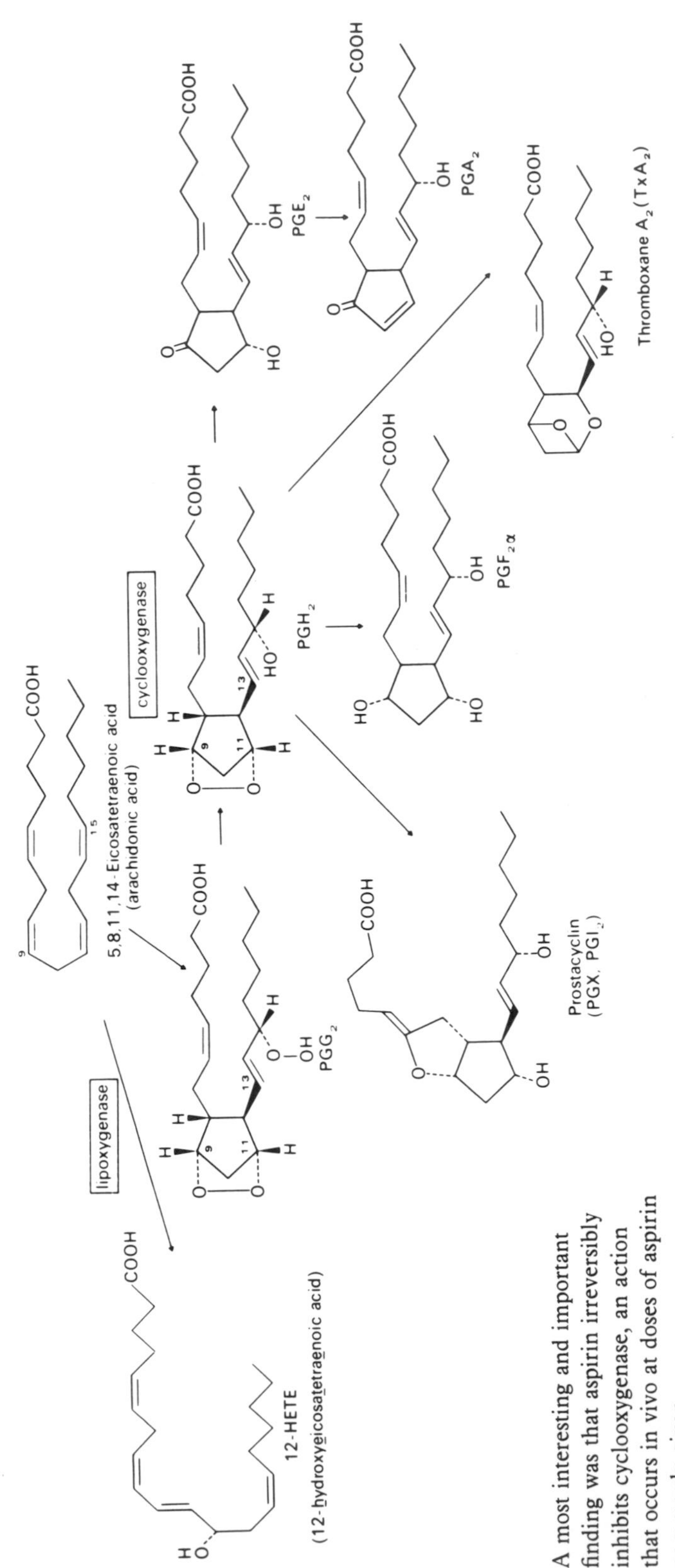

A most interesting and important finding was that aspirin irreversibly inhibits cyclooxygenase, an action that occurs in vivo at doses of aspirin commonly given.

Structure of prostaglandins and related compounds

COMMON FEATURES	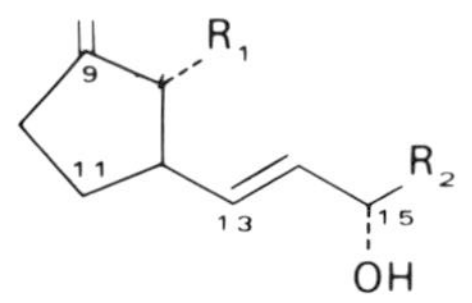 9, 11, 13, 15; R_1, R_2, OH
DIFFERENT CLASSES	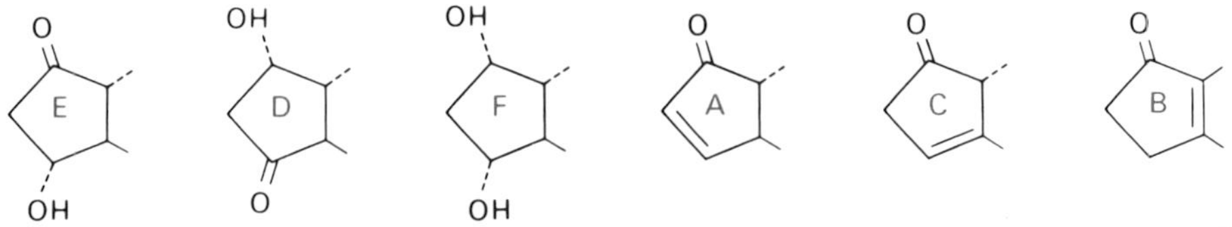 E (O, OH); D (OH, O); F (OH, OH); A (O); C (O); B (O)
DIFFERENT SERIES	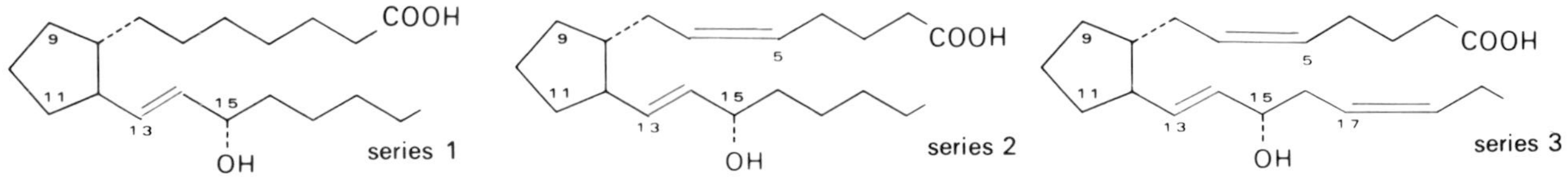 9, 11, 13, 15, COOH, OH — series 1; 9, 11, 13, 15, 5, COOH, OH — series 2; 9, 11, 13, 15, 5, 17, COOH, OH — series 3
ENDOPEROXIDES	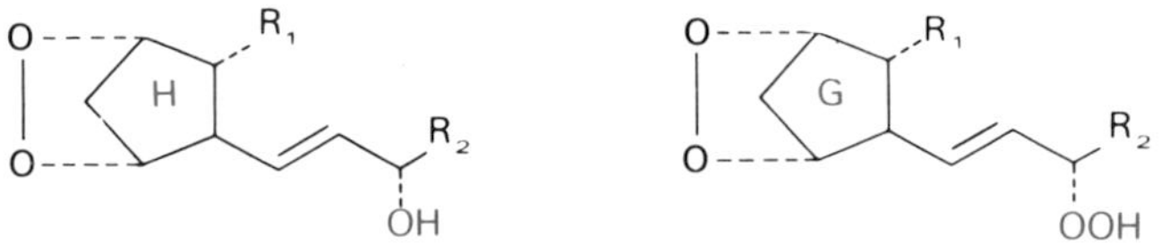 H (O—O, R_1, R_2, OH); G (O—O, R_1, R_2, OOH)
THROMBOXANES (TX)	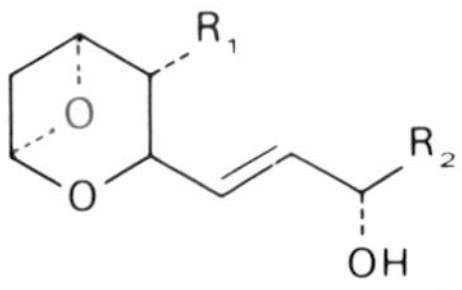 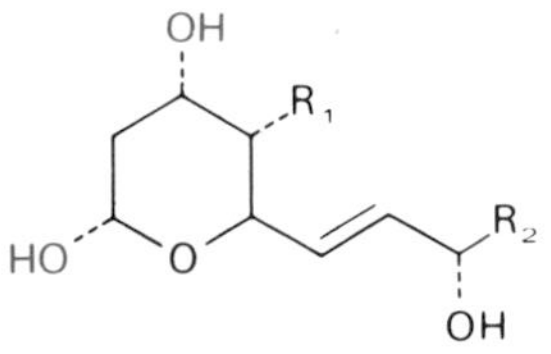 O, O, R_1, R_2, OH — TXA; OH, HO, O, R_1, R_2, OH — TXB
PROSTACYCLIN	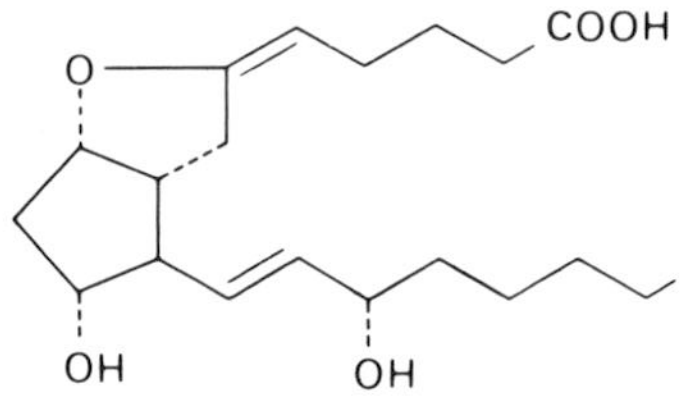 O, COOH, OH, OH

A

Linoleic acid can be elongated and desaturated to dihomo-γ-linolenic acid which gives rise to the '1' series of prostaglandins. Metabolism of α-linolenic acid gives rise to the '3' series of prostaglandins.

Linoleic acid (18:2,n-6,9) —elongation / desaturation→ C20:3 n-6,9,12 —oxygenation→ PGE_1

α-Linolenic acid (18:3,n-3,6,9) —elongation / desaturation→ C20:5 n-3,6,9,12,15 —oxygenation→ PGE_3

The most commonly found series is the '2' series, synthesized from arachidonic acid.

B **Leukotrienes**

The *leukotrienes* are arachidonic acid metabolites.

When conjugated to glutathione these are known as the peptidoleukotrienes. However, many polyunsaturated fatty acids undergo oxidation by different lipoxygenase enzymes to form hydroxyl derivatives, which appear to exercise a regulatory action in various cell functions. The structure of leukotriene B_4 is shown. This is chemotactic.

A recent development has been the elucidation of the structures of a series of compounds that have been termed the leukotrienes. These are derived from arachidonic acid by the action of lipoxygenases, which yield hydroxyeicosatetraenoic acids. These are conjugated to glutathione via its sulphydryl group. Slow reacting substance of anaphylaxis (SRS-A) has been shown to be such a compound, the search for its structure leading to the delineation of the leukotriene series. These promise to be a widely distributed class of compound, with actions in the inflammatory response and related systems.

The structures of some leukotrienes are shown.

Leukotriene is abbreviated LT, thus LTC_3 denotes leukotriene C_3.

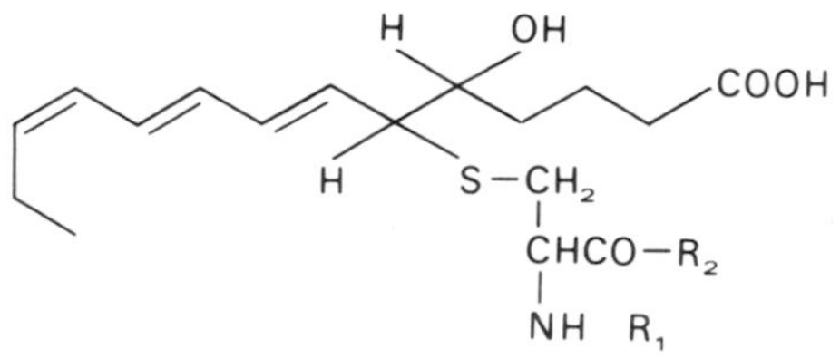

H OH
COOH
H S−CH₂
CHCO−R₂
NH R₁

	R_1	R_2	R_3
LTC_3	Glu	Gly	C_7H_{15}
LTC_4	Glu	Gly	C_7H_{13}
LTD_3	H	Gly	C_7H_{15}
LTD_4	H	Gly	C_7H_{13}

LTC_4 and LTD_4 derive from arachidonic acid, whilst LTC_3 and LTD_3 derive from C20 : 3 ω 9,12,15 which lacks the double bond in this region.

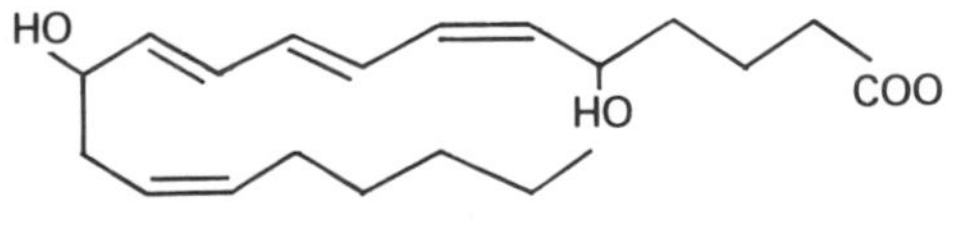

HO
HO
COO

Leukotriene B_4

A

Structures of complex carbohydrates

Complex carbohydrates (*oligosaccharides*) contain a wide variety of monosaccharides, including those that are acetylated, and are often bound to protein, to give the glycoproteins. They may also be bound to *ceramide* to give *gangliosides* and other complex glycolipids. These are often found at the outer surfaces of cells.

Submaxillary mucin (porcine A blood-group speficity)

GalNAc(α1⟶3)Gal(β1⟶3)GalNAc-O-Ser {or Thr}, with Fuc (α1⟶2) on Gal and NeuNAc (α2⟶6) on GalNAc

Gal NAc = N-acetylgalactosamine	Fuc = Fucose	Glc = Glucose
Gal = Galactose	NeuNAc = Sialic acid	Cer = Ceramide

The linkages between the sugars are shown in the brackets. The blood group substances are an example of the complex carbohydrates. The antigenic activity resides in the carbohydrate moiety, and is exhibited in both the glycoprotein and glycolipid forms of the substance.

In some cases, the carbohydrate chain is linked to protein via the amide nitrogen of asparagine, *N*-linked glycoprotein. In other cases, the carbohydrate chain is linked to protein via the OH group of serine or threonine, *O*-linked glycoprotein.

B

Fucose and *sialic acid* are important monosaccharides in complex carbohydrate structures.

α-L-Fucose

N-acetylneuraminic acid (sialic acid)

C

A typical and important ganglioside is termed G_{M1}. Other gangliosides are formed by the addition of further sialic acid residues (NeuNAc) to this structure.

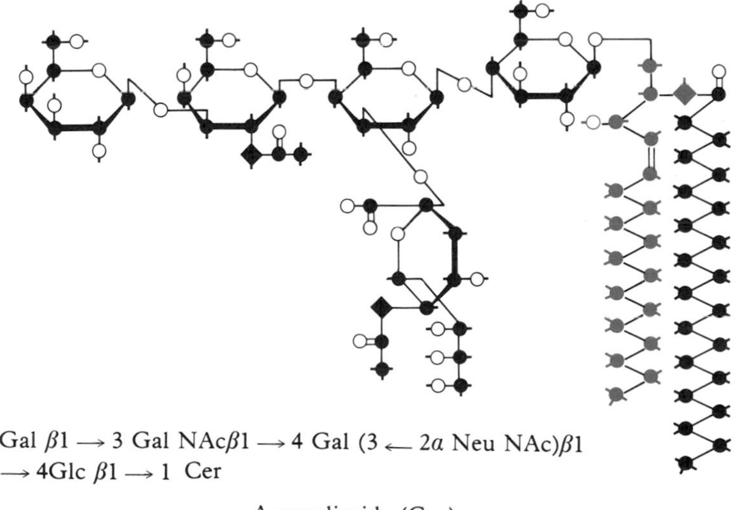

Gal β1 ⟶ 3 Gal NAcβ1 ⟶ 4 Gal (3 ⟵ 2α Neu NAc)β1 ⟶ 4Glc β1 ⟶ 1 Cer

A ganglioside (G_{M1})

A

Glycoproteins

Many proteins carry carbohydrate residues, usually in oligosaccharide chains attached to an asparagine (N-glycosylation) or to a serine or threonine (O-glycosylation)

The monosaccharide unit linked to asparagine is always β-N-acetylglucosamine.

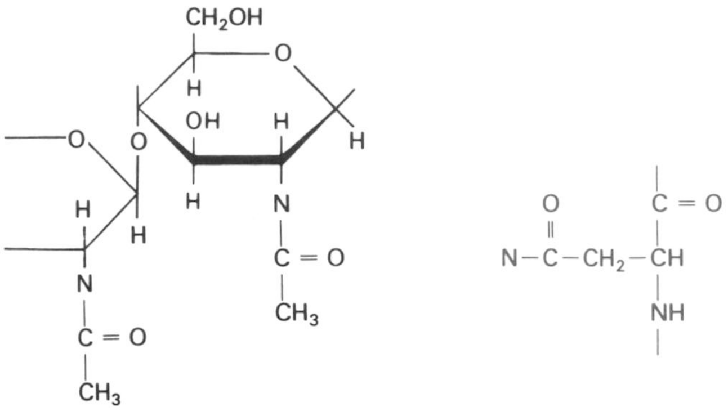

B

Intercarbohydrate links.

Two molecules of similar 6-carbon sugars can be linked in many different ways, depending on whether the link between them is α1,4, α1,3, β1,4 and so on. Three molecules thus can be linked in an even greater diversity of ways. When the monosaccharides themselves differ, the number of possible variations increases.

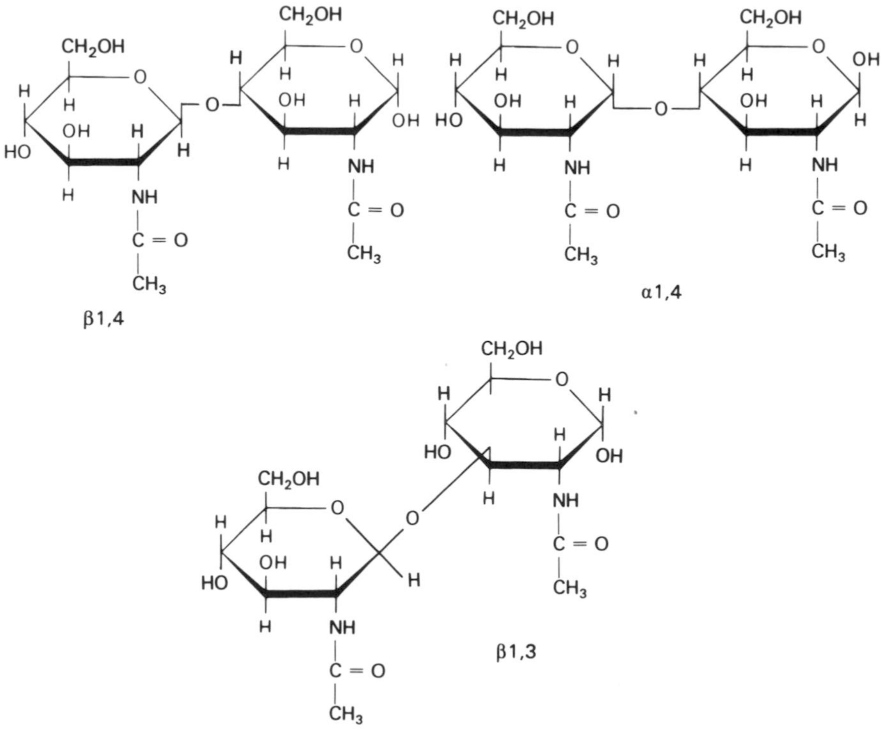

β1,4

α1,4

β1,3

The 3-dimensional structure will be profoundly affected by the type of linkage between molecules. Thus the number of shapes that can be made by even a trisaccharide is very great, and one role that the carbohydrates on proteins may play is that of recognition, for which they are ideally suited due to the great number of different identities they can assume.

A

The synthesis of complex polysaccharides involves *dolichol phosphate*, a lipid molecule.

$$(CH_2=\overset{\displaystyle CH_3}{\overset{|}{C}}-CH=CH)_n-CH_2-\overset{\displaystyle CH_3}{\overset{|}{C}H}-CH_2-CH_2-O-\overset{\displaystyle O}{\overset{\|}{\underset{\displaystyle O^-}{\underset{|}{P}}}}-O^-$$

Dolichol phosphate

Glycoprotein biosynthesis.

B

Chemistry

In the initial steps of synthesis, the oligosaccharide chain is built up on dolichol, embedded in the membrane of the endoplasmic reticulum, and involves nucleotide-linked or dolichol-linked monosaccharides. The oligosaccharide chain is then (see C below) transferred to the amido group of asparagine in a protein, after transfer from the endoplasmic reticulum to the Golgi, where further carbohydrate processing takes place.

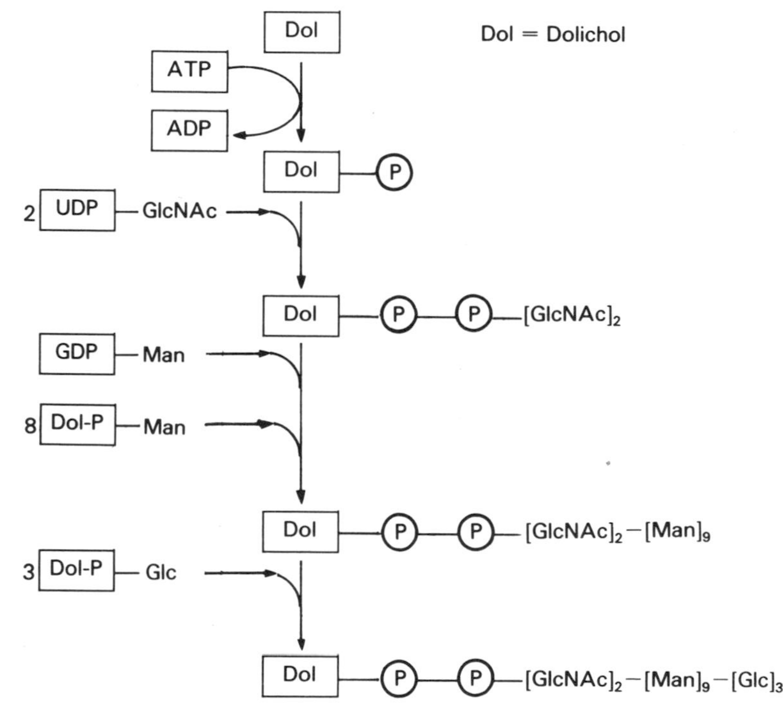

C

Morphology

The oligosaccharide is linked to the growing peptide chain in the rough surfaced endoplasmic reticulum (rough ER).

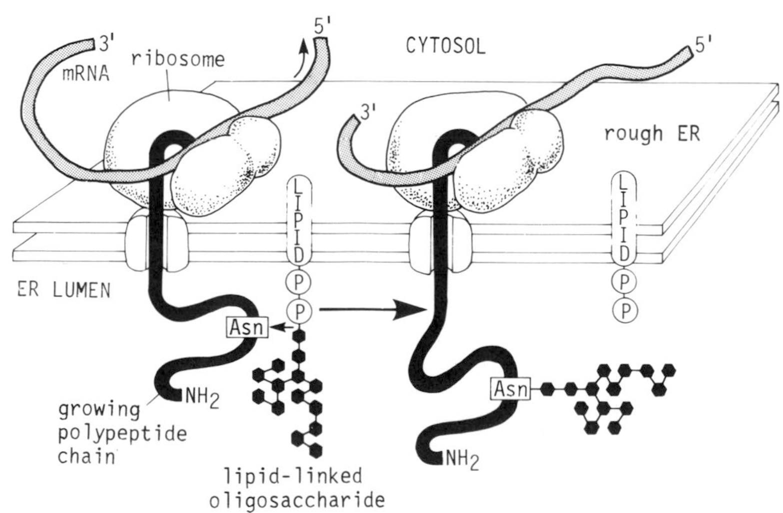

A

The core oligosaccharide.

As the nascent polypeptide passes through the Golgi some of the outer units are cleaved to leave a core, shown in red. N-glycosylation is inhibited specifically by *tunicamycin*, which prevents the dolichol-CHO complex being added to the asparagine.

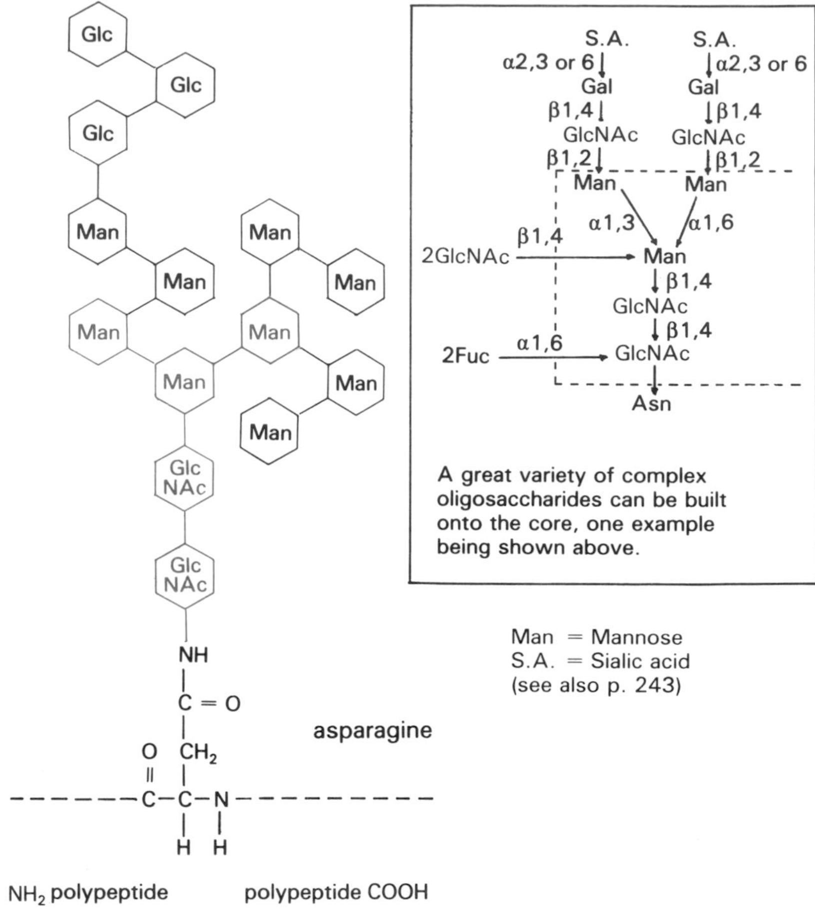

B

Lactose synthesis

An enzyme that plays a key role in glycoprotein synthesis is also involved in the synthesis of lactose

A key enzyme in the modification of the carbohydrate chains in the Golgi apparatus is N-acetylglucosamine galactosyl transferase. This ubiquitous enzyme is found in many types of cell bound to the membranes of the Golgi apparatus and is a marker enzyme for this organelle. It adds galactose in β 1,4-linkage to the terminal residues of both core oligosaccharide branches.

$$\text{UDP-galactose} + \text{N-acetylglucosamine} \longrightarrow \text{N-acetyllactosamine} + \text{UDP}$$

Lactose is synthesized by lactose synthase which consists of two proteins A and B. The A protein is N-acetylglucosamine galactosyl transferase and the B protein is α-lactalbumin, a milk protein. B protein modifies the substrate specificity of A protein from N-acetylglucosamine to glucose so that we have:

$$\text{UDP-galactose} + \text{glucose} \xrightarrow{A + B} \text{lactose} + \text{UDP}$$

B protein interacts with A protein which is bound to the membranes of the Golgi and this event is the initiation of lactation. The B protein (α-lactalbumin) is excreted in the milk.

The protein secretory pathway.

The Golgi apparatus is highly specialized for a variety of functions, including protein glycosylation and subsequent transport of glycoproteins to their destinations.

Three compartments have been identified in the Golgi stack, and each has a characteristic set of enzymes. These are termed the *cis*, *medial* and *trans* compartments. Each of these compartments is involved in a different stage of carbohydrate processing.

Note the complexity of the mechanisms that must exist for the budding off of vesicles and their progress to different destinations such as lysosomes, secretory storage granules or plasma membrane.

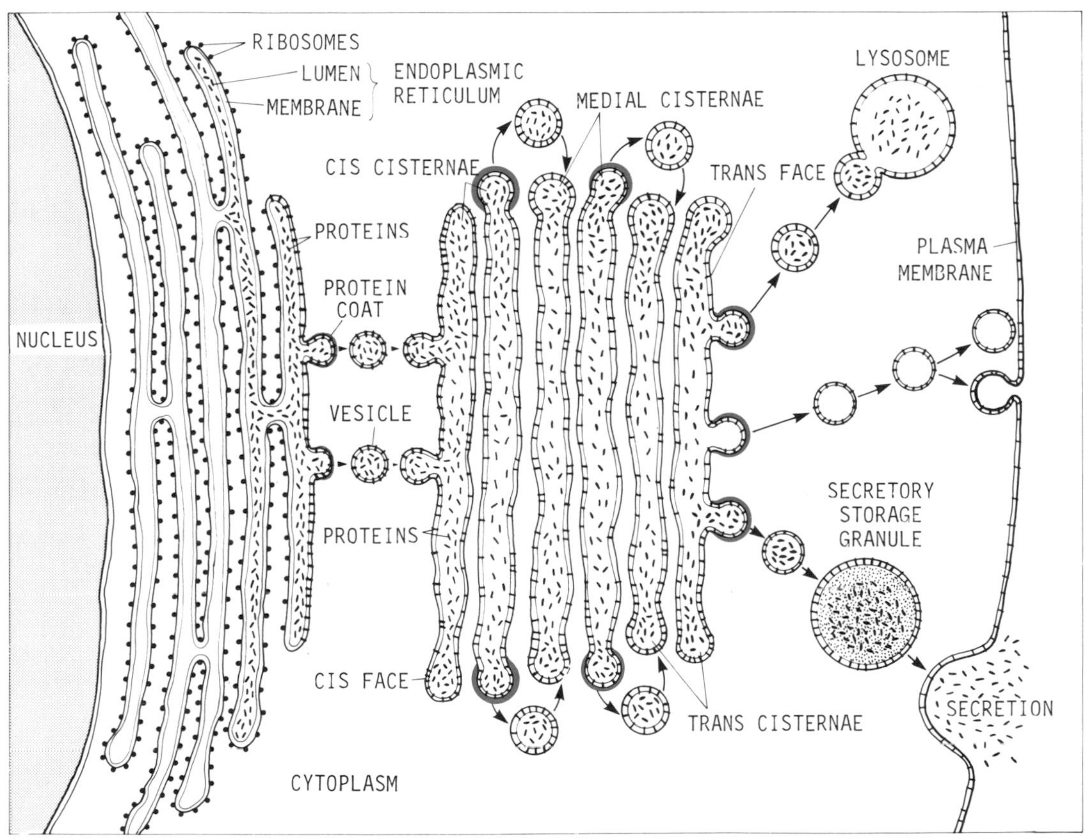

The primary structure of lysozyme and α-lactalbumin

Two proteins involved in the breakage and formation of β1,4 glycosidic bonds are derived from a common ancestral gene

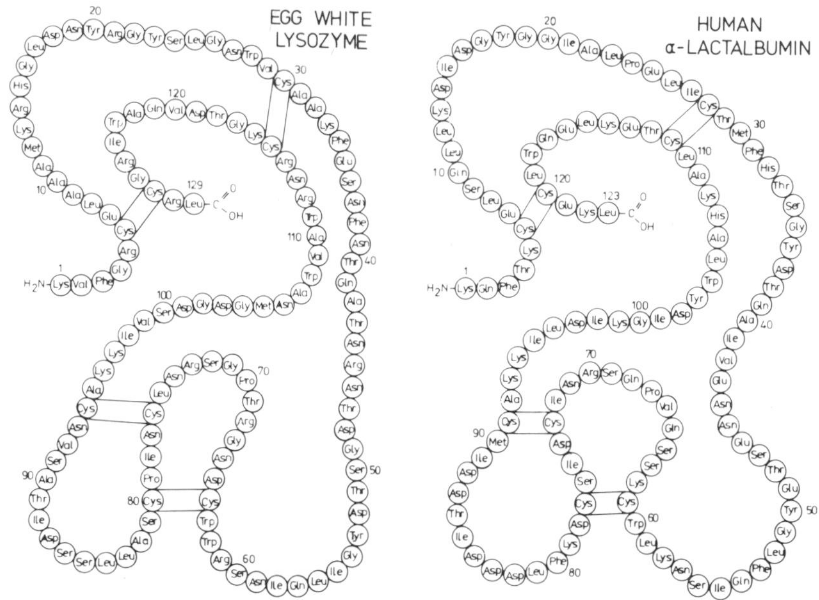

Lysozyme is a naturally-occurring antibiotic, being found in virtually all tissue secretions. It hydrolyses the β-glycosidic bond between C-1 of N-acetylmuramic acid and C-4 of N-acetylglucosamine and thereby degrades the cell wall of the bacteria. Chitin, a polysaccharide in the shell of crustaceae is also a substrate for lysozyme. Chitin consists only of N-acetylglucosamine residues joined by β (1⟶4) glycosidic links. As explained on page 246 α-lactalbumin plays a part in the formation of lactose, a β-1⟶4 disaccharide.

Above is shown the primary structure of lysozyme and α-lactalbumin. They are strikingly similar in structure in terms of both 1^{ary} and 3^{ary} structure. Also the gene structure with respect to location of introns and exons is strikingly similar. This supports the concept that the two proteins arose by duplication of a common ancestral gene. During the course of evolution the properties of the two proteins have diverged and α-lactalbumin has no lysozyme activity nor has lysozyme activity in respect to lactose synthesis.

Many residues play a part in the action of lysozyme but particularly Glu 35 and Asp 52. In α-lactalbumin its Ca^{2+} binding properties are important and Asp 82, 83, 87, 88 are particularly important in this respect.

9

MEMBRANE STRUCTURE AND FUNCTION

A

The phospholipid bilayer

Detergents, fatty acids and similar compounds form spherical micelles when added to water.

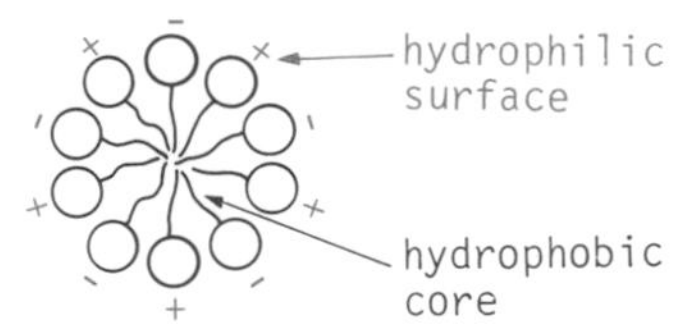

Mixture of positively and negatively charged detergents.

B

Phospholipids have a hydrophilic head group and a hydrophobic fatty acyl region.

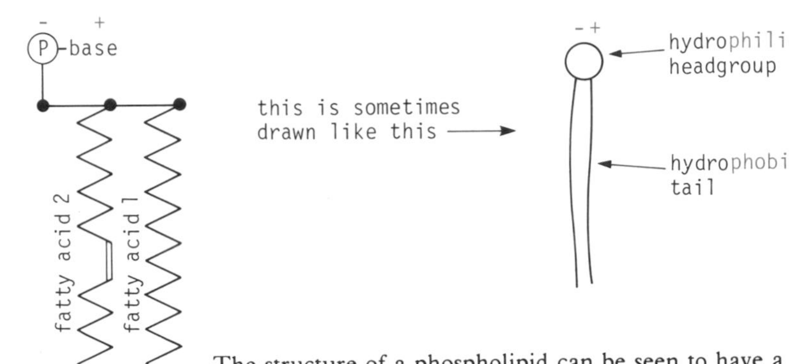

The structure of a phospholipid can be seen to have a hydrophilic region, composed of the headgroup phosphate and base and the hydrophobic region of the alkyl chains of the fatty acids.

C

Because of the geometry of phospholipid structure, some phospholipids form sheets, or *bilayers*, rather than micelles in aqueous solution.

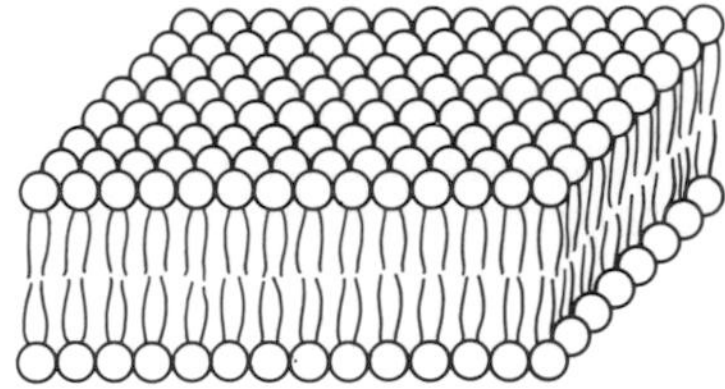

These two-dimensional arrays of phospholipid play an important role in membrane structure.

D

Not all phospholipids form bilayer structures. Phosphatidylcholine is important in membranes because it readily forms bilayer structures and promotes bilayer formation in the presence of other phospholipids.

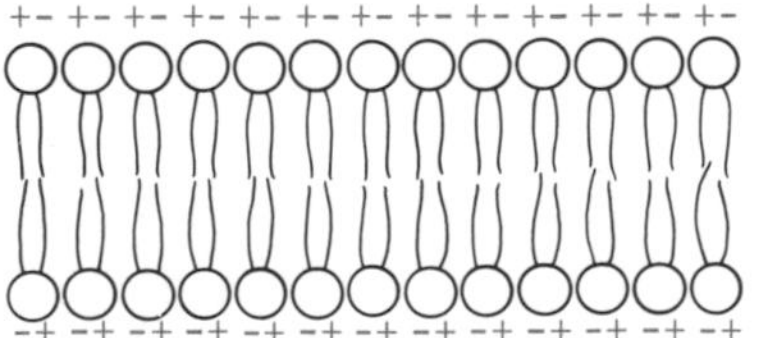

phospholipid bilayer

The outside surfaces of bilayers form a sheet of charged molecules. The inner core of the bilayer is highly hydrophobic.

A

Membrane proteins

1. *Intrinsic*, or *integral*, *proteins* have a hydrophobic moiety which penetrates into the bilayer.

2. Other proteins attach more loosely by electrostatic forces.

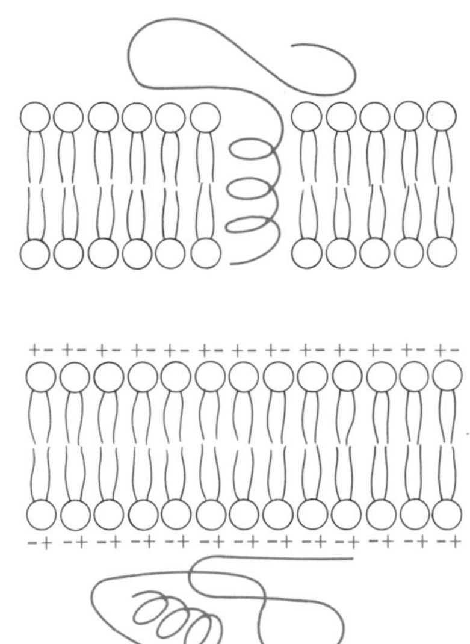

B

Membranes are composed of protein, lipid and carbohydrate.

The lipid is in the bilayer, and the carbohydrate is either attached to protein (glycoprotein) or lipid (glycolipids, such as gangliosides).

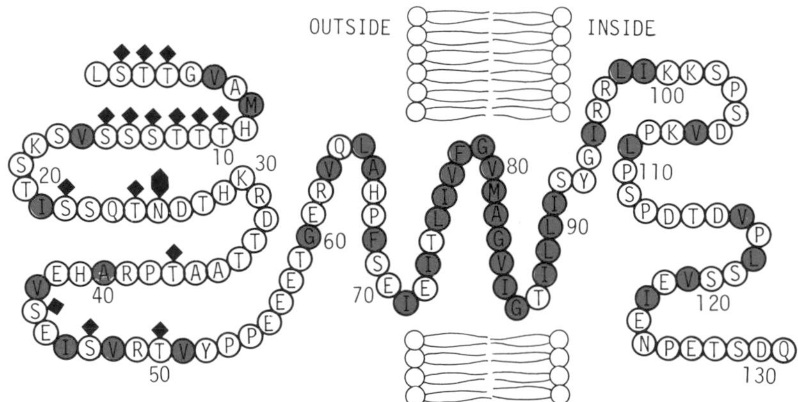

The complete amino acid sequence of many membrane proteins is known. That of glycophorin, an important protein of the red blood cell which bears some (AB & MN) of the blood group specific carbohydrates, reveals that hydrophobic side chains (shown in red) are clustered in the area of the molecule believed to lie in the core of the bilayer. The oligosaccharide side chains (shown as squares, or a hexagon) are on the outer surface of the membrane. Residue 1 is the amino terminus, and residue 131 the carboxyl terminus. Those oligosaccharides shown as squares are *O*-linked to serine or threonine residues and that shown as a hexagon is *N*-linked to asparagine.

A

Fluid nature of cell membranes

Fluidity of the lipid bilayer allows proteins to move laterally in the plane of the bilayer, e.g. after binding of antibodies to antigens on the cell surface.

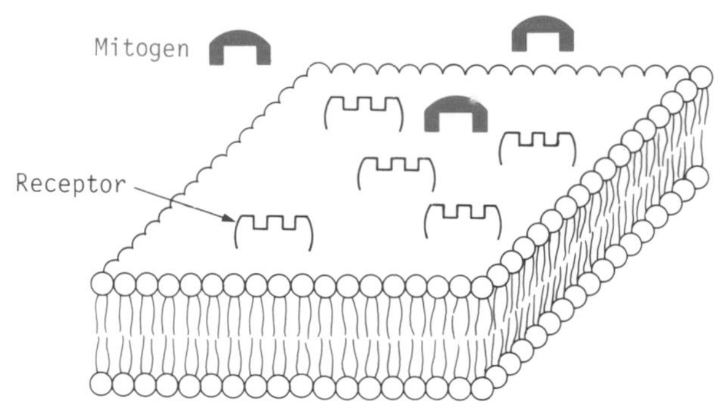

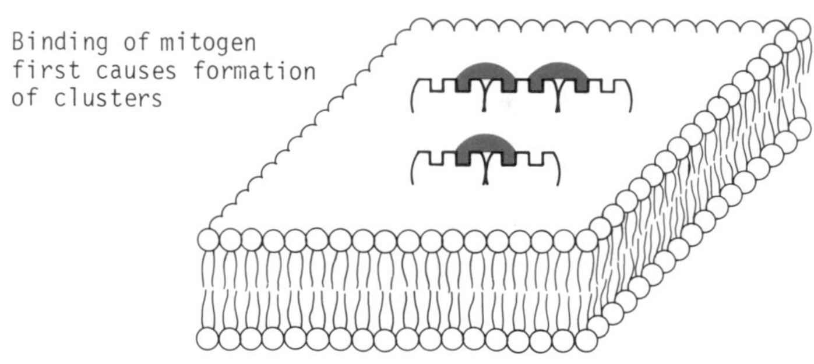

B

The *clustering* or *capping* of plasma membrane proteins is believed to be an important phenomenon associated with certain events such as cell stimulation by mitogens (substances that stimulate the cell to grow and enter mitosis).

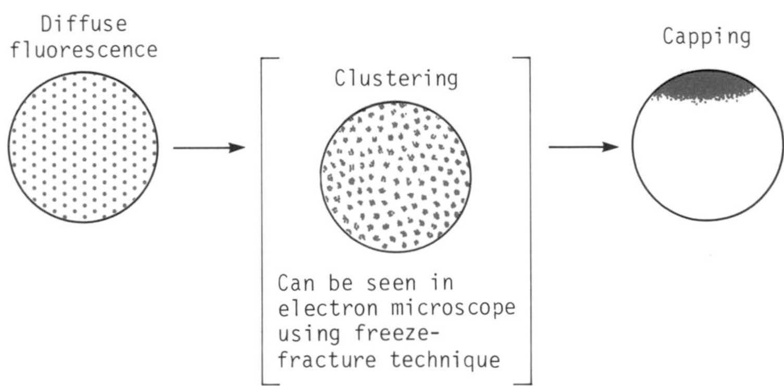

If the antibody or mitogen is 'labelled' with a fluorescent dye, it can be seen that after about 30 min the label has migrated into a 'cap' at one end of the lymphocyte.

Such experiments demonstrate that the proteins are free to migrate in the plane of the membrane.

A

Proteins can move laterally in the plane of the membrane, and can rotate around an axis vertical to the plane of the membrane. However, they cannot 'tumble' through the plane of the membrane.

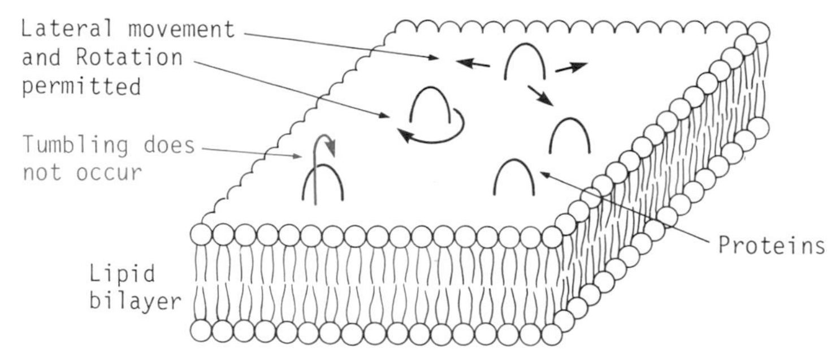

B

Proteins not actually integral to the membrane can exert an influence on the membrane and its proteins. *Spectrin*, a red cell protein, is an example of proteins which control membrane shape and movement.

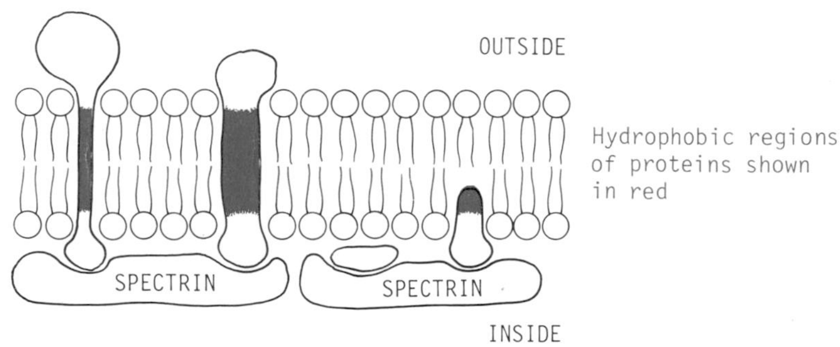

C

The cytoskeleton

A model has been proposed for the arrangement of the red cell cytoskeleton.

Spectrin has α- and β-chains and is thought to interact through assemblies of actin and tropomyosin. The protein ankyrin links spectrin to the plasma membrane. The band 4.1 protein (so named because of its position on gels of red cell proteins) may also be involved in linkage to the plasma membrane.

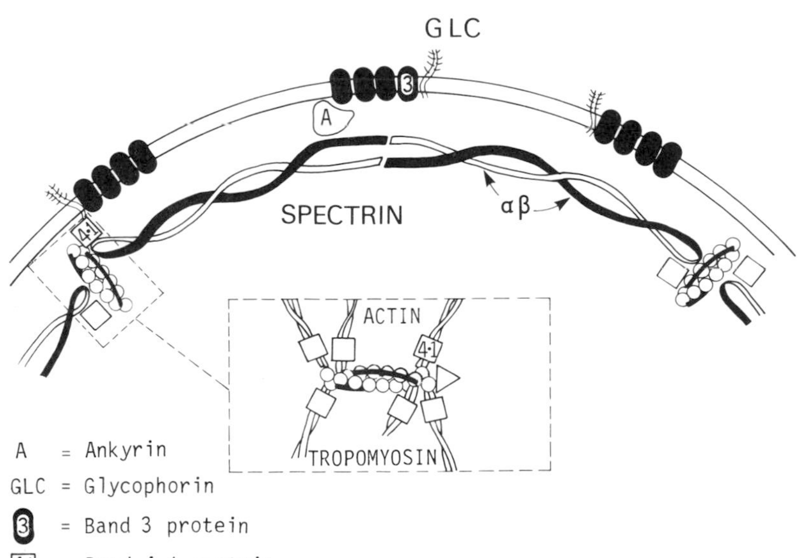

A

In cells other than the red cell, a cytoskeleton of microtubules and microfilaments exists.

Microtubules are formed from the protein *tubulin*, whilst *microfilaments* are formed from *actin*

The protein tubulin can polymerize and depolymerize for microtubules to form and dissociate. Tubulin consists of dimers, composed of α tubulin and β tubulin monomers. Microtubules are cylindrical polymers of tubulin, shown below (α tubulin shown in pink. β tubulin white).

A. Cross section B. Surface view

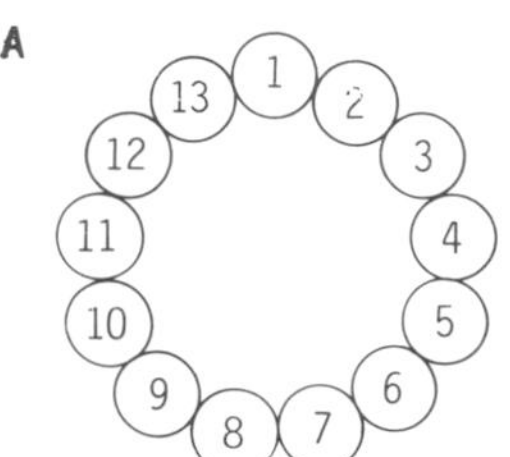

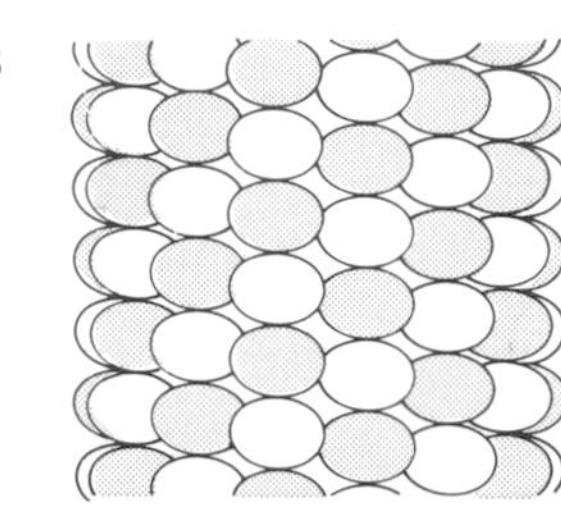

B

By using fluorescent antibodies against tubulin, the cytoskeleton can be visualized.

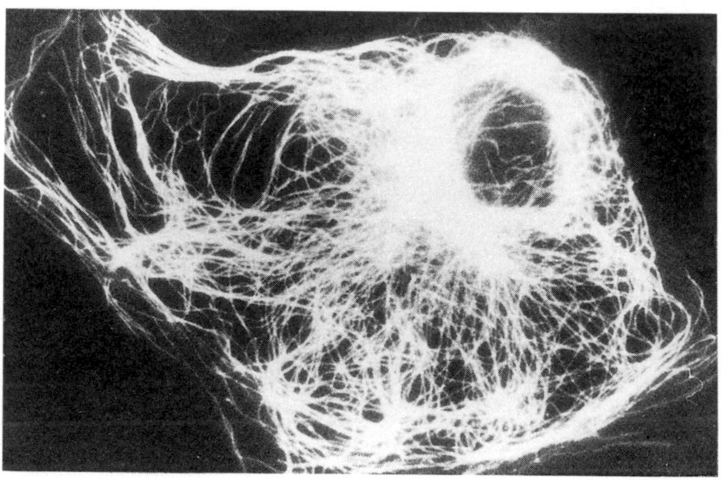

Microtubules visualized in a cell. The most dense staining surrounds the nucleus.

C

Membrane transport

Transport across cell membranes.

It is obvious that communication systems must exist in the membranes to transport substances from one compartment to another, otherwise all of the compartments would be totally closed.

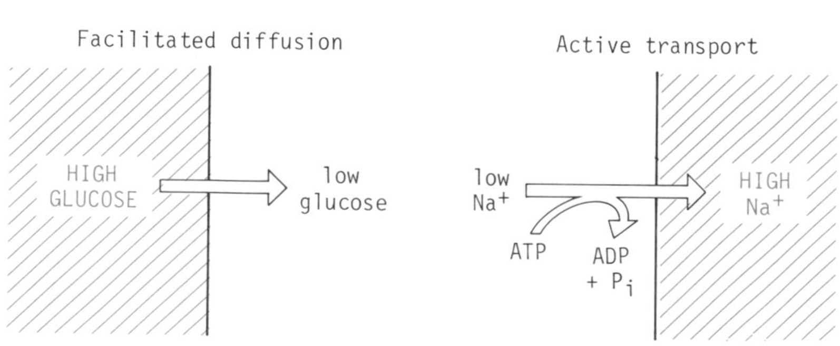

Facilitated diffusion is energy independent and is in the direction of the concentration gradient (i.e. from high concentration to low concentration). Active transport requires the provision of energy (often involving ATP hydrolysis) and can occur against the concentration gradient.

A

Active transport is a process whereby molecules are transported against the concentration gradient. In some cases, this requires ATP. In other cases, a molecule can be transported against its concentration gradient by virtue of the fact that another molecule is simultaneously transported in the direction of its own concentration gradient.

Transport against the concentration gradient (i.e. from a compartment of low concentration into a compartment of higher concentration) can occur, but requires some form of energy compensation. This can be provided by hydrolysis of ATP, as in the sodium pump, that maintains sodium and potassium concentrations in nerve cells.

Action of sodium pump

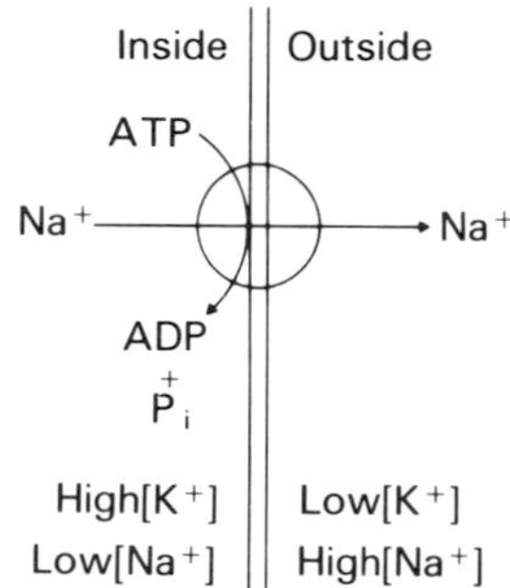

Another way of describing the action of the pump

$$\text{ATP} \xrightarrow[\text{Na}^+\ \text{K}^+]{\text{Mg}^{2+}} \text{ADP} + \text{P}_i$$

Isolated preparations of the enzyme hydrolyse ATP when stimulated by both Na^+ and K^+. As with other enzymes utilizing ATP, Mg^{2+} is also required.

Transport driven by an ion gradient

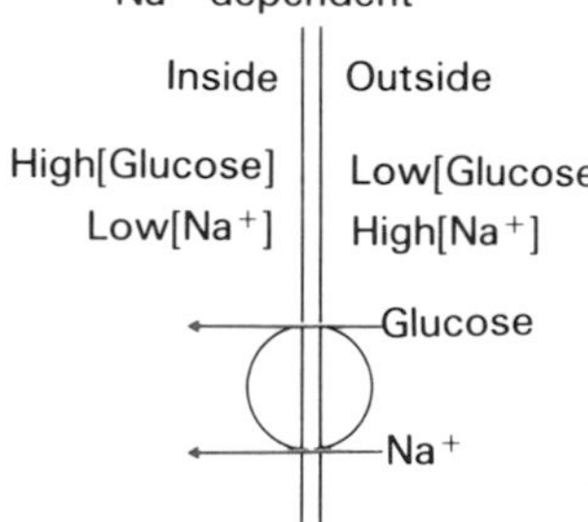

This is termed *symport* i.e. both molecules moving in the same direction. When the molecules move in opposite directions the term *antiport* is used.

B

Transport systems show saturation kinetics, demonstrating that a transport site exists which can be occupied by only a limited number of molecules. Inhibition can occur and a K_m can be measured.

This is found both for facilitated diffusion and for active transport.

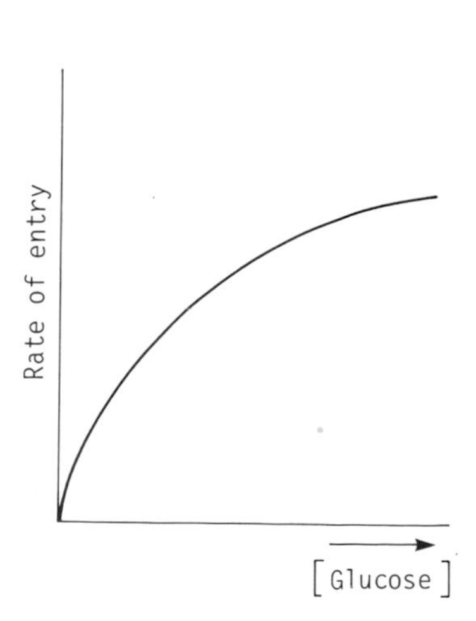

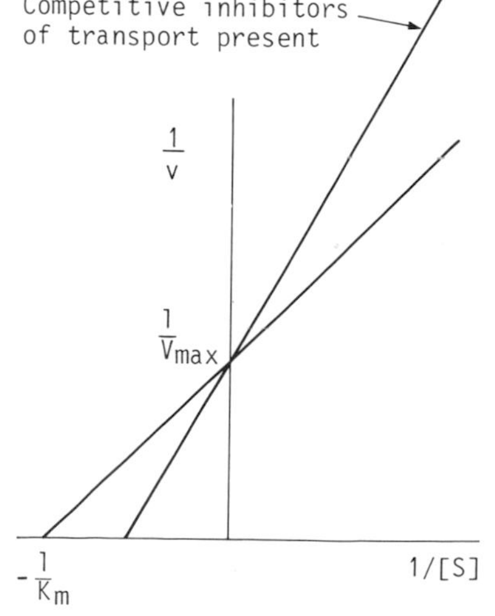

A

Components of a transport system

	Diffusion through bilayer	Passive transport	Gated transport	Active transport
Facilitation	-	+	+	+
Selection	-	-	+	+
Energy transformation	-	-	-	+

The components of a transport system may contain any or all of the following: a selective gate (denoted ○), a non-selective channel through the membrane, and an energy-coupling system (denoted □).

Many of the possible arrangements of these components probably occur, e.g. the gate and/or the energy-coupler may be on the inside of the membrane.

B

In the *nerve axon*, the *action potential* is regulated by the influx of Na^+ through a gated channel. K^+ ions then leave by other channels to return the membrane potential to normal. Ionic concentrations are eventually restored by the *sodium pump*.

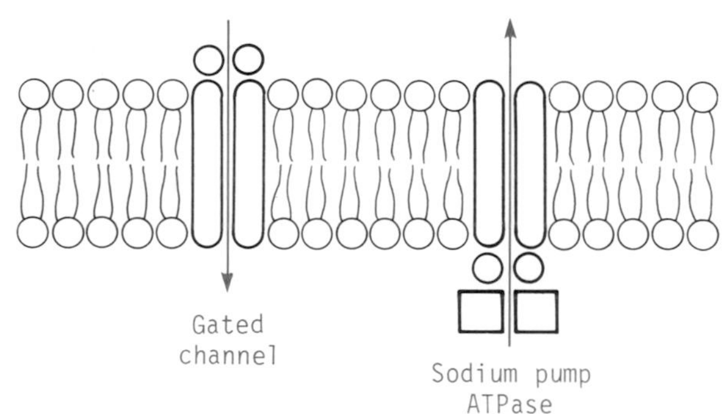

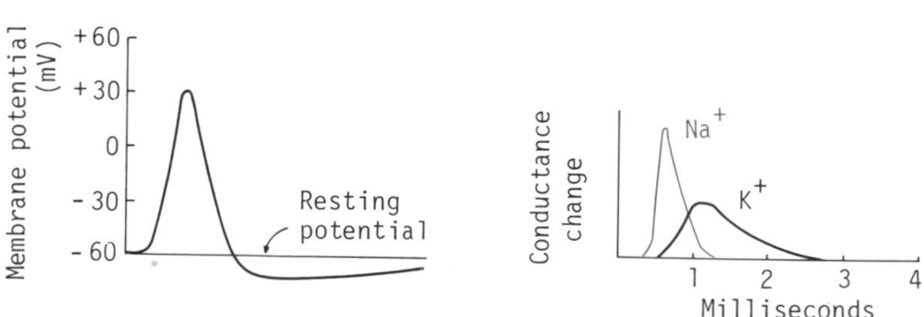

Events during the propagation of the action potential

A

Nerve transmitter substances

The architecture of *transmission.*

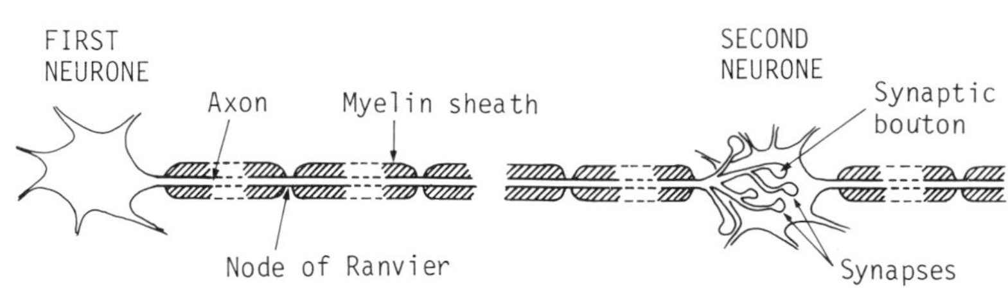

B

Neurotransmission

Transmission involves chemical transmitter substances that mediate the passage of the action potential across the synapse from one axon to the next.

Propagation of a nerve impulse involves two processes. The first, which is termed *conduction*, involves the movement of the *action potential* along the axon and requires the passage of Na^+ and K^+ through gates in the membrane and subsequent pumping of these cations by the 'sodium pump' to restore their concentration to the normal level (see p. 255). The second, termed *transmission*, is the communication of the impulse across a synapse and is achieved by release of transmitter substances from one neurone to the next.

The nature of some transmitter substances

Compounds that are thought to be transmitter substances include acetylcholine, noradrenaline, dopamine, 5-hydroxytryptamine (serotonin), ATP, and γ-aminobutyrate. Acetylcholine is destroyed, after excitation of the post synaptic membrane, by the enzyme acetylcholinesterase (see next page), which hydrolyses it to choline and acetic acid. Acetylcholine is synthesized by the reaction:

$$\text{choline} + \text{acetyl CoA} \longrightarrow \text{acetylcholine} + \text{CoA}$$

More recently, a number of peptides, collectively called the endorphins, have been thought to influence transmission in certain areas of the brain. These bind to receptors which are known to bind opiates (such as morphine) especially in the mid brain and thalamus where pain-conducting tracts collect.

A

The sequence of events involved in transmission at a cholinergic synapse

Acetylcholine is released into the synaptic cleft.

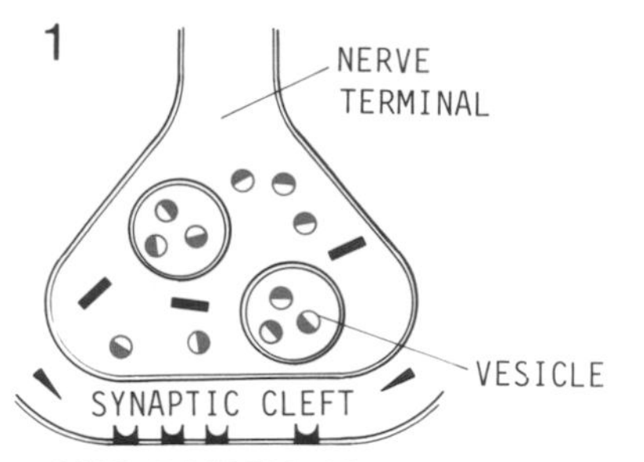

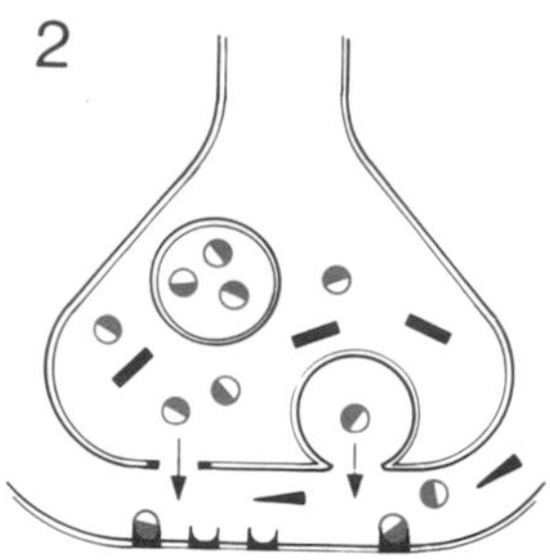

1. The nerve terminal contains acetylcholine in vesicles and in the cytosol.
2. The arrival of an action potential causes release of acetylcholine into the synaptic cleft, where it binds to the receptor in the membrane of the postsynaptic cell. (There is some discussion whether vesicular or cytosolic acetylcholine is released into the synaptic cleft). Activation of the postsynaptic cell causes the action potential to be propagated in the next neurone.

B

After transmission of the action potential, acetylcholine is degraded by *acetylcholinesterase*.

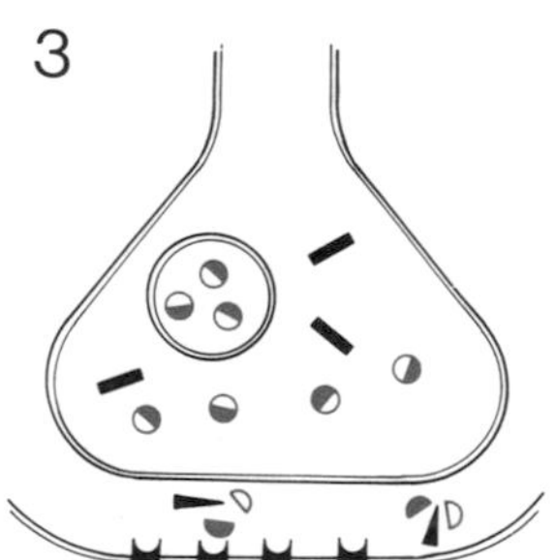

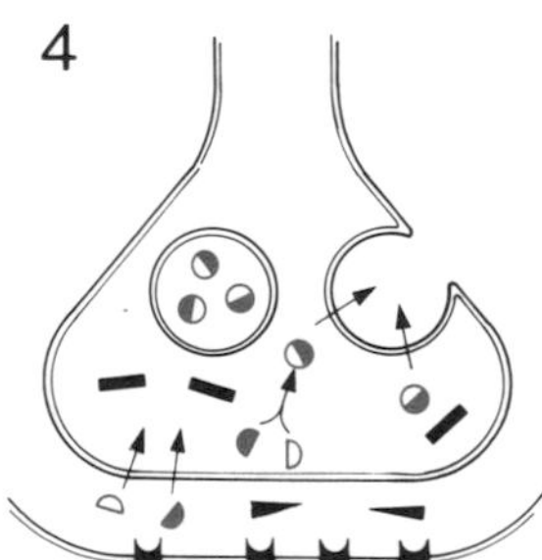

3. The enzyme acetylcholinesterase cleaves the acetylcholine to acetate and choline to prevent persistent stimulation of the postsynaptic cell.

4. Acetate and choline re-enter the presynaptic cell where acetylcholine is resynthesized by choline acetyltransferase (acetyl CoA + choline react to form acetylcholine and CoA). Endocytosis may be involved in the formation of new vesicles.

C

The nerve terminal resynthesizes acetylcholine.

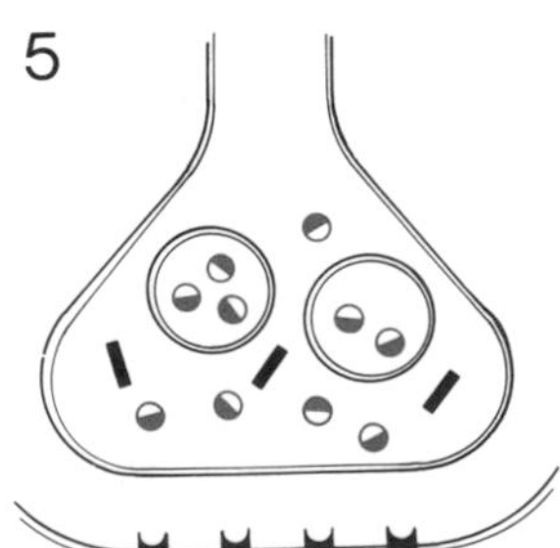

5. The nerve terminal is restored to a state in which it can await a further action potential.

Receptor in post-synaptic cell membrane
Choline acetyltransferase
Acetylcholinesterase
Acetylcholine

Calcium plays an important role in the sequence. The action potential causes Ca^{2+} ions to enter the nerve terminal, and the resulting elevation of cytosolic Ca^{2+} triggers the release of the acetylcholine.

A

Coupling of receptors to adenylate cyclase

The action of many hormones and other agonists is mediated through activation of adenylate cyclase (see p. 190). The coupling of the hormone-receptor complex to adenylate cyclase involves a *G-protein*

Examples of agonists which act in this way are glucagon, adrenaline, TSH, ACTH and many growth factors.

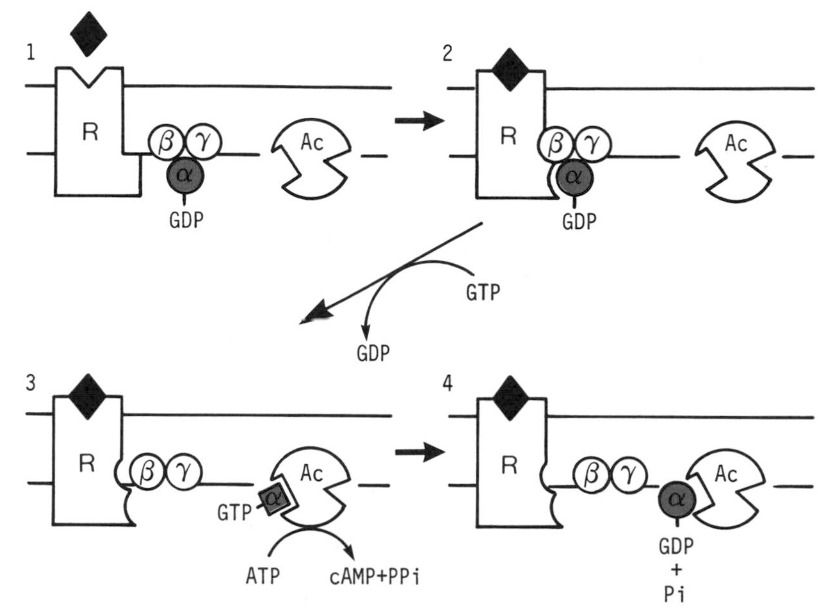

1. The receptor-adenylate cyclase system consists of a receptor (R), an enzyme (AC, the catalytic unit of adenylate cyclase) and a modulator protein, termed the G (or N) protein (G signifies guanylate, N signifies nucleotide) so-called because the binding of a guanine nucleotide is involved in its action. The G protein consists of α, β, γ subunits, the α subunit carrying the guanine nucleotide binding site.

2. Binding of an agonist, such as a hormone (♦), to its receptor causes the G protein to bind to the hormone-receptor complex, in which form it can accept GTP in place of GDP.

3. The GTP-α subunit dissociates and binds to AC, activating it so that cyclic AMP is formed from ATP.

4. The α-subunit has GTP-ase activity, and hydrolyses GTP to GDP and Pi. This hydrolysis causes inactivation of AC, and re-formation of the complex as in 2, so that the sequence 2 $\longrightarrow$ 4 then repeats.

A

Inhibitory receptors

Activation of some receptors causes inhibition of adenylate cyclase and this is mediated by a distinct G-protein. The stimulatory G-protein is designated G_s, and the inhibitory G-protein G_i. The β and γ subunits of G_i are the same as those of G_s, but the α subunit is distinctive for G_i.

(see also page 259A)

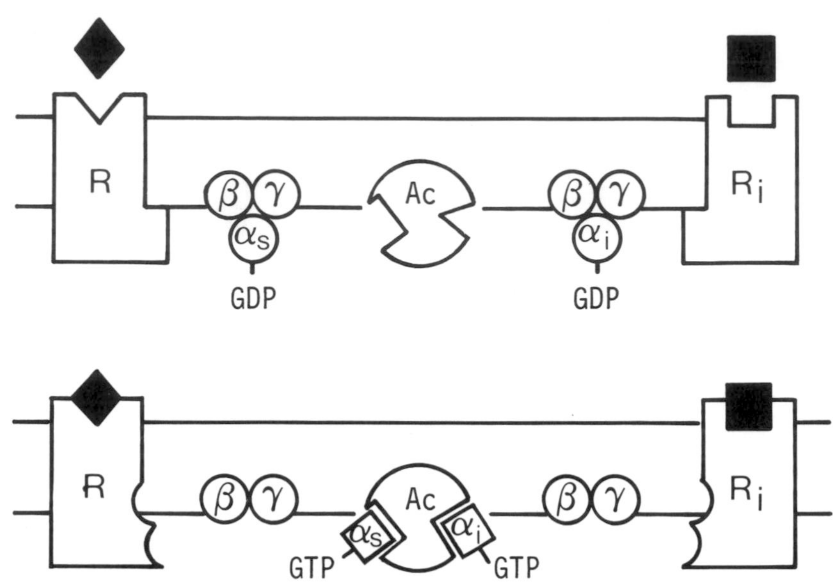

B

Toxins produced by certain bacteria interact with the adenylate cyclase system by causing ADP-ribosylation. *Vibrio cholerae*, the causative agent for cholera, produces a toxin which causes ADP-ribosylation of α_s, and pertussis toxin (from *Bordetella pertussis*, the causative agent of whooping cough) causes ADP-ribosylation of α_i.

ADP-ribosylation of α_s by cholera toxin prevents association of α_s with $\beta\gamma$ subunits, and thus persistently activates adenylate cyclase. ADP-ribosylation of α_i by pertussis toxin also activates the enzyme.

ADP-ribosylation has been studied in the nucleus, where it is brought about by the enzyme ADP-ribosyl transferase.

Protein—C(=O)—O—1′ Rib5′ —O—P(O⁻)(=O)—O—P(O⁻)(=O)—O—5′Rib1′—Ade
(2′—O—) 1′ Rib5′ —O—P(O⁻)(=O)—O—P(O⁻)(=O)—O—5′Rib1′—Ade
(2′—O—) 1′ Rib5′ —O—P(O⁻)(=O)—O—P(O⁻)(=O)—O—5′Rib1′—Ade (2′—O)
1′ Rib5′ —O—P(O⁻)(=O)—O—P(O⁻)(=O)—O—5′Rib1′—Ade (2′—O—~)

The structure of poly (ADP-ribose) attached to a protein carboxyl group. Rib, ribose; Ade, adenine.

Receptor structure and function

A

Receptors are mostly found in the plasma membrane and consist of protein, often with carbohydrates attached to the extracellular peptide chain.

Apo B (LDL) receptor

Transferrin receptor

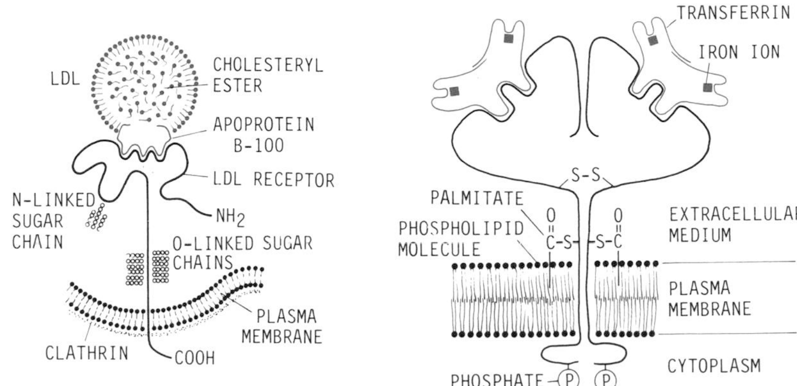

A number of the receptors that have been purified have been found to contain a hydrophobic region which anchors the receptor in the lipid bilayer. Models of this type for the LDL and transferrin receptors have been proposed.

B

Not all receptors are thought to penetrate the bilayer in a single span. The *acetylcholine receptor* (AChR) appears to re-cross the bilayer several times.

The *α*-subunit of the receptor consists of five membrane-spanning helical regions. As explained on pages 267 and 268, it has been possible to bring about amino acid substitutions in different parts of the peptide chain. In some cases these substitutions have no effect on receptor function (labelled N), others alter gating or permeation (labelled P), whilst others affect the binding of acetylcholine (labelled B).

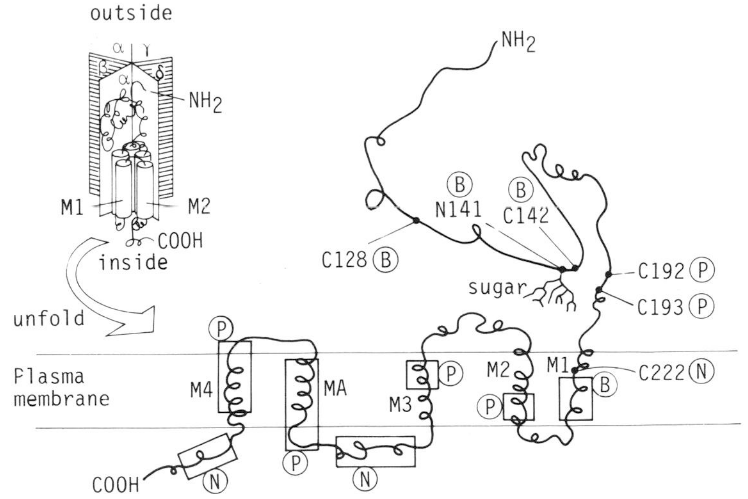

A

A comparison of the stuctures of various receptor proteins

As shown on the previous page, the structure of an increasing number of membrane receptor proteins is now known, and the orientation and mode of insertion in the membrane lipid bilayer deduced. A considerable variety is found in the number of polypeptide chains, and their orientation in and through the bilayer. These include the acetylcholine receptor ACh, which loops several times through the bilayer, the insulin receptor which consists of four polypeptide chains, only two of which penetrate the bilayer, the transferrin receptor consisting of two bilayer-penetrating chains linked by a disulphide bond, and the single bilayer-penetrating chain of the low density lipoprotein (LDL) and other receptors such as epidermal growth factor (EGF) or immunoglobulins (IgA and IgM).

If the NH_2 terminal is extracellular (on the outer surface of the plasma membrane) then it is usual for the polypeptide to be synthesized with a signal peptide (see p. 108). If the NH_2 terminal is on the inside of the membrane, then a cleavable signal peptide at the NH_2 terminus is not formed.

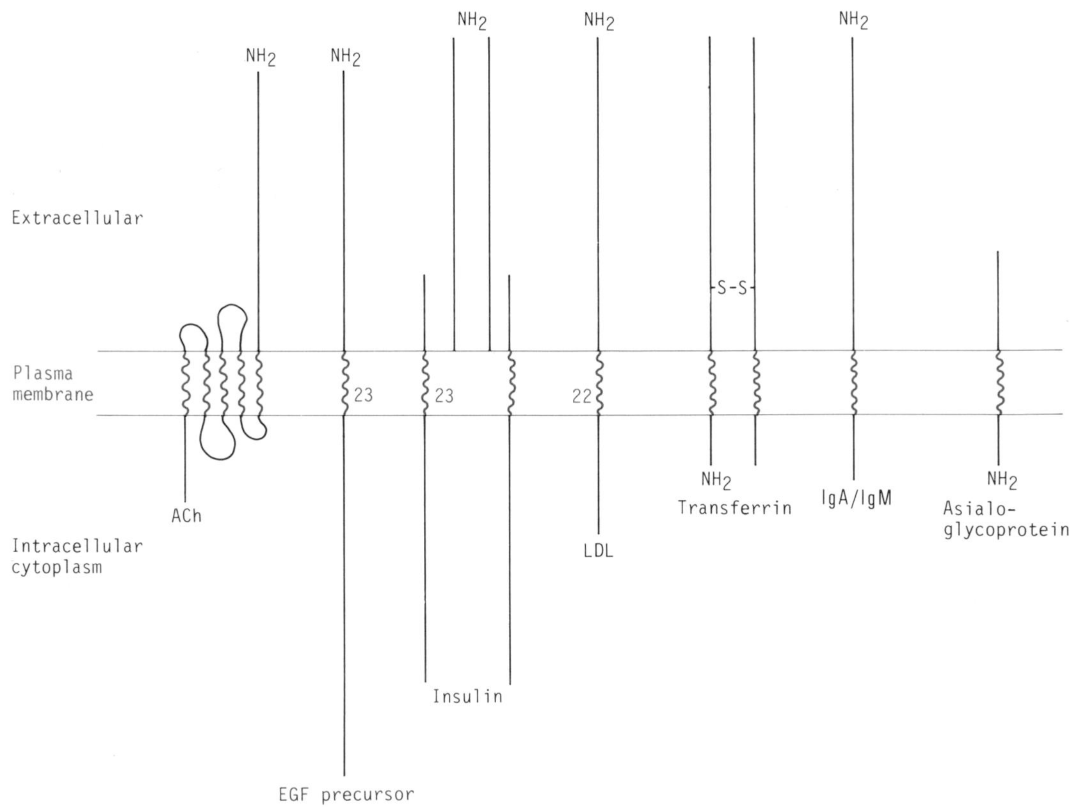

Details of the LDL and Transferrin receptors are shown on page 261A.

A

Internalization of receptors

Receptors are constantly being internalized, and, when ligand is bound, carry it into the cell. After internalization, ligand and receptor separate, the ligand being degraded and the receptor being recycled back to the plasma membrane.

The coated pits invaginate and seal off as coated vesicles. The *clathrin* is then lost, and the vesicles fuse to form endosomes. A proton pump is probably responsible for a lowering of the pH in the endosome. There is good evidence that the pH within the endosome is about 5. This lowering of the pH has effects on the ligands and receptors, often causing them to dissociate. The endosome, or CURL (*C*ompartment for *U*ncoupling of *R*eceptor and *L*igand), then pinches off to yield a vesicle ('tubular portion') containing the receptors. This recycles back to the plasma membrane. The remainder of the endosome (or CURL) then fuses with a lysosome, an organelle rich in degradative enzymes active at acid pH, and the remaining contents of the CURL are degraded.

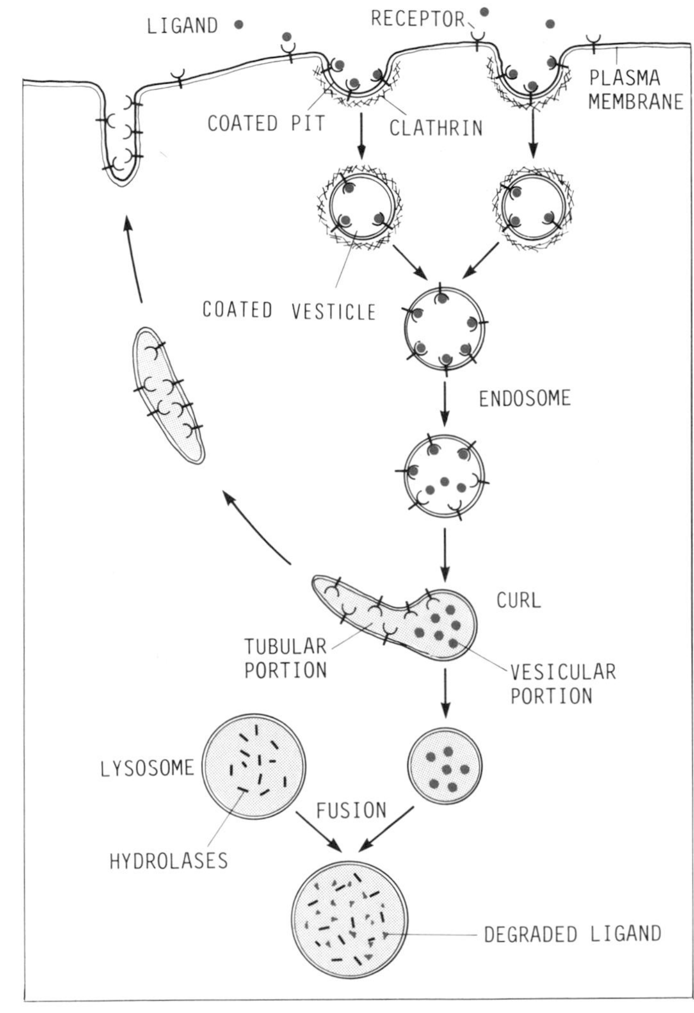

A

There are several modes of processing receptor/ligand complexes during receptor-mediated endocytosis (more strictly pinocytosis rather than phagocytosis).

After endocytosis, the vesicle develops into an endosome, or CURL and becomes more acidic. The lowering of the pH may cause dissociation of receptor and ligand. In other cases (e.g. transferrin) ligand and receptor remain attached, but transferrin releases its iron, and recycles to the plasma membrane with its receptor thus conserving the transferrin. In other cases, such as epidermal growth factor, receptor and ligand do not dissociate until the endosome has fused with a lysosome. In the case of immunological receptors, in at least one case, where the F_c receptor binds immune complex, only unoccupied receptors recycle to the cell surface. Occupied receptors are degraded.

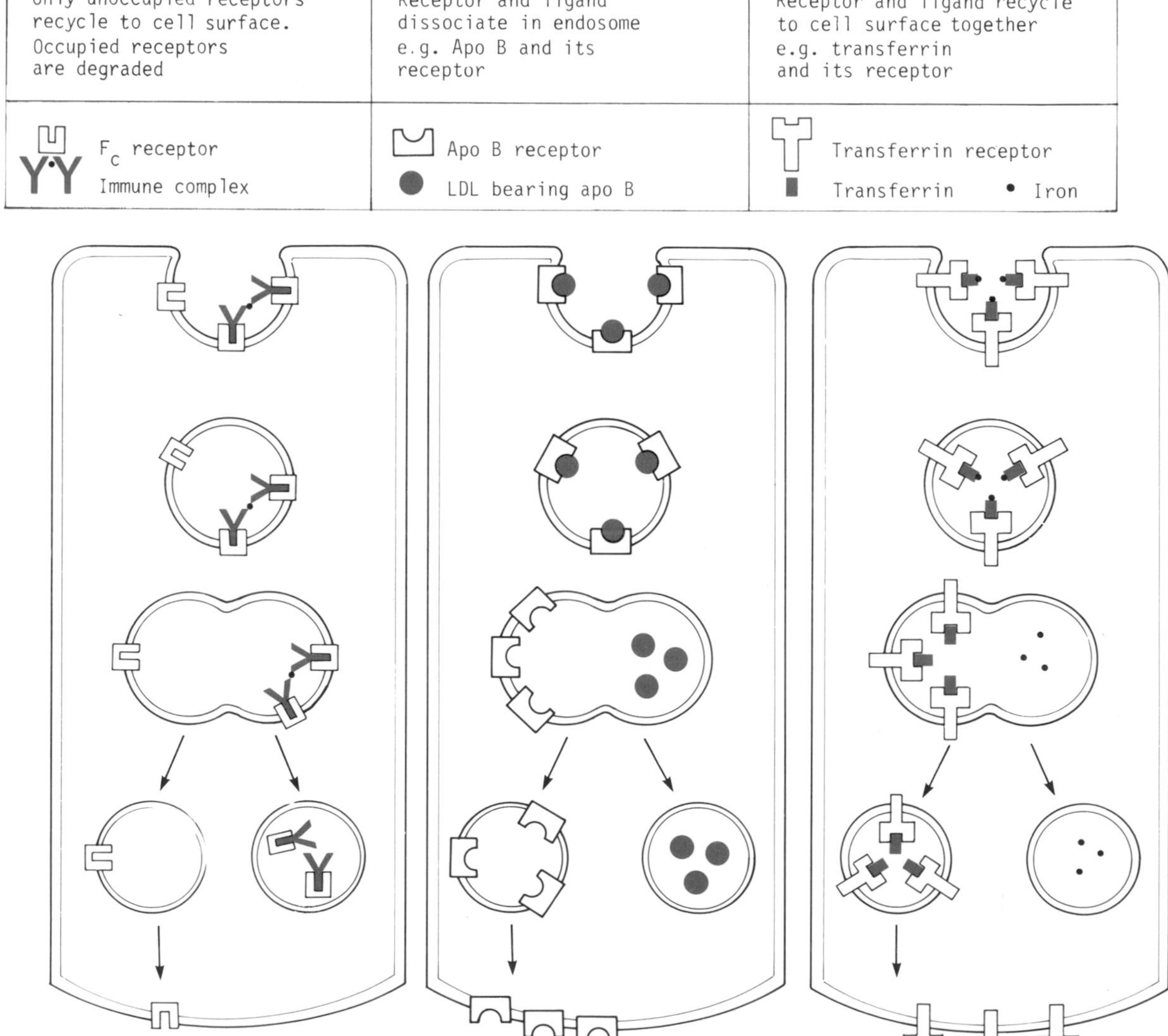

Note: Epidermal growth factor is a protein which stimulates growth of a variety of cells.

A

The role of clathrin

Clathrin, the protein that controls coated pit and coated vesicle formation exists in a conformation known as a triskelion, a trimeric, three-legged structure.

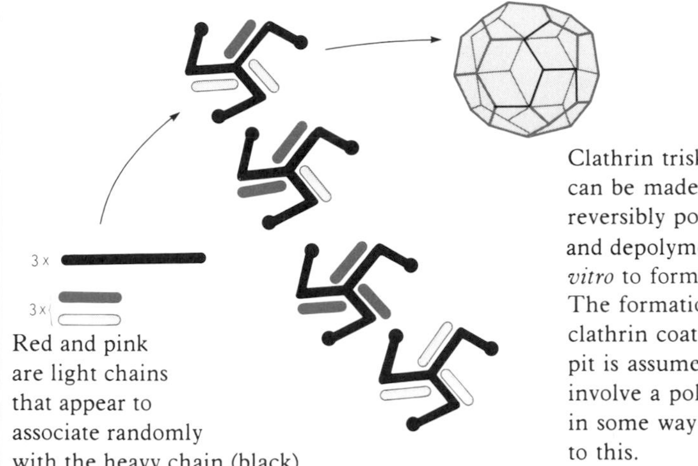

Red and pink are light chains that appear to associate randomly with the heavy chain (black).

Clathrin triskelions can be made to reversibly polymerize and depolymerize *in vitro* to form 'baskets'. The formation of the clathrin coat of a coated pit is assumed to involve a polymerization in some way similar to this.

B

Triskelions (1) and isolated coated vesicles (2) have been visualized by electron microcscopy.

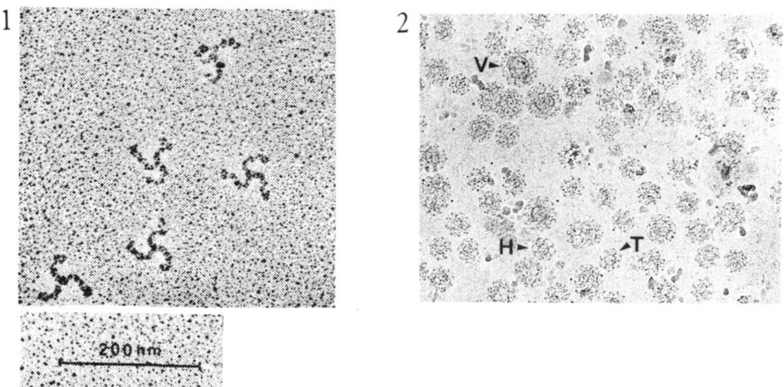

1. Individual triskelions

2. Unstained placental coated vesicles. Hexagonal barrels (H), so-called tennis ball structures (T) and larger coats containing vesicles (V) can be seen.

3.

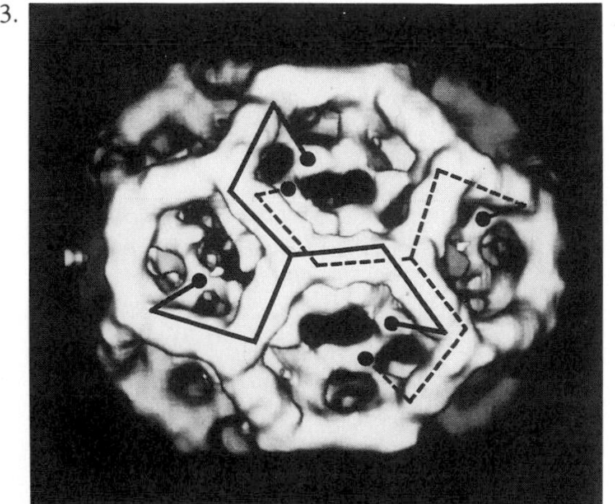

3. 3D-reconstruction of a clathrin cage. This cage is a 'hexagonal barrel' which has a hexagon at the top and the bottom, six hexagons around the equator and two rings of six pentagons joining them. Two clathrin triskelions centered at the vertices between neighbouring equatorial hexagons have been superimposed diagrammatically on the reconstruction. Each leg of a triskelion runs from one vertex, along two neighbouring polygonal edges and then turns inwards with its terminal domain forming the inner shell of density seen in the reconstruction.

A

The PI response

Hydrolysis of phosphatidylinositol or its phosphorylated derivatives is an early step in cell stimulation by many agents. One consequence of this hydrolysis is release of Ca^{2+} from intracellular stores. This reaction is known as 'the phosphoinositide response' or the '*PI response*'.

An early step in this response involves activation of phospholipase C to hydrolyse PIP_2 to IP_3 and diacylglycerol

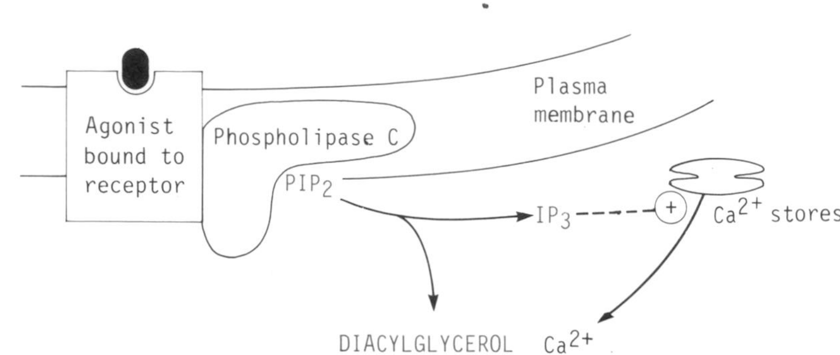

Ca^{2+} is considered to be an important second messenger in the action of many hormones, growth factors, neurotransmitters and other agonists that stimulate excitable cells. It was originally proposed that hydrolysis of phosphatidylinositol led to the opening of Ca^{2+} gates in the plasma membrane. It is now thought that it is in fact release of inositol trisphosphate* (IP_3) from phosphatidylinositol 4, 5-bisphosphate (PIP_2) that causes the intracellular Ca^{2+} concentration to rise. IP_3 has been shown to release Ca^{2+} from intracellular stores and may thus be regarded as a second messenger.

*. Inositol 1,4,5-trisphosphate see page 232.

B

Concerted action by the diacylglycerol (DG) and Ca^{2+} released by the PI response activates protein kinases.

Protein kinase C acts on a substrate in platelets with an M_r of about 40 000. Ca^{2+}/calmodulin activates the kinase for myosin light chain, an important protein of the cytoskeleton.

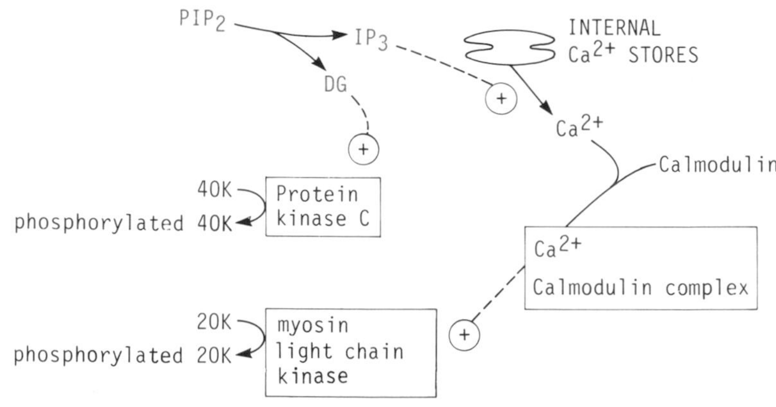

K above indicates the M_r of the proteins $\times 10^{-3}$.

Much of our understanding of the connection between the PI response and protein phosphorylation has been gained from experiments with platelets. In the platelet, exogenously-added diacyglycerol leads to activation of protein kinase C with phosphorylation of its substrate. Addition of a Ca^{2+} ionophore leads to phosphorylation of myosin light chain. Individually, at low concentrations, exogenous diacylglycerol or Ca^{2+} ionophore bring about slow aggregation of platelets but when added together at these low concentrations they initiate rapid and complete aggregation.

A

The cloning of the acetylcholine receptor cDNA

An example of how protein structure and function can be probed using *site-directed mutagenesis* (see below).

The genes for the four subunits of AChR of *Torpedo californica* (an electric fish) have been cloned and the cDNA corresponding to each subunit was transcribed *in vitro*. The four mRNAs synthesized were injected into Xenopus oocytes (which do not normally possess an acetylcholine receptor). The mRNA was translated by the oocytes, which were incubated in the presence of [^{35}S]methionine.

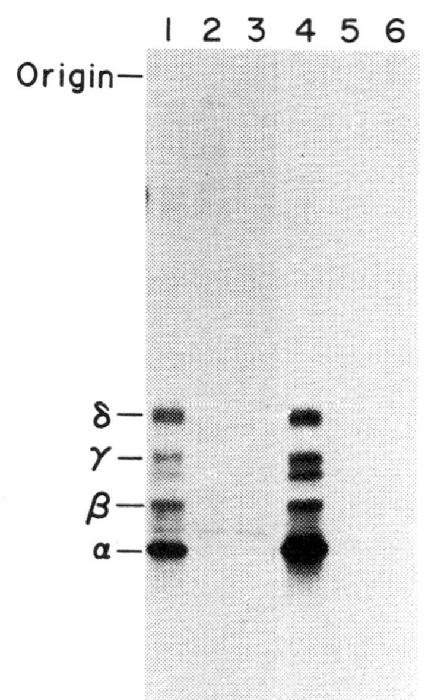

SDS 10% PAGE of synthesized subunits

After incubation of the oocytes with [^{35}S]methionine, the AChR was precipitated by rabbit antiserum to the *T. californica* AChR and was electrophoresed in the presence of SDS on 10% polyacrylamide gel. Lane 1-Immunoprecipitate from Xenopus injected with *T. californica* total mRNA Lane 4-Immunoprecipitate from Xenopus injected with mRNAs synthesized *in vitro* from cDNA. Lanes 2, 3, 5 + 6 are controls to verify that non-specific precipitation of protein by the rabbit antiserum used did not occur.

B

The membrane potential-dependence of acetylcholine (ACh) responses was recorded from a Xenopus oocyte injected with the four AChR mRNAs and then incubated 3 days.

The oocytes that had been injected with the mRNAs and then incubated for 3 days responded to ACh by causing a current, recorded under voltage clamp, at different levels of applied membrane potential, indicating that a functional AChR had been synthesized by the occytes from the *T. californica* mRNA and incorporated into the membrane.

C

Protein Engineering

The technique which allows proteins of any desired structure to be synthesized by modification of a cloned DNA is commonly referred to as *protein engineering*. The DNA can either be modified at specific sites by means of mutagens or by the replacement of stretches of nucleotides.

A

Protein Engineering, an example.

The cloning of the cDNA of the acetylcholine receptor has permitted site-directed mutagenesis for the introduction of deletions or specific amino acid substitutions in the subunit.

The effect of these changes on function could be tested. Each of the α-subunit-specific mRNAs synthesized with the cDNA templates with internal deletions, combined with the wild-type and subunit-specific mRNAs was injected into Xenopus oocytes. The oocytes were examined for the content of AChR subunit polypeptides and for the functional properties of the AChR formed. Binding was tested by incubation with [^{125}I]-bungarotoxin, which is related to ACh binding, and is inhibited specifically by carbamylcholine. ACh potentials were recorded at a holding potential of −60 mV.

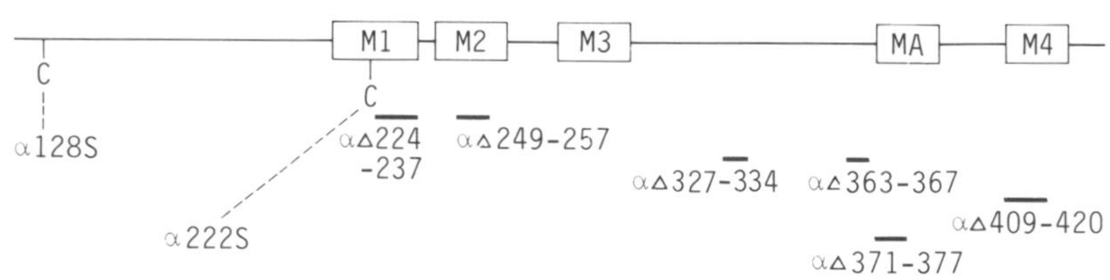

Subunit	Residue From	Change* To	[^{125}I]-bungarotoxin binding Activity %	% Inhibition by carbamylcholine	Response to ACh(%)
Wild type			100	80	100
αΔ224-237	LFSFLTGL VFYLPT	PSS	0.5	NT	ND
αΔ249-257	VLLSLTV FL	SS	17	87	ND
αΔ327-334	STMKRASK	RAR	80	82	70
αΔ363-367	QTPLI	PSS	83	77	101
αΔ371-377	DUKSAIE	A	44	81	0.2
αΔ409-420	ILLCVFML ICII	EL	23	77	ND
α128S	C	S	0.08	NT	ND
α222S	C	S	79	83	107

*using the single letter code for amino acids

Deletion mutations

The bungarotoxin binding capacity of most of the deletion mutants was comparable with the wild type except for αΔ224-237. Segment M1 is near to the extracellular amino terminal section of the subunit, so this deletion may result in a conformational change near the binding site. Deletions in M1, M2, MA or M4 resulted in loss of AChR response to ACh.

Substitutions

These explored the role of cysteines which had been replaced with serines. Replacement of 128 abolished binding and AChR response, whilst 222 replacement had little effect.

These experiments provide a fine illustration of the power of these techniques for exploring the relationship between protein structure and function.

A

Oncogenes

A number of viruses that have the ability to transform normal cells into tumour cells contain genes that have been identified as coding for products that are responsible for providing a continuous growth stimulus for the host cell. These genes have been termed *oncogenes*

Oncogene products probably mimic different elements of normal systems by which a variety of proteins normally stimulate cell growth. One of these is released by stimulated platelets and is named *platelet-derived growth factor* (PDGF). It stimulates the growth of endothelial cells, in situations where damage to the wall of blood vessels has occurred, thereby accelerating the repair process.

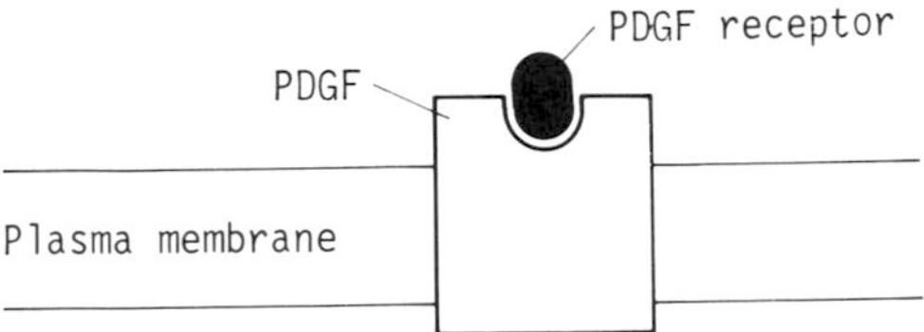

When the PDGF receptor is occupied, protein tyrosine kinase is activated and this stimulates cell growth.

B

There are at least three ways in which oncogene products can provide this continuous growth stimulus.

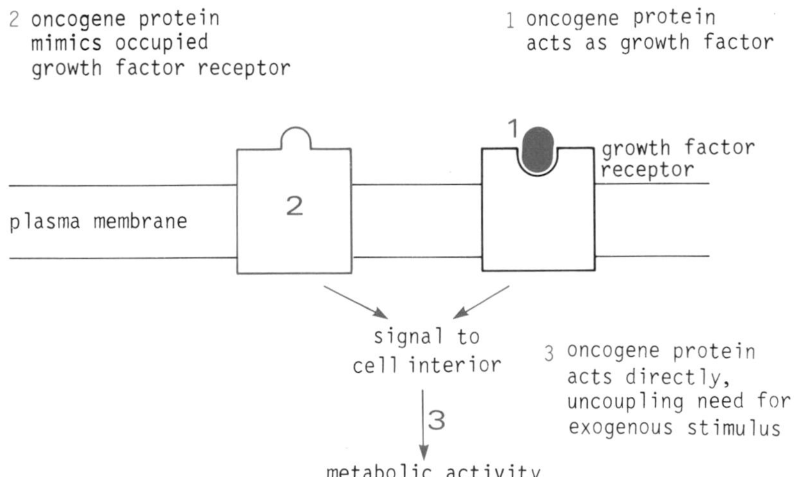

There are oncogene proteins for (1), (2) and (3).

1. An example is PDGF and *v-sis* of Simian sarcoma virus (SSV).
2. Epidermal growth factor (EGF) receptor is mimicked by *v-erb*-B oncogene product of avian erythroblastosis virus (AEV). Protein tyrosine kinase activity is increased when EGF binds. The oncogene product is the truncated form of the EGF receptor.
3. Oncogenes of the *src* family encode proteins with tyrosine kinase activity. There are other oncogenes whose products are nuclear, e.g. *myc*, *myb*, and these may play a role in activating other genes responsible for the growth response.

INDEX